MICROWAVE ENGINEERING USING MICROSTRIP CIRCUITS

To my wife, Marilyn,
and sons, Andrew and Stuart

Henry

Mano brangiems Tėveliams

Ramutis

MICROWAVE ENGINEERING USING MICROSTRIP CIRCUITS

Dr E.H. Fooks
Dr R.A. Zakarevičius

PRENTICE HALL

New York London Toronto Sydney Tokyo Singapore

1 2 3 4 5 94 93 92 91 90

Printed and bound in Australia by
Brown Prior Anderson, Burwood, Victoria.
Cover design by Cathy Hoare.

0 7248 0915 5 (paperback)
0-13-691650-3 (hardback)

National Library of Australia
Cataloguing-in-Publication Data

Fooks, E. H.
  Microwave engineering using microstrip circuits

  Includes index.
  ISBN: 0 13 691650 3.
  ISBN: 0 7248 0915 5 (pbk.).

  1. Microwaves. 2. Microwave wiring. 3. Microwave circuits.
  I. Zakarevičius, R. A. II. Title.

621.38132

Library of Congress
Cataloguing-in-Publication Data

Fooks, E.H.
  Microwave engineering using microstrip circuits / E. H. Fooks,
  R. A. Zakarevičius.
      p.   cm.
  Includes bibliographical references.
  ISBN: 0-13-691650-3

  1. Microwave integrated circuits. 2. Strip transmission lines.
  I. Zakarevičius, R. A. II. Title.

TK7876.F66 1989   89-26619
621.381'32--dc20   CIP

Prentice Hall, Inc., *Englewood Cliffs, New Jersey*
Prentice Hall Canada, Inc., *Toronto*
Prentice Hall Hispanoamericana, SA, *Mexico*
Prentice Hall of India Private Ltd, *New Delhi*
Prentice Hall International, Inc., *London*
Prentice Hall of Japan, Inc., *Tokyo*
Prentice Hall of Southeast Asia Pty Ltd, *Singapore*
Editora Prentice Hall do Brasil Ltda, *Rio de Janeiro*

PRENTICE HALL

A division of Simon & Schuster

Contents

Preface

Microstrip circuits for microwave engineering have progressed through the research and development stages and are now found as an integral part of many small-signal microwave systems. This form of circuit is also becoming increasingly important in baseband circuits for optical communications and has implications for interstage communication in high-speed digital networks.

This book has been written primarily as a textbook for the student who is interested in microwave engineering, and is aimed at developing a practical understanding of microstrip components and systems. It presents a thorough grounding in the basics of microstrip components and for interfacing them with transistors and diodes in active circuits. The material makes the book suitable both as a textbook for a graduate course in microwave circuits over one or two semesters, and as a textbook or reference book for final year undergraduate courses on high frequency (or microwave) circuits and electronics. The book will also be useful for practicing engineers, physicists and other specialists requiring a reference for microstrip circuits. The chapters are, in the main, self-contained and include examples, making the book suitable for self study. Exercises and a list of key references are included at the end of most chapters, together with selected answers at the end of the book.

As this is a teaching text and not a treatise, there is an emphasis on fundamentals at the expense of peripheral detail. The treatment of the subject matter is approached in a manner that is suitable for teaching purposes, based on a number of years experience in teaching this type of material. The book emphasizes the physical point of view. Proofs of important results are always given and considerable effort has been made to find or develop simple, yet rigorous, proofs. Nevertheless, physical interpretations of mathematical results are stressed whenever possible. Many useful results, the proofs of which are reasonably simple, have been left to be solved as exercises. This helps to limit the size of the book without sacrificing rigor and, at the same time, provides pedagogical advantages.

Many new approaches and some new material are included. The scattering parameter treatment is from a physical point of view which emphasizes a traveling wave approach and studiously avoids calculations which are no more than manipulations of formulae and symbols. The proof of various power gain formulae follows a similar original approach. The concept of a zero length line has been found to be very useful in disentangling the traveling waves in a number of circumstances. Original formulations of various design formulae are also given. Charts for the line widths and separation for parallel-coupled lines in couplers and filters are presented in a manner that is suitable for circuit synthesis.

All major aspects of microstrip circuits, principles and design are covered. Assumed knowledge includes complex and matrix algebra, two-port parameter theory, elementary differential equations, the basics of lumped filter design, the fundamentals of electromagnetic theory, and elements of transistor and diode circuits. Elementary transmission line theory is so fundamental to much of the book, that the first chapter gives a thorough revision of the necessary material. Two-port networks are characterized mainly in terms of scattering parameters at microwave frequencies, and a thorough treatment of the characterization of these networks is included in Chapter 2. Nevertheless, when results are more elegantly derived with the help of other parameters, such parameters are used, as we do not believe in slavishly adhering to a scattering parameter treatment simply for the sake of consistency. The microstrip line is used in all the circuit components in this book. A low frequency analysis of the line itself is given in Chapter 3, with details of the numerical analysis techniques provided as appendix material, that will be of particular interest to those who wish to have an introduction to the field analysis of microstrip line configurations. In the analysis of components, it is desirable at first to separate the component design from the second-order effects of non-ideal lines. However, these effects are not to be ignored and are brought together in Chapters 4 and 5, where lines in a practical environment are discussed. The concept and use of the voltage reflection coefficient with its interpretation on the Smith Chart is important and, wherever possible, graphical interpretations and solutions on the chart are used.

Hybrid-line couplers are the first class of component to be considered and are dealt with in Chapter 7. Here, the notion of even and odd modes with respect to a plane of symmetry through the device is introduced, so that a four-port network may be analyzed in terms of equivalent two-port networks. Even- and odd-mode analysis is extended to parallel-coupled lines for directional couplers in Chapter 8. Filter analysis, in particular for band-pass and band-stop filters, follows directly from the concepts that have been developed for coupling between parallel lines in directional couplers. Filter design and analysis is now such a broad and detailed field, that it has been necessary to limit the material to such topics that provide a useful coverage of the basic filter types together with their relationship with other components. A collection of components (some of great importance such as those for launching microwaves into a microstrip line) that do not fit neatly into other chapters are described as miscellaneous components in Chapter 10.

Active circuit characterization is included in a self-contained manner in Chapter 11, to allow a sensible design to be achieved when microstrip lines are interfaced to active circuits. Some examples of practical systems that include active components are given in Chapter 12, which thus consolidates earlier material. In the final chapter on measurements, a number of laboratory experiments associated with the basic line and the effects of discontinuities are described, and may be carried out to complement the lectures.

In order to keep the book within sensible bounds, some related topics are omitted. Other types of planar circuits, such as slot lines and coplanar waveguides, are only briefly alluded to, as they are seen to be peripheral to the main thrust of this book. Even though much of the practical realization of microstrip circuits is in the monolithic microwave integrated circuit (MMIC) form, we have left out any specific treatment of MMICs as not being as fundamental as a thorough grounding in microstrip basics. For similar reasons, we do not deal with aspects of the manufacturing technology for thick and thin film circuits.

The omission of CAD techniques requires special mention. Again, space limitations play their part but, more fundamentally, we consider that an intelligent use of CAD techniques requires a thorough understanding of the modeling assumptions that underpin them. Without such an understanding, the use of CAD is pedagogically unsound, as the user of a package will neither appreciate its limitations nor notice when obvious errors are being produced. We feel that the reader who has thoroughly grasped the principles of microstrip circuits expounded in this book will quickly learn how to apply any specific CAD package.

Several diagrams have been redrawn from a number of key journals serving the microwave community. For permission to do so, we thank the IEEE, the IEE, the publishers of the *Microwave Journal* and of *Microwaves and RF*, and Mrs Anita M. Smith for supplying Figure 6.3 (The Smith Chart) through the Analog Instruments Company.

This book would not have been written without the inspiration of and interaction with many people. We thank our students, both graduate and undergraduate, for the opportunity and the challenge of teaching this material. The enthusiasm of Ted Gannan provided the spark that set the project on its way. To our manuscript reviewers and the staff of Prentice Hall Australia, and in particular to Fiona Marcar and Andrew Binnie, our thanks for their help and tolerance. Finally, our thanks go to our colleagues whose interest in the progress of the book has always been an encouragement that deserves its due recognition.

Symbols

Symbols are described and indexed to an early instance of their major use. Units, standard mathematical symbols, symbols used in a local context or derivation, and coordinate component parts of variables are not included. The subscript, n, is used as a general variable for a range of values.

A, B, C, D	The transmission or ABCD-matrix elements	2.2.1
A_n, B_n, C_n	Power series coefficients	
a_n	The normalized input wave into the nth port	2.1.1
a_S	A (normalized) wave launched by a source into a matched load	2.1.4
$\mathbf{B}$	Magnetic flux density	
b	Normalized susceptance, as in $y = g + jb$	6.5
b_n	Susceptance of the nth normalized admittance	6.5
b_n	The normalized reflected wave from the nth port	2.1.1
b_S	As a_S, but from the output of a two-port network	11.5.2
C	Capacitance (also an ABCD-parameter element)	
C_c	Corner capacitance	5.3
C_{even}, C_{odd}	Even- and odd-mode capacitance	5.6
C_F	Open-circuit fringing capacitance	5.2
C_L	End-capacitance in a transmission line Π-equivalent circuit	9.3.5
C_S	Step transition capacitance	5.4
C_T	T-junction discontinuity capacitance	5.5
C	Coupling coefficient in dB	7.1
C_L, C_S	Load and source stability circle centers	11.4
$\boldsymbol{C}$	Transmission line capacitance per meter	1.2
$\boldsymbol{C_w}$	Capacitance per meter for a microstrip line, width w	5.4
c	Velocity of e.m. waves in free space $= 2.997925 \times 10^8$ m.s^{-1}	1.2.3
c	Voltage coupling coefficient for $\theta = 90°$	7.3
D	Coupler directivity in dB	7.1
$\mathbf{D}$	Electric flux density	
d	Transmission line length	
d	Voltage coupling coefficient for arbitrary θ	8.4
$\mathbf{E}$	Electric field	
e_{i1}, e_{i2}	Incident wave voltage components at ports 1 and 2	2.1.1
e_{r1}, e_{r2}	Reflected wave voltage components at ports 1 and 2	2.1.1
e_S	The voltage wave launched by a source into a matched line	2.1.4
f	Frequency in Hz	
f_0	Design or center frequency	7.4
f_0	Quasi-static approximation frequency limit	4.4.1
G_a	Available power gain	11.2
G_p	(Ordinary) power gain	11.2
G_t	Transducer power gain	11.2

G_{au}, G_{pu}, G_{tu}	Unilateral power gains	11.5.3
G	Transmission line conductance per meter	1.2
g	Normalized conductance, as in $y = g + jb$	6.5
g_i, g_r, g_f, g_o	The real parts of the y-parameters	11.5.4
g_n	Conductance of the nth normalized admittance	6.5
H	Magnetic field	
h	Substrate height	3.1.3
I	Current	
I_f, I_r, I_τ	Forward, reverse and transmitted traveling wave currents	1.2
I_L	Load current	1.3.1
I_n	Total current at port n	2.1.3
I	Insertion loss in dB	7.1
i_{i1}, i_{i2}	Incident wave current components at ports 1 and 2	2.1.1
i_{r1}, i_{r2}	Reflected wave current components at ports 1 and 2	2.1.1
J	Immittance inverter parameter	9.4.3
J'	Normalized admittance inverter parameter, J/Y_0	9.4.5
K	Stability factor	11.4.2
K	Flux reflection coefficient	A2
K(k)	Elliptic integral	3.1.2
k_0	Phase coefficient for a plane wave in free space	4.5.2
L, L_n	Inductance; the nth inductance	
L_c	Corner inductance	5.3
L_c	End-inductance in a transmission line T-equivalent circuit	9.3.5
L_s	Step transition inductance	5.4
L	Filter attenuation in dB	9.3.2
L_r	Passband ripple attenuation in dB	9.3.3
L	Transmission line inductance per meter	1.2
L_w	Inductance per meter for a transmission line, width w	5.4
l	Transmission line length	
m	Corner miter percentage	5.3
m	Step transition correction factor	5.4
P	Power	
P_{av}	Available power	2.1.5
P_{av}	Average power flow	4.4.2
P_{av_o}	Available power at the output of a two-port network	11.2
P_{av_s}	Available power of a source	9.3.1
P_{in}, P_{out}	Input and output power	11.2
P_L	Load power	11.5.2
P	Power-split ratio in dB	7.3
p	Power-split ratio as a ratio of voltages	7.3
Q	Quality factor	9.5
q	Effective filling fraction	3.3.2
q_n	Point charge at node n	3.2.3
q	Column matrix of point charges	3.2.3
R	Resistance	
R_s	Surface resistance	4.2.2
R_1, R_2, R_3	Reference planes	5.5
R_L, R_s	Load and source stability circle radii	11.4
R	Transmission line resistance per meter	1.2

r	Normalized resistance, R/Z_0	6.2
r, θ, z	Cylindrical coordinates	
S	Voltage standing wave ratio (V.S.W.R.)	1.3.2
s	Gap or parallel line separation	
s_i, s_r, s_f, s_o	Scattering parameters	2.1.1
$s_i^{(e,o)}, s_f^{(e,o)}$	Even- and odd-mode scattering parameters	7.3
s_{mn}	General multi-port scattering parameter	2.1.1
T	Transmission coefficient	1.3.4
$[T]$	Transmission (ABCD) matrix	2.2.1
T_n	Noise temperature	11.6
t	Microstrip line thickness	4.2.1
t	Time	
t	Voltage transmission coefficient for arbitrary θ	8.4
U	Mason's U-function	11.5.4
u	Unilateral approximation check parameter	11.5.3
u	Surface wave transverse decay coefficient	4.5.2
u	Real part of Γ for the Smith Chart derivation	6.2
V	Voltage	
V_f, V_r, V_τ	Forward, reverse and transmitted traveling wave voltages	1.2
V_i, V_o	Input and output voltages on a transmission line	8.2
$V_i^{(e,o)}, V_o^{(o)}$	Even- and odd-mode input and output voltages	8.2
V_L	Load voltage	1.3.1
V_n	Voltage at port n	2.1.3
V_s	Open-circuit source voltage	2.1.4
$\mathbf{V}$	Column matrix of point potentials	3.2.3
v	Imaginary part of Γ for the Smith Chart derivation	6.2
v_{ph}, v_{phase}	Phase velocity	1.2.3
w	Microstrip line width	3.1.3
$w^{(50)}$	Width of a 50Ω microstrip line	8.3
X, x	Reactance and its normalized value	6.2
x, y, z	Cartesian coordinates	
Y_0	Transmission line characteristic admittance	1.4
Y_{0n}, Y_n	Characteristic admittance of line n	6.4
Y_{in}, Y_{out}	Input and output admittance	1.4
Y_L, y_L	Load admittance and its normalized value	1.4
Y_n, y_n	The nth admittance and its normalized value	7.4
Y_T	Transformer characteristic admittance	9.4.3
y_i, y_r, y_f, y_o	Admittance or y-parameters	2.2.4
Z_0	Transmission line characteristic impedance	1.2
Z_{0n}, Z_n	Characteristic impedance of line n	6.4
Z_{0e}, Z_{0o}	Even- and odd-mode characteristic impedances	8.2
Z_{fs}	Characteristic impedance of air-filled (free space) transmission line	3.3.2
Z_H, Z_{HIGH}	Characteristic impedance of a high impedance line	6.8.1
Z_{in}, z_{in}	Input impedance and its normalized value	1.4
Z_L, z_L	Load impedance and its normalized value	1.3.1
Z_L, Z_{LOW}	Characteristic impedance of a low impedance line	6.8.1
Z_n, z_n	The nth impedance and its normalized value	6.10
Z_{out}, z_{out}	Output impedance and its normalized value	2.1.2
Z_s	Source impedance	2.1.4

Z_S	Surface impedance	4.2.2
Z_{series}, Z_{shunt}	Series and shunt line characteristic impedances	7.3
z_{series}, z_{shunt}	Z_{series} and Z_{shunt} normalized to the feed line impedance	7.6.3
Z_{stub}, z_{stub}	Stub line impedance and its normalized value	7.6.3
Z_T	Transformer characteristic impedance	6.6.1
α	Attenuation coefficient, neper.m^{-1}	1.2
α_c	Attenuation due to conductor loss	4.2.2
α_d	Attenuation due to dielectric loss	4.2.2
β	Phase coefficient, radian.m^{-1}	1.2
Γ	Voltage reflection coefficient	1.3.4
Γ_0	Optimum source reflection coefficient for minimum noise	11.6
Γ_e, Γ_o	Even- and odd-mode voltage reflection coefficients	8.2
$\Gamma_{in}, \Gamma_{out}$	Input and output voltage reflection coefficients	2.1.6
Γ_L, Γ_s	Load and source voltage reflection coefficients	2.1.5
$\Gamma_L(opt), \Gamma_s(opt)$	Γ_L and Γ_s for simultaneous conjugate matching	11.3.1
Γ_n	The nth reflection coefficient	6.6.2
γ	Propagation coefficient, $\gamma = \alpha + j\beta$	1.2
γ	A function of Γ	8.4
Δ	Determinant of a matrix	
Δ	Surface roughness	4.2.2
Δn	An increment of any variable, n	
$\Delta x_{(C)}, \Delta x_{(L)}$	Line correction for capacitance and inductance	5.3
δ	Skin depth	4.2.2
δ	Angle of the dielectric loss tangent, $\tan\delta$	4.2.2
ε_0	Permittivity of free space $= 8.854 \times 10^{-12}$ F.m^{-1}	
ε_{eff}	Effective relative permittivity	3.3.2
$\varepsilon_{eff}^{(e)}, \varepsilon_{eff}^{(o)}$	Even- and odd-mode effective relative permittivity	8.3
ε_r	Relative permittivity	1.2.3
η_0	Intrinsic impedance of free space	4.5.1
$\Theta_{e,o}$	Even- and odd-mode transmission coefficient phase angles	7.3
θ	Coupling length in radians	8.1
θ, ϕ	A general phase angle or electrical length	
κ_n	Low-pass filter prototype value	9.3
λ	Wavelength; the microstrip line wavelength	1.2.3
λ_0	Free space wavelength	1.2.3
λ_c	Free space wavelength at the waveguide cut-off frequency	4.5.4
λ_H, λ_L	Wavelength in high and low impedance lines	9.3.5
λ_s	Wavelength for a plane wave in the substrate material	4.4.1
μ	Permeability	4.2.2
μ_0	Permeability of free space $= 4\pi \times 10^{-7}$ H.m^{-1}	
ρ	Charge density	3.2.3
σ	Conductivity	4.2.2
ϕ_n	Discrete potential at node n	3.2.2
Ω	Fractional bandwidth	9.1
ω	Angular frequency in radian.s^{-1}	1.2
ω_0	Center frequency	9.1
ω_1, ω_2	Band-edge frequencies	9.1
ω_1'	Band-edge frequency in the prototype filter	9.1

1 Transmission line theory

1.1 INTRODUCTION

The basic elements of transmission line theory that are applicable to a wide variety of wave propagation problems are reviewed in this chapter. This basic theory will be built on in later chapters to describe the propagation properties of microstrip lines. A microstrip line is illustrated in Figure 1.1. The generation and processing of signals at microwave frequencies, say from below 1.0 to beyond 30 GHz, these days are typically achieved in microstrip circuits using elements constructed from microstrip transmission lines and combined with semiconductor components. The line widths and substrate thickness are small while the circuit line lengths are generally a significant fraction of a wavelength. The planar geometry is particularly suited to production using photographic techniques to give the line patterns on a copper-clad substrate. At the upper end of the microwave range, complete subsystems including the active devices may be fabricated onto a single semiconductor slice. Within this book, however, emphasis will be placed at the lower frequencies to give a solid framework for the understanding of circuit design and operation.

In this chapter on the basic theory of transmission lines, a two-conductor transmission line with a uniform dielectric material is considered. The complexities of a transmission line with propagation in a mixed air/dielectric environment as shown in Figure 1.1 will be described later in Chapter 3.

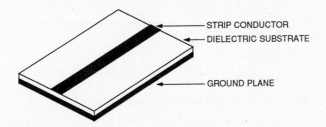

STRIP CONDUCTOR
DIELECTRIC SUBSTRATE
GROUND PLANE

Figure 1.1 A microstrip transmission line

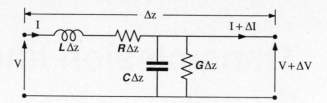

Figure 1.2   The equivalent circuit for a short length of transmission line

1.2   PRIMARY AND SECONDARY PARAMETERS

Consider a transmission line that has the distributed parameters for a short line length, Δz, as illustrated in Figure 1.2. This incremental model is valid, provided the following conditions and assumptions are true:

i)   The line is a distributed system. Each infinitesimal element of the line is identical to all other similar length elements.

ii) At any point along the line, the voltage V and current I are meaningful quantities. This is satisfied for a two-conductor transmission line that has a uniform dielectric material and is operating in a mode where all the fields are in the transverse plane (i.e. a transverse electromagnetic or TEM-mode). The voltage $V = -\int E \cdot dl$ may be evaluated along any path between the two conductors. The current is the total longitudinal current flow in each conductor.

iii) The inductance, capacitance, resistance and conductance per unit length (**L, C, R** and **G**) exist as the primary parameters of the line. The propagation coefficient γ and characteristic impedance Z_0, as secondary parameters for the line, will be derived in terms of these primary parameters.

iv) All time variations will be expressed as $e^{j\omega t}$ with $d(\cdot)/dt \equiv j\omega(\cdot)$.

The incremental voltage change for the short line length due to the current flow through the series components is given by

$$\Delta V = -\left\{ RI + L\frac{\partial I}{\partial t} \right\} \Delta z \tag{1.1}$$

The incremental current change, ignoring second order small quantities and assuming a voltage V across the shunt components, is

$$\Delta I = -\left\{ GV + C\frac{\partial V}{\partial t} \right\} \Delta z \tag{1.2}$$

In the limit as $\Delta z \to 0$, $\partial V/\partial z$ and $\partial I/\partial z$ may be obtained.

The wave equation is an equation that gives the time and spatial dependence of either V or I. It is obtained for V by eliminating the current from these two first-order differential equations. Having used $\partial/\partial t \equiv j\omega$, time is no longer a variable and $\partial/\partial z$ becomes d/dz, giving

$$\frac{d^2 V}{dz^2} = (R + j\omega L)(G + j\omega C)\, V \tag{1.3}$$

An identical equation to (1.3) with the current, I, in place of the voltage may also be obtained. A solution to the wave equation (1.3) for the voltage along the line is given by

$$V = (V_f e^{-\gamma z} + V_r e^{\gamma z}) e^{j\omega t} \tag{1.4}$$

where
$$\gamma = \left\{(R + j\omega L)(G + j\omega C)\right\}^{\frac{1}{2}} \tag{1.5}$$

Likewise
$$I = (I_f e^{-\gamma z} + I_r e^{\gamma z}) e^{j\omega t} \tag{1.6}$$

In general, the propagation coefficient γ is complex, i.e. $\gamma = \alpha + j\beta$, with α the attenuation coefficient in neper.m^{-1}, and β the phase coefficient in radian.m^{-1}.

The term *coefficient* is used in preference to *constant* for α, β and γ as it will soon be realized that, although these quantities may be constant with respect to time, they are generally frequency, material and geometry dependent.

The neper for attenuation is based on using voltage, current or field strength ratios expressed in terms of natural logarithms. For example

$$\alpha = ln\left\{\frac{V(z = 1\ meter)}{V(z = 0)}\right\} \quad neper.m^{-1} \tag{1.7}$$

In practice, however, attenuation is frequently expressed in decibels (dB), the bel being a power ratio expressed in terms of logarithms to the base ten. Thus, in decibels

$$\alpha = 10\,log_{10}\left\{\frac{P(z = 1\ meter)}{P(z = 0)}\right\} \quad dB.m^{-1} \tag{1.8}$$

Provided that the voltages in (1.7) are expressed at the same impedance level, then $8.686\,dB \equiv 1.0\,neper$.

The solution (1.4) for the voltage from the second order differential equation has two independent parts, with coefficients V_f and V_r. The first term, $V_f e^{j\omega t} e^{-\gamma z}$, represents the forward wave that is traveling in the +z direction. Moving in the +z direction along a transmission line, the magnitude of the wave decays as $V_f e^{-\alpha z}$ and, at any instant, the phase of the wave progressively lags by $e^{-j\beta z}$ with increasing z. The second term, $V_r e^{j\omega t} e^{\gamma z}$, is the reverse wave traveling in the −z direction and is attenuated as z decreases.

From (1.1)

$$I = -\frac{1}{R + j\omega L} \times \frac{dV}{dz} \tag{1.9}$$

i.e.
$$I = -\frac{1}{R + j\omega L}\left\{-\gamma V_f e^{-\gamma z} + \gamma V_r e^{\gamma z}\right\} e^{j\omega t} \tag{1.10}$$

Substituting for γ from (1.5) gives

$$I = \left[\frac{G + j\omega C}{R + j\omega L}\right]^{\frac{1}{2}} \times \left\{V_f e^{-\gamma z} - V_r e^{\gamma z}\right\} e^{j\omega t} \tag{1.11}$$

Considering either the forward or the reverse traveling wave, then the ratio of voltage to current for the wave is a constant, known as the characteristic impedance. When

(1.11) is compared with the solution of the wave equation in terms of the current (1.6), the characteristic impedance of the line is seen to be

$$Z_0 = \left[\frac{R + j\omega L}{G + j\omega C}\right]^{\frac{1}{2}} \quad \Omega$$

(1.12)

with

$$Z_0 = \frac{V_f}{I_f} \quad \text{or} \quad -\frac{V_r}{I_r}$$

forward   reverse
wave   wave

(1.13)

Note that a backward wave, as opposed to a reverse wave, is one that travels and decays in the +z direction while having a negative phase coefficient. Such a wave will not occur in the context of distributed microstrip transmission lines.

1.2.1 Lossless transmission lines

A lossless transmission line has both perfect conductors (i.e. $R = 0$) and a perfect dielectric material with no conduction currents flowing between the two conductors (i.e. $G = 0$). Therefore

$$\gamma = \alpha + j\beta = \left\{(j\omega L)(j\omega C)\right\}^{\frac{1}{2}}$$

(1.14)

giving $\alpha = 0$ neper.m^{-1}

(1.15)

and $\beta = \omega\sqrt{LC}$ radian.m^{-1}

(1.16)

The characteristic impedance for the lossless transmission line from (1.12) is

$$Z_0 = \left[\frac{L}{C}\right]^{\frac{1}{2}} \quad \Omega$$

(1.17)

1.2.2 Low-loss transmission lines

It is usual for a practical transmission line to be considered as a low-loss line. The *mathematical* requirement is that for the series impedance elements, $R \ll \omega L$, and for the shunt admittance elements, $G \ll \omega C$. Inequality ratios of $1 : 100$ will give approximately a 1% error in the attenuation and in the change of the phase coefficient. Since the resistance per unit length of a transmission line is dependent upon the depth of penetration of the current in the conductor, and this depth is found in §4.2.2 to be proportional to the square root of frequency, then the mathematical requirement for a low-loss line becomes more easy to achieve with increasing frequency. For a dielectric material, the ratio of the shunt current components, $G/\omega C$, is known as the loss tangent for the material and is substantially constant over a very wide frequency range for commonly used dielectric materials. Typical ratios for low-loss dielectric materials are less than 0.001.

Under the low-loss conditions, the following expressions are obtained for the secondary parameters. Rearranging (1.5) such that the loss ratios appear as small quantities compared with unity, gives the propagation coefficient as

$$\gamma = j\omega\sqrt{LC}\left\{\left[1 - j\frac{R}{\omega L}\right]\left[1 - j\frac{G}{\omega C}\right]\right\}^{\frac{1}{2}}$$

(1.18)

Ignoring the second-order small quantities, and thus taking only the first two terms of the series expansion for the square root, gives

$$\gamma \approx j\omega\sqrt{LC}\left\{1 - j\left[\frac{R}{2\omega L} + \frac{G}{2\omega C}\right]\right\}$$ (1.19)

In this expression, with no first-order small quantity to be added to β

$$\beta_{\text{low-loss}} \approx \beta_{\text{lossless}} = \omega\sqrt{LC} \quad \text{radian.m}^{-1}$$ (1.20)

For the attenuation

$$\alpha \approx \omega\sqrt{LC}\left\{\frac{R}{2\omega L} + \frac{G}{2\omega C}\right\}$$ (1.21)

i.e. $$\alpha = \frac{1}{2}\left\{\frac{R}{Z_0} + GZ_0\right\} \quad \text{neper.m}^{-1}$$ (1.22)

where the characteristic impedance is the lossless value given by (1.17).

The effects of loss on the characteristic impedance of the line are found from (1.12), where using similar series expansions reduces (1.12) to

$$Z_0 = \left\{\frac{L}{C}\right\}^{\frac{1}{2}}\left\{1 - j\left[\frac{R}{2\omega L} - \frac{G}{2\omega C}\right]\right\} \quad \Omega$$ (1.23)

However, the characteristic impedance given by (1.17) is normally used, even for low-loss lines, since the perturbation terms allowing for the loss are insignificant at microwave frequencies.

1.2.3 Velocities and wavelengths

The phase velocity

To travel along the transmission line at the velocity of the forward voltage wave and maintain a constant phase requires

$$\omega t - \beta z = \text{constant}$$ (1.24)

This velocity is known as the phase velocity. Consider a point on the wave shown in Figure 1.3 that has traveled from z_1 at t_1 to z_2 at t_2. Keeping the phase constant

$$\omega t_1 - \beta z_1 = \omega t_2 - \beta z_2$$ (1.25)

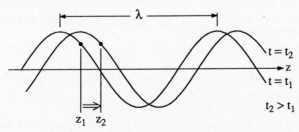

Figure 1.3 Evaluating the phase velocity

Now with the phase velocity given as the distance traveled by the constant phase front per unit time along the line then, from (1.25)

$$v_{phase} = \frac{z_2 - z_1}{t_2 - t_1} = \frac{\omega}{\beta}$$

(1.26)

For a low-loss transmission line with $\beta = \omega\sqrt{LC}$

$$v_{phase} = \frac{1}{\sqrt{LC}} \quad m.s^{-1}$$

(1.27)

The phase velocity in an air-filled transmission line is approximately equal to the velocity of light, c, in free space with $c = 2.997925 \times 10^8$ $m.s^{-1}$. The capacitance of a line that is uniformly filled with a dielectric material, relative permittivity ε_r, is proportional to ε_r. Thus

$$v_{phase} = \frac{c}{\sqrt{\varepsilon_r}} \quad m.s^{-1}$$

(1.28)

Substituting for L which is independent of the dielectric material, it follows from (1.27) and (1.17) that for any dielectric-filled transmission line

$$Z_0 = \frac{1}{v_{phase}C} \quad \Omega$$

(1.29)

where C is the capacitance per unit length of the line. However, for both transmission lines with losses and for waveguides modes where the fields are not entirely in the transverse plane, the phase coefficient β is not proportional to ω. Therefore, in these cases, the phase velocity will be frequency dependent and dispersion will occur.

The group velocity

It is necessary for a band of frequencies to be transmitted if information is to be conveyed in a signal. Consider the case of a wave represented by the two components

$$e^{j((\omega + d\omega)t - (\beta + d\beta)z)} + e^{j((\omega - d\omega)t - (\beta - d\beta)z)} = 2\cos(d\omega.t - d\beta.z)\, e^{j(\omega t - \beta z)}$$

(1.30)

The envelope of the wave is given by $\cos(d\omega.t - d\beta.z)$. Maintaining the phase of the envelope constant leads to the envelope or group velocity, namely

$$v_{group} = \frac{d\omega}{d\beta}$$

(1.31)

Any variation of the slope $(d\omega/d\beta)$ with frequency across the spectrum of the signal that is being propagated will result in "group delay distortion" and the information content of the signal will be corrupted.

The line wavelength

For a forward traveling wave, the voltage along a lossless line is given from (1.4) by

$$V_f e^{j(\omega t - \beta z)} = V_f e^{j\phi}$$

(1.32)

At any time t, two points on a transmission line that are separated by one wavelength have a phase difference of 2π. Therefore, at z and $(z + \lambda)$, with a phase difference of $\phi = 2\pi$

$$\omega t - \beta z = \omega t - \beta(z + \lambda) + 2\pi$$

(1.33)

giving $\qquad \beta = \dfrac{2\pi}{\lambda}$ or $\lambda = \dfrac{2\pi}{\beta}$ $\qquad\qquad\qquad\qquad$ (1.34)

where λ is the wavelength along the transmission line at the frequency of operation. In terms of the free space wavelength λ_0 and the relative permittivity ε_r of the uniform dielectric material

$$\beta = \frac{2\pi\sqrt{\varepsilon_r}}{\lambda_0} \quad \text{radian.m}^{-1} \qquad\qquad\qquad (1.35)$$

and $\qquad \lambda = \lambda_0/\sqrt{\varepsilon_r}$ m $\qquad\qquad\qquad\qquad\qquad$ (1.36)

These two equations are applicable to the transverse electromagnetic (TEM) mode of propagation only. However, they will also be used for microstrip transmission lines under quasi-static approximations by using an effective relative permittivity ε_{eff} to replace ε_r.

Example 1.1

A coaxial transmission line, filled with a uniform dielectric medium, has the following primary parameters at 1.0GHz

$$L = 250\text{nH.m}^{-1},\ C = 95\text{pF.m}^{-1},\ R = 1.6\Omega.\text{m}^{-1} \text{ and } G = 600\mu\text{S.m}^{-1}$$

i)   Verify that the line may be treated as a low-loss transmission line.
ii) Calculate the characteristic impedance and the attenuation and phase coefficients at 1.0GHz.
iii) From the phase coefficient, calculate the wavelength along the line and the relative permittivity of the uniform dielectric medium.

Solution:

i)   The line is low-loss at 1.0GHz if $R \ll \omega L$ and $G \ll \omega C$. Now with

$$R : \omega L = 1.6 : 2\pi \times 10^9 \times 250 \times 10^{-9} \qquad = 1 : 982$$

$$G : \omega C = 6 \times 10^{-4} : 2\pi \times 10^9 \times 95 \times 10^{-12} \quad = 1 : 995$$

it is seen that the low-loss inequality is well and truly satisfied.

ii) From (1.17)

$$Z_0 = \left[\frac{L}{C}\right]^{\frac{1}{2}} = \left[\frac{250 \times 10^{-9}}{95 \times 10^{-12}}\right]^{\frac{1}{2}} = 51.3\,\Omega$$

From (1.22)

$$\alpha = \frac{1}{2}\left\{\frac{1.6}{51.3} + 6 \times 10^{-4} \times 51.3\right\}$$

i.e. $\qquad \alpha = 0.031\ \text{neper.m}^{-1} \equiv 0.27\ \text{dB.m}^{-1}$

From (1.20)

$$\beta = \omega\sqrt{LC} = 2\pi \times 10^9 \left\{250 \times 10^{-9} \times 95 \times 10^{-12}\right\}^{\frac{1}{2}}$$

i.e. $\qquad \beta = 30.62\ \text{radian.m}^{-1}$

iii) The wavelength along the line at 1.0GHz

$$\lambda = \frac{2\pi}{\beta} = \frac{2\pi}{30.62} = 0.205 \text{ m}$$

and is seen to be less than the free space wavelength of 0.30m. Finally, from (1.36), the relative permittivity of the dielectric medium

$$\varepsilon_r = \left\{ \frac{0.30}{0.205} \right\}^2 = 2.14$$

1.3   TRANSMISSION LINES WITH A LOAD TERMINATION

The equations for the total voltage and current on a transmission line are

$$V = V_f e^{-\gamma z} + V_r e^{\gamma z} \tag{1.37}$$

and

$$I = I_f e^{-\gamma z} + I_r e^{\gamma z} = \frac{V_f e^{-\gamma z} - V_r e^{\gamma z}}{Z_0} \tag{1.38}$$

In (1.37) and (1.38), a time dependence of $e^{j\omega t}$ is implied. The forward wave represents transmission along the line from the source to the termination or load at the far end of the line. At the point where the load is connected to the transmission line, a reflected wave may be set up on the line and will propagate back towards the source.

Let a load impedance Z_L be placed on a line of characteristic impedance Z_0. It is convenient to choose the load plane at $z = 0$, as it is at this plane that the relationship between the reverse and forward waves will be evaluated in terms of the voltage reflection coefficient, Γ_L. Thus, for a line of length l, the input to the line will be at the plane $z = -l$, at which plane the ratio of total voltage to current will give the input impedance while the ratio between reverse and forward waves will give an equivalent input reflection coefficient. These parameters are illustrated in Figure 1.4.

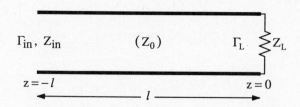

Figure 1.4   A transmission line with a load termination

1.3.1 Reflection coefficients

At the load plane where $z = 0$, the load voltage and current are

$$V_L = V_f + V_r \tag{1.39}$$

and $\qquad I_L = I_f + I_r \quad$ or $\quad I_L = \dfrac{V_f - V_r}{Z_0} \tag{1.40}$

Therefore, at the load plane, the load impedance

$$Z_L = \frac{V_L}{I_L} = \frac{V_f + V_r}{V_f - V_r} \times Z_0 \tag{1.41}$$

This equation may be written as

$$z_L = \frac{Z_L}{Z_0} = \frac{1 + (V_r/V_f)}{1 - (V_r/V_f)} \tag{1.42}$$

where z_L is the load impedance, normalized to the characteristic impedance of the transmission line. Substituting for the voltage reflection coefficient at the load, $\Gamma_L = V_r/V_f$

$$z_L = \frac{1 + \Gamma_L}{1 - \Gamma_L} \tag{1.43}$$

while rearranging gives the reflection coefficient at the load

$$\Gamma_L = \frac{Z_L - Z_0}{Z_L + Z_0} = \frac{z_L - 1}{z_L + 1} \tag{1.44}$$

The reflection coefficient at any point on the line and, in particular, at the input to the line at $z = -l$ is the ratio of the reverse to forward wave voltages at that point. In general, this value will be a complex quantity. Thus, from (1.37)

$$\Gamma_{in} = \frac{V_r e^{\gamma z}}{V_f e^{-\gamma z}} = \frac{V_r e^{-\gamma l}}{V_f e^{\gamma l}} \tag{1.45}$$

giving $\qquad \Gamma_{in} = \Gamma_L e^{-2\gamma l} \quad$ at the input to the line. $\tag{1.46}$

1.3.2 The voltage standing wave ratio

The voltage standing wave ratio (V.S.W.R.) is the ratio of the maximum to minimum voltages of a standing wave on a transmission line.

$$\text{The maximum voltage} \quad = \quad |V_f| + |V_r|$$

$$\text{The minimum voltage} \quad = \quad |V_f| - |V_r|$$

Hence, the V.S.W.R.

$$S = \frac{|V_f| + |V_r|}{|V_f| - |V_r|} \tag{1.47}$$

or $\qquad S = \dfrac{1 + |\Gamma_L|}{1 - |\Gamma_L|} \tag{1.48}$

When the line is terminated with a matched load, i.e. $Z_L = Z_0$, then from (1.44) it is seen that there are no reflections, $\Gamma_L = 0$, giving $S = 1$. Please note that in some publications, the V.S.W.R. is defined as the ratio of the minimum to maximum voltages, giving $S \le 1$. In comparing (1.43) and (1.48), it is seen that when Γ_L is real and positive at a voltage maximum, then $z_L = S$. Thus the input impedance at a voltage maximum on a transmission line is

$$Z_{in} = S Z_0 \qquad\qquad (1.49)$$

Example 1.2

A $75\,\Omega$ characteristic impedance line is terminated with a load that has an impedance of $(68 - j\,12)\,\Omega$. Calculate

i)   the voltage reflection coefficient at the load,
ii) the V.S.W.R. along the line,
iii) the position of the first voltage minimum from the load,
iv) the impedance at the plane of the voltage minimum.

Solution:

i)   From (1.44), the reflection coefficient at the load is given by

$$\Gamma_L = \frac{(68 - j\,12) - 75}{(68 - j\,12) + 75} = \frac{13.89\,/-120.3°}{143.5\,/-4.8°}$$

i.e. $\Gamma_L = 0.097\,/-115.5°$

ii) From (1.48), the V.S.W.R.

$$S = \frac{1 + 0.097}{1 - 0.097} = 1.215$$

iii) The magnitudes of the forward and reverse traveling waves remain constant along a lossless transmission line. At a voltage minimum the reflection coefficient is real and negative, i.e. $/\pm 180°$. If l is the distance of the voltage minimum plane from the load, then Γ_{in} from (1.46) gives the reflection coefficient at that plane, Γ.

Thus $\Gamma = \Gamma_L e^{-j2\beta l}$

or $0.097\,/\pm 180° = 0.097\,/-115.5° \times e^{-j2\beta l}$

giving $-2\beta l = \dfrac{2\pi}{360}(-180 + 115.5) = -1.126$

Substituting $\beta = 2\pi/\lambda$, then

$$l = \frac{1.126\,\lambda}{4\pi} \approx 0.09\,\lambda$$

iv) At the voltage minimum, from (1.43)

$$Z = Z_0 \frac{1 + \Gamma}{1 - \Gamma} \quad \text{with } \Gamma = -0.097$$

giving $Z = 61.7\,\Omega$

1.3.3 Related parameters

The effects of a given voltage reflection coefficient on a transmission line may be interpreted in several ways. Taking a linear scale of $|\Gamma|$ as the fundamental quantity, Figure 1.5 illustrates its relationship to other quantities.

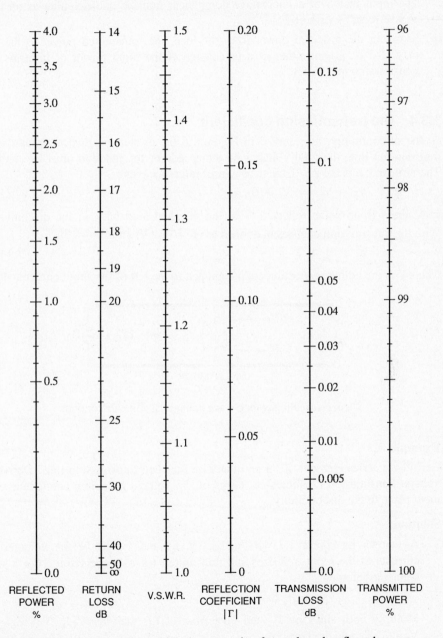

Figure 1.5 The various relationships between the forward and reflected waves on a transmission line, given by drawing an appropriate horizontal line on the chart

i)   The reflection coefficient is associated with a standing wave created by the forward and reverse traveling voltage waves as seen in (1.48).

ii) The reflection coefficient is associated with power being reflected back from the load. In particular, this reflected power may be expressed as a return loss in dB. The input match of a microwave component may be quoted either as an input V.S.W.R. or as a return loss.

iii) Knowing the reflected power from the load, the transmitted power to the load may also be given, either as a percentage of the input power or in terms of a transmission loss in dB.

1.3.4   The transmission coefficient

At the discontinuity, represented in Figure 1.6 by an element in parallel across the transmission line, the total voltages on either side of the junction must be identical. Therefore at the plane $z = 0$, taken as a local reference plane

$$V_f + V_r = V_t + 0 \tag{1.50}$$

since there is no wave reflected from the matched load back to the discontinuity. With the transmission coefficient defined as $T = V_t/V_f$, it follows that

$$T = 1 + \Gamma \tag{1.51}$$

where Γ is the voltage reflection coefficient just to the left of the shunt component.

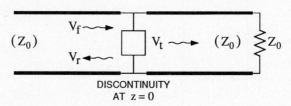

DISCONTINUITY
AT $z = 0$

Figure 1.6   Propagating wave voltages at a line discontinuity

Example 1.3

Consider a series element, Z, in an otherwise matched transmission line. Derive the voltage transmission coefficient in terms of the voltage reflection coefficient at the input plane to the discontinuity.

Solution:

As shown in Figure 1.7, let (V_f, I_f), (V_r, I_r) and (V_t, I_t) be the voltages and currents of the input, reflected and transmitted waves respectively. For a series connected element

$$I_t = I_f + I_r$$

Multiply both sides by Z_0 and note from (1.13) that $V_t = Z_0 I_t$, $V_f = Z_0 I_f$ and $V_r = -Z_0 I_r$, which gives

$$V_t = V_f - V_r$$

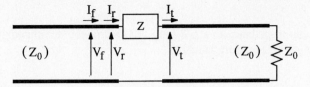

Figure 1.7 A series impedance in a matched transmission line

and
$$\frac{V_t}{V_f} = 1 - \frac{V_r}{V_f}$$

Thus
$$T = 1 - \Gamma \quad \text{for a series-connected element.}$$

1.4 THE INPUT IMPEDANCE

The input impedance Z_{in} for a lossless transmission line depends on three parameters: the characteristic impedance of the line Z_0, the electrical length of the line βl, and the load impedance Z_L. A knowledge of the variation of the voltage reflection coefficient along a transmission line is used to derive the input impedance in three basic steps. They are:

i) the reflection coefficient at the load is given in terms of the load impedance, using (1.44),

ii) the reflection coefficient at the input is given in terms of the reflection coefficient at the load, using (1.46), and

iii) the input impedance is given in terms of the reflection coefficient at the input. The general relationship between the impedance at any plane along a transmission line is related to the reflection coefficient at that plane, in the same way as occurs at the load plane in (1.43).

Commencing at the input plane with step (iii) and progressing towards the load through steps (ii) and (i), the following expressions are obtained

$$Z_{in} = Z_0 \left\{ \frac{1 + \Gamma_{in}}{1 - \Gamma_{in}} \right\} = Z_0 \left\{ \frac{1 + \Gamma_L e^{-j2\beta l}}{1 - \Gamma_L e^{-j2\beta l}} \right\} \tag{1.52}$$

$$= Z_0 \left\{ \frac{(Z_L + Z_0)e^{j\beta l} + (Z_L - Z_0)e^{-j\beta l}}{(Z_L + Z_0)e^{j\beta l} - (Z_L - Z_0)e^{-j\beta l}} \right\} \tag{1.53}$$

giving
$$Z_{in} = Z_0 \left\{ \frac{Z_L \cos(\beta l) + j Z_0 \sin(\beta l)}{Z_0 \cos(\beta l) + j Z_L \sin(\beta l)} \right\} \tag{1.54}$$

For admittances, take the reciprocal of each side of (1.54), rewrite each impedance with the appropriate $Z = 1/Y$ and rearrange to give

$$Y_{in} = Y_0 \left\{ \frac{Y_L \cos(\beta l) + j Y_0 \sin(\beta l)}{Y_0 \cos(\beta l) + j Y_L \sin(\beta l)} \right\} \tag{1.55}$$

Example 1.4

Calculate the input impedance to a $\lambda/8$ long 50Ω characteristic impedance line that is terminated with a 100Ω load impedance.

Solution:

For a $\lambda/8$ line, $\beta l = \dfrac{2\pi}{\lambda} \times \dfrac{\lambda}{8} = \dfrac{\pi}{4}$, giving $\cos(\beta l) = \sin(\beta l)$. Thus, from (1.54) for this particular case

$$Z_{in} = Z_0 \left[\frac{Z_L + jZ_0}{Z_0 + jZ_L} \right]$$

$$= 50 \times \frac{100 + j50}{50 + j100}$$

giving $Z_{in} = 50\underline{/-36.87°} \equiv (40 - j30)\,\Omega$

EXERCISES

1.1   A lossless air-spaced transmission line has uniform, but unusual, conductor shapes and a capacitance of 100pF.m^{-1}. What is the characteristic impedance of the transmission line?

1.2   Derive equations that relate the following terms to the reflection coefficient magnitude, $|\Gamma|$:
   i)   V.S.W.R.,
   ii)   percentage power reflected,
   iii) return loss, in dB,
   iv)   percentage power transmitted,
   v)   transmission loss, in dB.
   Evaluate each of these quantities for $|\Gamma| = 0.01, 0.1, 0.3$ and 1.0.

1.3   i)   Calculate the input V.S.W.R. at the input of a perfect 10dB attenuator, when its output is connected to a line terminated in a short circuit.
   ii)   If the attenuator is not perfect but has an input V.S.W.R. of 1.05 when it is terminated by a matched load, what is the possible range of input V.S.W.R. when the attenuator is terminated by a variable position short circuit?

1.4   A transmission line termination in the form of a non-ideal short circuit gives a V.S.W.R. of 200. What is the equivalent power loss compared with a perfect short circuit? If there is an additional 0.1dB distributed attenuation between the short-circuit termination and the measurement plane, what will be the measured V.S.W.R.?

1.5   From (1.54), deduce Z_{in} for the following special cases of terminated transmission lines:
   i)   Find Z_{in}, if $Z_L = Z_0$ for all values of l.
   ii)   Find Z_{in}, if $Z_L = \infty$ for $l = \lambda/4$ and $\lambda/2$.
   iii) Find Z_{in}, if $Z_L = 0$ for $l = \lambda/4$ and $\lambda/2$.
   iv)   Find Z_{in}, if $Z_L = 0$ and $l = \lambda/8$. Is the input impedance capacitive or inductive in this case?
   v)   Find Z_{in}, if Z_L is real and $l = \lambda/4$.

1.6 i) Write individual equations for the three steps of §1.4, to include the effects of losses along the line.

ii) Repeat Example 1.4, using the same parameters but also including a loss of 3.0dB per wavelength along the line.

iii) For the general case where $\gamma = \alpha + j\beta$, deduce the transmission line equation that will allow for the losses along the line.

1.7 A lossless air-spaced coaxial line has a 50Ω characteristic impedance and is terminated with a load impedance of $(100 + j100)\ \Omega$. Calculate the voltage reflection coefficient at the load, the V.S.W.R. on the line and the position of the voltage minimum nearest the load at a frequency of 1.0GHz.

1.8 Calculate the input impedance of a transmission line 0.3λ long with a characteristic impedance of 75Ω and a load impedance $Z_L = 37.5 + j52.5\,\Omega$. Consider the two cases with the attenuation taken as (i) zero and (ii) $3.83\,\mathrm{dB}.\lambda^{-1}$.

1.9 Consider an arbitrary length of line with characteristic impedance Z_0. On the line, a load Z_{L1} with a reflection coefficient Γ_{L1} gives an input impedance Z_{in1}. Likewise $Z_{L2} \rightarrow \Gamma_{L2} \rightarrow Z_{in2}$. If $Z_{L1} Z_{L2} = Z_0^2$, prove that $\Gamma_{L1} = -\Gamma_{L2}$ and $Z_{in1} Z_{in2} = Z_0^2$.

1.10 V_1, I_1 and V_2, I_2 are the total voltages and currents at the input and output respectively of a lossless $\lambda/4$ line with characteristic impedance Z_0.

i) Show that for all source and load combinations, $V_1 = 0 \implies I_2 = 0$.

ii) More generally, show that

$$\left| \frac{V_1}{I_2} \right| = Z_0$$

2 Two-port parameters

2.1 THE SCATTERING MATRIX

In high frequency work, two-port networks are best characterized in terms of scattering parameters, rather than in terms of the admittance or hybrid parameter representations that interrelate the actual port voltages and currents in a simple manner and are used at lower frequencies. Scattering parameters are defined in terms of traveling waves, which are the natural variables to be used in a transmission line environment, and are popular because:

i) Matched loads are used in their determination. At microwave frequencies matched loads are relatively easy to realize, while the short and open circuits required for the traditional low frequency parameters are much more difficult to achieve and, furthermore, are more likely to make an active device unstable.

ii) When only the magnitudes of the scattering parameters are required, it is not necessary to be concerned with the position of the reference planes, that is, the planes at which the device under test begins and the connecting test network ends. The position of the reference planes only affects the phase of the scattering parameters.

For these reasons and the fact that instruments to achieve the required measurements are readily available, scattering parameters are used almost exclusively at microwave frequencies to characterize both active and passive networks. In this section, scattering parameters are defined and their evaluation is described. The scattering parameters are introduced via traveling waves in transmission lines connected to the network. The definition is then extended in §2.1.3 to include the more general situation, where transmission lines do not necessarily exist. An active source is described in traveling wave terms in §2.1.4, leading to the evaluation of the available power of a source in §2.1.5.

2.1.1 Traveling waves and scattering parameters

Consider a two-port network with transmission lines connected to it as illustrated in Figure 2.1. The traveling waves in these transmission lines will be the variables that are used to characterize the two-port network.

For a linear two-port network, there are linear relations among the incident and reflected wave variables that may be expressed in matrix form.

$$\begin{bmatrix} e_{r1} \\ e_{r2} \end{bmatrix} = \begin{bmatrix} s_{11} & s_{12} \\ s_{21} & s_{22} \end{bmatrix} \begin{bmatrix} e_{i1} \\ e_{i2} \end{bmatrix} \qquad (2.1)$$

Figure 2.1 Incident and reflected waves at the input and output of a two-port network

The resulting matrix is known as the scattering matrix and the elements as the scattering parameters of the two-port network. Using an alternative notation, the equation may be written as

$$\begin{bmatrix} e_{r1} \\ e_{r2} \end{bmatrix} = \begin{bmatrix} s_i & s_r \\ s_f & s_o \end{bmatrix} \begin{bmatrix} e_{i1} \\ e_{i2} \end{bmatrix} \tag{2.2}$$

In (2.1) and (2.2) the following four points should be noted as they are important in the understanding of scattering parameters:

i)   The convention is to take the waves traveling *towards* each port of the two-port network as the incident waves.

ii) A traveling wave, whether incident or reflected, has both voltage and current components, as illustrated in Figure 2.2. For the incident and reflected waves

$$\frac{e_{i1}}{i_{i1}} = \frac{e_{r1}}{i_{r1}} = Z_0 \tag{2.3}$$

where Z_0 is the characteristic impedance of the transmission line. Note that the currents are taken as positive in the direction of wave propagation. The currents i_{i1} etc., as traveling waves, may just as readily be used as the variables to obtain the identical scattering parameters. In fact, the variables that are often used are the a's and b's defined as

$$a_1 = \frac{e_{i1}}{\sqrt{Z_0}} \quad \left\{ = i_{i1}\sqrt{Z_0} \right\}$$

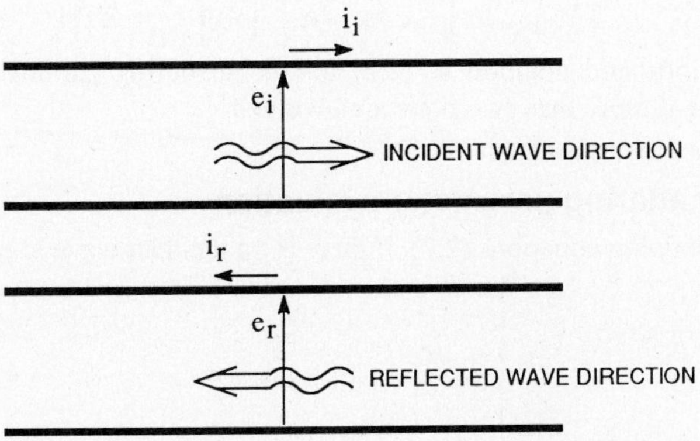

i_i

e_i

INCIDENT WAVE DIRECTION

i_r

e_r

REFLECTED WAVE DIRECTION

Figure 2.2 The voltage and current components of traveling waves

$$a_2 = \frac{e_{i2}}{\sqrt{Z_0}} \left\{ = i_{i2}\sqrt{Z_0} \right\}$$

$$b_1 = \frac{e_{r1}}{\sqrt{Z_0}} \left\{ = i_{r1}\sqrt{Z_0} \right\}$$

and $\qquad b_2 = \frac{e_{r2}}{\sqrt{Z_0}} \left\{ = i_{r2}\sqrt{Z_0} \right\}$

$$(2.4)$$

Thus, in terms of the scattering parameters

$$\begin{bmatrix} b_1 \\ b_2 \end{bmatrix} = \begin{bmatrix} s_i & s_r \\ s_f & s_o \end{bmatrix} \begin{bmatrix} a_1 \\ a_2 \end{bmatrix}$$

$$(2.5)$$

Choosing these variables is convenient, because the expressions for power flow are simplified. For example, in the incident direction

$$\text{power flow} = |a|^2 - |b|^2$$

$$= \frac{|e_i|^2}{Z_0} - \frac{|e_r|^2}{Z_0}$$

$$(2.6)$$

The variables a and b are called "power waves".

iii) Even though the concept of traveling waves is considered, it must be remembered that they are waves at some particular point, e.g. waves at the input plane. In order to evaluate the wave variables at other planes along the connecting transmission lines, the variables are multiplied by phase factors.

iv) Even though the discussion has been in terms of a two-port network, scattering parameters can readily be defined for networks with more than two ports; see Exercises 2.1(vii), 2.2 and 2.3. For example, for a three-port network

$$\begin{bmatrix} b_1 \\ b_2 \\ b_3 \end{bmatrix} = \begin{bmatrix} s_{11} & s_{12} & s_{13} \\ s_{21} & s_{22} & s_{23} \\ s_{31} & s_{32} & s_{33} \end{bmatrix} \begin{bmatrix} a_1 \\ a_2 \\ a_3 \end{bmatrix}$$

$$(2.7)$$

The short-hand notation of (2.2) for the scattering parameter subscripts is not suitable if more than two ports are involved.

2.1.2 Scattering parameter evaluation

From the defining equations (2.2), if there is no incident wave at port 2, then

$$s_i = \left. \frac{e_{r1}}{e_{i1}} \right|_{e_{i2}=0}$$

and $\qquad s_f = \left. \frac{e_{r2}}{e_{i1}} \right|_{e_{i2}=0}$

$$(2.8)$$

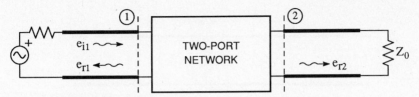

Figure 2.3 The traveling waves when the ouput is matched

Now, $e_{i2} = 0$ is ensured by having a matched load at the output shown in Figure 2.3, where all the connecting transmission lines have the characteristic impedance Z_0.

Thus $\quad\quad s_i$ = the input reflection coefficient with the output matched
and $\quad\quad s_f$ = the forward transmission coefficient with the output matched.

Similarly, to obtain s_r and s_o, the input and output ports are reversed for test purposes to give

$\quad\quad\quad\quad s_o$ = the output reflection coefficient with the input matched
and $\quad\quad s_r$ = the reverse transmission coefficient with the input matched.

Example 2.1

Calculate the scattering parameters of each of the following two-port networks, all normalized to Z_0, as illustrated in Figure 2.4. For each case, it may thus be assumed that the two-port networks are connected at both the input and output to transmission lines of characteristic impedance Z_0.

Solution:

Circuit (a)

In the case of circuit (a), the phase change due to the propagation delay through the 0.6λ length of transmission line with no reflections for the input or output

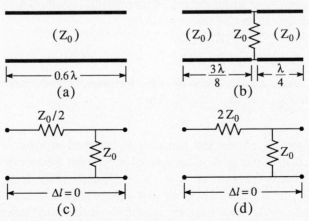

Figure 2.4 Two-port networks, the scattering parameters of which are to be evaluated

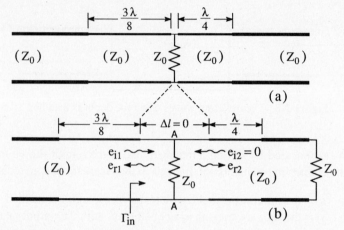

Figure 2.5 Evaluating the scattering parameters of Figure 2.4b. (a) Figure 2.4b embedded into two transmission lines of characteristic impedance Z_0. (b) Evaluating the scattering parameters of the central part of Figure 2.5a.

ports, i.e. connected to Z_0 lines, is 1.2π radians. This is represented by $e^{-j1.2\pi}$ in the scattering matrix

$$\begin{bmatrix} 0 & e^{-j1.2\pi} \\ e^{-j1.2\pi} & 0 \end{bmatrix}$$

This case is so simple that a formal solution is not necessary.

Circuit (b)

To calculate the scattering parameters for circuit (b), connect transmission lines of characteristic impedance Z_0 to it as in Figure 2.5a. First of all, calculate the scattering parameters of just the section containing the resistor Z_0 connected across the line, redrawn as a line of zero length in Figure 2.5b. For this section of zero length, s_i is given by the input reflection coefficient Γ_{in} with the output matched, i.e. Γ_{in} in this case is given by

$$\Gamma_{in} = \frac{Z_{in} - Z_0}{Z_{in} + Z_0} \tag{2.9}$$

In this case, Z_{in} is given by the resistance, Z_0, in parallel with the load impedance Z_0, i.e. $Z_{in} = Z_0/2$. Thus

$$s_i = \Gamma_{in} = -\tfrac{1}{3}$$

To evaluate s_f, an incident wave e_{i1} is launched in Figure 2.5b and the total voltage that results at A-A, as a result of e_{i1} and the reflected wave e_{r1} being simultaneously present at A-A, is evaluated. This total voltage is thus $(1 + s_i)e_{i1} = (2/3)e_{i1}$. Looking at A-A from the output side, the total voltage at A-A is also equal to $e_{r2} + e_{i2}$, the sum of the incident and reflected waves at A-A' at the output side. Since $e_{i2} = 0$, there being a matched load on the output side, it follows that $e_{r2} = (2/3)e_{i1}$. Thus $s_f = 2/3$. By virtue of the symmetry of input to output, s_o and s_r in Figure 2.5b are identical to s_i and s_f respectively.

The scattering parameters of the zero length section in Figure 2.5b are thus

$$\begin{bmatrix} -\frac{1}{3} & \frac{2}{3} \\ \frac{2}{3} & -\frac{1}{3} \end{bmatrix}$$

To get the scattering parameters of the full circuit in Figure 2.5a, the scattering parameters for the resistor in a zero-length of line are modified by the addition of appropriate phase factors. The phase shift introduced by a $3\lambda/8$ length of line is $135°$. Comparing the e_{i1}'s and e_{r1}'s for the two-port networks in (a) and (b), the e_{i1} in (a) leads the e_{i1} in (b) by $135°$, while the e_{r1} in (a) lags the e_{r1} in (b) by $135°$, making the $s_i = -(1/3)e^{-j270°}$. Similarly, the additional phase shift introduced to s_f (and s_r) is $-(135° + 90°)$ and to s_o is $2 \times (-90°)$, making the scattering parameters of the complete two-port network in Figure 2.4b

$$\begin{bmatrix} -\frac{1}{3}e^{-j270°} & \frac{2}{3}e^{-j225°} \\ \frac{2}{3}e^{-j225°} & -\frac{1}{3}e^{-j180°} \end{bmatrix}$$

Circuit (c)

The scattering parameters of Figure 2.4c will now be evaluated. The electrical length of this two-port network is zero (i.e. $\Delta l = 0$). Again, connect transmission lines of characteristic impedance Z_0 to the two-port network and match the output line. Now, s_i is given by Γ_{in} in terms of Z_{in}, where Z_{in} is equal to $Z_0/2$ in series with another $Z_0/2$, the latter being due to the parallel combination of the Z_0

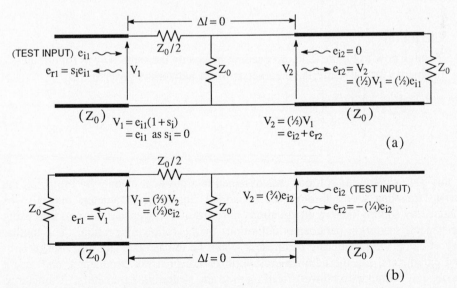

Figure 2.6 The circuits for evaluating the scattering parameters of Figure 2.4c, showing the circuit (a) for s_i and s_f, and (b) for s_r and s_o

across the line and the matched line. This makes $Z_{in} = Z_0$, leading to $s_i = 0$. Referring now to Figure 2.6, since $s_i = 0$, it can be seen that the total voltage at the input, V_1, becomes equal to e_{i1}. As $Z_0/2$ and $Z_0 \| Z_0$ form a potential divider network, this in turn makes the total voltage at the output

$$V_2 = \left[\frac{Z_0 \| Z_0}{Z_0/2 + Z_0 \| Z_0} \right] e_{i1} = \tfrac{1}{2} e_{i1}$$

The matched load condition ensures that $e_{i2} = 0$, making $e_{r2} = V_2$. Thus $s_f = e_{r2}/e_{i1}$ gives $s_f = 1/2$.

To determine s_o and s_r, apply an e_{i2} test incident wave to the output and terminate the input in a matched load as shown in Figure 2.6b. Thus the output has become the input for test purposes. Proceeding as before, the output impedance in Figure 2.6b is determined to be

$$Z_{out} = (Z_0) \| (Z_0 + Z_0/2) = 0.6 Z_0$$

giving $$s_o = \Gamma_{out} = \frac{Z_{out} - Z_0}{Z_{out} + Z_0} = -0.25$$

Thus, e_{i2} produces a total output voltage

$$V_2 = (1 + s_o) e_{i2} = 0.75 e_{i2}$$

With the potential divider network $Z_0/2$ and Z_0 at the input, the total voltage at the input $V_1 = (1/2) e_{i2}$. Since again $e_{r1} = V_1$, this makes $s_r = 1/2$. Hence the scattering matrix for Figure 2.4c is

$$\begin{bmatrix} 0 & \tfrac{1}{2} \\ \tfrac{1}{2} & -\tfrac{1}{4} \end{bmatrix}$$

Circuit (d)

Turning now to Figure 2.4d, proceeding in exactly the same way as for circuit (c), the matrix for the scattering parameters of the network is given as

$$\begin{bmatrix} \tfrac{3}{7} & \tfrac{2}{7} \\ \tfrac{2}{7} & -\tfrac{1}{7} \end{bmatrix}$$

In this last example the solution was obtained in a physical manner by following the various traveling waves through the networks. Scattering parameters may also be obtained by solving the circuit equations in terms of voltages and currents and using the general scattering parameters definitions of the following section. Yet another way of calculating scattering parameters would be to calculate any other set of two-port parameters, say the admittance, hybrid or transmission parameters, and to use conversion formulae to convert from them to the scattering parameters. Tabulations of such conversion formulae exist (e.g. Gonzalez [2.2]).

However, the approach that has been adopted in this example is recommended for the majority of cases, because it is quite simple and provides considerable physical insight. Other techniques of calculating scattering parameters are just mathematical manipulations, and do not provide such physical insight.

2.1.3 General definition of scattering parameters

As a starting point, the total voltage V_1 and the total current I_1 at the input are evaluated in terms of the traveling wave components at the input. The power wave variables, and thus the scattering parameters, are then expressed in terms of these port voltages and currents. In Figure 2.1, while only the voltage components of the traveling waves are labeled, it must be remembered that the waves have both voltage and current components, as illustrated in Figure 2.7. The voltage and current at the input plane are given by

$$V_1 = e_{i1} + e_{r1}$$

and

$$I_1 = i_{i1} - i_{r1} \tag{2.10}$$

with a similar pair of equations for the output plane. Solving for the traveling wave voltages and currents and using (2.4), it is found that the power wave variables are given by

$$a_1 = \frac{V_1 + Z_0 I_1}{2\sqrt{Z_0}}$$

$$b_1 = \frac{V_1 - Z_0 I_1}{2\sqrt{Z_0}}$$

$$a_2 = \frac{V_2 + Z_0 I_2}{2\sqrt{Z_0}}$$

$$b_2 = \frac{V_2 - Z_0 I_2}{2\sqrt{Z_0}} \tag{2.11}$$

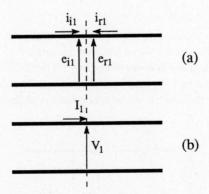

Figure 2.7 (a) The traveling wave components at a point on a line and (b) the total voltage and current at that point

These equations may be regarded purely as a transformation of variables, i.e. given the total voltages, V, and currents, I, the a and b variables are obtained and used to define the scattering parameters using (2.5). This may be done even at d.c., but then the interpretation in terms of traveling waves is not as immediately obvious, though such an interpretation is still possible. It is important to note that this transformation of variables involves a normalizing impedance Z_0. In principle, this impedance can be any impedance, but a physical interpretation will only be possible for certain selected impedances. In general, Z_0 may be complex, in which case the definition of the power wave variables is as follows

$$a_1 = \frac{V_1 + Z_0 I_1}{2\sqrt{Re\,Z_0}}$$

$$b_1 = \frac{V_1 - Z_0^* I_1}{2\sqrt{Re\,Z_0}}$$

$$a_2 = \frac{V_2 + Z_0 I_2}{2\sqrt{Re\,Z_0}}$$

$$b_2 = \frac{V_2 - Z_0^* I_2}{2\sqrt{Re\,Z_0}} \tag{2.12}$$

There is no requirement in (2.12) that the normalizing impedances must be the same at both the input and output.

To obtain a physical interpretation in the general case, the following result, to be proved as Exercise 2.6, is used. If V and I are the voltage and current at the *terminals* of a source with an internal impedance Z_S, as illustrated in Figure 2.8, and

$$a = \frac{V + Z_S I}{2\sqrt{Re\,Z_S}}$$

$$b = \frac{V - Z_S^* I}{2\sqrt{Re\,Z_S}} \tag{2.13}$$

then $$Re(VI^*) = |a|^2 - |b|^2 \tag{2.14}$$

$Re(VI^*)$ is the power delivered by the source terminals. This power becomes the available power from the source when the load connected to the terminals is equal to Z_S^*, in which case $b = 0$. Thus it is seen that $|a|^2$ is the available power of the source. $|a|^2$ is the power launched by the source towards the load and, if the load is conjugately matched to the source, all of this power is absorbed and none is reflected.

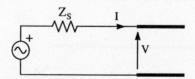

Figure 2.8   An active source

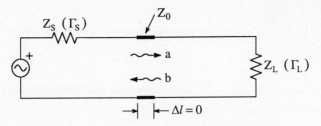

Figure 2.9 The traveling waves in a line of zero length

For other loads, some power, i.e. $|b|^2$, is reflected back and the net power absorbed by the load is $|a|^2 - |b|^2$.

To summarize, the a's and b's are defined in terms of terminal voltages and currents at the input and output with the scattering parameters being defined in terms of the a's and b's. A physical meaning to the a's and b's may be attached when the normalizing impedance Z_0 is chosen in specific ways:

i)   When Z_0 is chosen to be the characteristic impedance of a transmission line, then the a's and b's are the incident and reflected waves in the transmission lines connected to the two-port network.

ii) When Z_0 is chosen to be the impedance of a source, then the a's and b's may be interpreted as power waves in the manner discussed above.

iii) When Z_0 is some other impedance, the a's and b's still have meaning, if a transmission line of infinitesimal length and characteristic impedance Z_0 is thought of as connecting the source to the two-port network. The a's and b's are then the traveling waves in this infinitesimally short length of line, as shown in Figure 2.9.

This concept of waves in a line of zero length will be very useful on a number of occasions in this and following chapters.

2.1.4 Active source representation

Take an active source with an open-circuit voltage V_S and internal impedance Z_S connected to a line of characteristic impedance Z_0 and producing at its terminals the voltage V and the current I. The a and b waves that result in the line are a function of the parameters of the source and also of the termination at the other end of the line. The source and the load each provide a linear equation connecting a and b. The solution of these two equations gives the a,b that result.

The relation between the a and b waves imposed by the parameters of the source is now obtained (i.e. the source is characterized in traveling wave terms). Referring to Figure 2.10, solve for a and b as before and further impose the condition $V_S = V + Z_S I$. This yields

$$a = a_S + \Gamma_S b \qquad\qquad (2.15)$$

where $\qquad a_S = \dfrac{e_S}{\sqrt{Z_0}}, \qquad e_S = V_S \dfrac{Z_0}{Z_0 + Z_S} \qquad\qquad (2.16)$

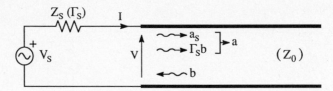

Figure 2.10   The traveling waves at the terminals of an active source

and Γ_S is the reflection coefficient of Z_S with respect to a line of characteristic impedance Z_0.

The wave a_S is the wave that is launched into a line of characteristic impedance Z_0 by a source V_S, Z_S. The wave a_S is the only wave that travels in the line if the line is matched at the other end. If there is a mismatch at the load end of the line, then the wave b is reflected back towards the source. In turn, b is reflected at the source to produce the wave $\Gamma_S b$. The wave $\Gamma_S b$ combines with a_S to give the resultant wave "a". It is important to note that the (sinusoidal) steady state situation is being considered and the traveling waves that are being referred to are waves that have resulted in the steady state. Transient conditions are not being dealt with here. The reflected wave b is produced by the total wave "a" and not just a_S.

2.1.5 Available power

Take an active source Z_S, characterized alternatively by a reflection coefficient Γ_S, connected to a load Z_L (Γ_L). Imagine the connection between the two as consisting of a line of infinitesimal length Δl and characteristic impedance Z_0 and supporting the traveling waves a,b as illustrated in Figure 2.9. The source will deliver its available power, P_{av}, when $Z_S = Z_L^*$ (or equivalently when $\Gamma_S = \Gamma_L^*$).

Now
$$a = a_S + \Gamma_S b \tag{2.17}$$

and
$$b = \Gamma_L a \tag{2.18}$$

which gives
$$a = \frac{a_S}{1 - \Gamma_S \Gamma_L} \tag{2.19}$$

and   the power into load
$$= |a|^2 - |b|^2 \tag{2.20}$$

$$= |a_S|^2 \left\{ \frac{1 - |\Gamma_L|^2}{|1 - \Gamma_S \Gamma_L|^2} \right\} \tag{2.21}$$

Setting $\Gamma_S = \Gamma_L^*$ together with $\Delta l = 0$ gives
$$P_{av} = \frac{|a_S|^2}{1 - |\Gamma_S|^2} \tag{2.22}$$

Note
The word "matched" may be used in two different senses: (a) matching for no reflection, or (b) conjugate matching for maximum power transfer. Sometimes

matching in both senses is accomplished simultaneously. In general, however, one or the other but not both may be achieved.

Further insight on this point is provided by the following result that is to be proved as Exercise 11.1. For a *lossless reciprocal* two-port network

$$\Gamma_S = \Gamma_{in}^* \iff \Gamma_L = \Gamma_{out}^* \tag{2.23}$$

This result shows that if, say, one has a source with an internal impedance Z_0, that is to be conjugately matched to some load Z_L then, if a lossless network that is inserted between Z_0 and Z_L achieves conjugate matching at the Z_L end, conjugate matching will also result at the Z_0 end. Further, since Z_0 is real, this means that the source sees Z_0 and the input is matched for no reflections as well. One thus can achieve matching for no reflections and conjugate matching at the same time if a lossless network is used for the matching.

2.1.6 Some comments and a few useful results

i) The reader will have noted that the scattering parameters of the circuits in Figures 2.4c and 2.4d, as worked out in Example 2.1, give a symmetrical scattering matrix with $s_f = s_r$, even though the two circuits are decidedly non-symmetrical. This is because both these circuits are *reciprocal*, i.e. circuits that obey the reciprocity theorem. Any circuit consisting of resistors, capacitors, inductors, transformers and lengths of transmission line will be reciprocal. Any circuit that happens to be symmetrical, whether reciprocal or not, will of course have a symmetrical matrix, but the converse does not hold.

On the other hand, an active network such as a transistor will not be reciprocal. In fact, for a transistor it is desirable for s_f to be as large as possible and for s_r to be as small as possible. A two-port network for which $s_r = 0$ is known as a *unilateral* two-port. Thus a unilateral two-port network is one that works only in one direction, while a reciprocal network is one that, in a sense (and in a sense only), acts equally well in both directions. Unilateral and reciprocal two-port networks may be regarded as the extreme limits between which the action of most active two-port networks lies.

Sometimes deliberate steps are taken to unilateralize a transistor, i.e. making $s_r = 0$, for stability reasons. Unfortunately, in general this can only be achieved at a spot frequency and unilateralization over a broad band is not possible. One then settles for the next best thing, namely the minimization of s_r. This is known as *neutralization*. Neutralization can thus be regarded as imperfect unilateralization. Finally, one may just simply assume $s_r = 0$ for design purposes. More will be said about this *unilateral approximation* in Chapter 11.

ii) [S] is a *unitary* matrix for a lossless network, as shown in Exercise 2.5.

iii) Input and output reflection coefficients of a two-port network, see Exercise 2.4(i), are given by

$$\Gamma_{in} = s_i - \frac{s_r s_f}{s_o - \dfrac{1}{\Gamma_L}}$$

and $\qquad \Gamma_{\text{out}} = s_o - \dfrac{s_r\, s_f}{s_i - \dfrac{1}{\Gamma_s}}$

$\qquad\qquad\qquad\qquad\qquad\qquad\qquad\qquad\qquad\qquad$ (2.24)

Thus $\Gamma_{\text{in}} = f(\Gamma_L)$ and $\Gamma_{\text{out}} = f(\Gamma_s)$, unless the network is unilateral, in which latter case Γ_{in} simply equals s_i and Γ_{out} equals s_o, irrespective of Γ_L and Γ_s. This result leads to a great deal of simplification in Chapter 11, when the unilateral approximation is used in maximizing the power gain.

2.2 TRANSMISSION (ABCD) AND y-PARAMETERS

Although scattering parameters are undoubtedly the most useful and the most commonly used parameters for characterizing a two-port network constructed with microstrip lines, they do not always present the simplest way of dealing with certain problems. Occasionally a more elegant description results when other parameters are used. The authors do not believe in slavishly sticking to the exclusive use of scattering parameters purely for the sake of consistency, when other parameters may give answers more succinctly.

In this section, transmission and y-parameters are revised on the assumption that the reader is acquainted with them. Transmission parameters are also used here, with the aid of examples, to derive useful equivalent circuits for transmission lines.

2.2.1 Transmission parameters

The transmission parameters are most useful when two-port networks are cascaded. Multiplying the matrices of the individual two-port networks (in the correct order) simply gives the transmission matrix for the combination.

Given the (total) voltages and currents of the two-port network in Figure 2.11, the transmission or chain matrix $[\,T\,]$ is given by

$$\begin{bmatrix} V_1 \\ I_1 \end{bmatrix} = \begin{bmatrix} T \end{bmatrix} \begin{bmatrix} V_2 \\ -I_2 \end{bmatrix}$$

$\qquad\qquad\qquad\qquad\qquad\qquad\qquad\qquad\qquad\qquad$ (2.25)

or in terms of the individual matrix elements

$$\begin{bmatrix} V_1 \\ I_1 \end{bmatrix} = \begin{bmatrix} A & B \\ C & D \end{bmatrix} \begin{bmatrix} V_2 \\ -I_2 \end{bmatrix}$$

$\qquad\qquad\qquad\qquad\qquad\qquad\qquad\qquad\qquad\qquad$ (2.26)

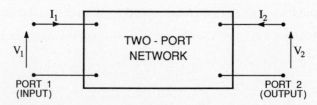

Figure 2.11 The voltages and currents for a two-port network

A, B, C and D are often used as the symbols for the individual matrix elements and thus the transmission matrix is also commonly known as the ABCD-matrix and the parameters as the ABCD-parameters.

It is important to note that the current variable on the port 2 (output) side is taken as $-I_2$. This is so as to make the output current in the same direction as the input current of the next stage in a cascade, thus making the matrix $[T]$ of a cascade, $[T_1]$ followed by $[T_2]$

$$\left[T\right] = \left[T_1\right]\left[T_2\right] \tag{2.27}$$

To determine individual A, B, C, D parameters, either V_2 or I_2 is set to zero in the appropriate defining equation to eliminate the parameter that is not required. Thus

$$A = \left.\frac{V_1}{V_2}\right|_{I_2=0} \qquad B = \left.\frac{V_1}{-I_2}\right|_{V_2=0}$$

$$C = \left.\frac{I_1}{V_2}\right|_{I_2=0} \qquad D = \left.\frac{I_1}{-I_2}\right|_{V_2=0} \tag{2.28}$$

Example 2.2

Derive the ABCD-parameters of a lossless transmission line of length l.

Solution:

The voltages and currents at any point z on a transmission line are given by (1.37) and (1.38) and are reproduced here for $\gamma = j\beta$.

$$V = V_f e^{-j\beta z} + V_r e^{j\beta z} \tag{2.29}$$

and $$I = I_f e^{-j\beta z} + I_r e^{j\beta z} \tag{2.30}$$

The resultant voltages and currents at the input and output ends ($z = -l$ and $z = 0$ respectively) are illustrated in Figure 2.12.

With the output open-circuited

$$I_f - I_r = -I_2 = 0$$

making $$I_f = I_r, \quad V_f = V_r$$

$$V_2 = V_f + V_r = 2V_f$$

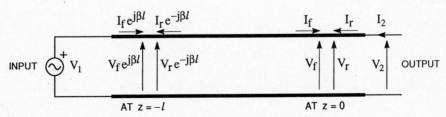

Figure 2.12 The forward and reverse voltages and currents at the two ends of a transmission line of length l

and $\qquad V_1 = V_f e^{j\beta l} + V_r e^{-j\beta l} = 2 V_f \cos\beta l$

Thus $\qquad A = \left. \dfrac{V_1}{V_2} \right|_{I_2 = 0}$

$\qquad\qquad = \cos\beta l$

Now with the output short-circuited

$$V_f + V_r = V_2 = 0$$

making $\qquad V_f = -V_r, \quad I_f = -I_r$

$$-I_2 = I_f - I_r = 2 I_f$$

and $\qquad V_1 = V_f(e^{j\beta l} - e^{-j\beta l}) = 2 j V_f \sin\beta l$

Thus $\qquad B = \left. \dfrac{V_1}{-I_2} \right|_{V_2 = 0}$

$\qquad\qquad = j \dfrac{V_f}{I_f} \sin\beta l = j Z_0 \sin\beta l$

C and D may be determined in a similar manner, but in this example it suffices to note that the two-port network is reciprocal and symmetrical so that $A = D$ and $AD - BC = 1$ (see Exercise 2.8), thus making $C = j Y_0 \sin\beta l$. The complete ABCD-matrix is thus

$$\begin{bmatrix} A & B \\ C & D \end{bmatrix} = \begin{bmatrix} \cos\beta l & j Z_0 \sin\beta l \\ j Y_0 \sin\beta l & \cos\beta l \end{bmatrix} \qquad (2.31)$$

Example 2.3

Derive expressions for the scattering parameters of a two-port network in terms of its ABCD-parameters.

Solution:

Referring to Figure 2.11, if there is a matched load at the output, $V_2 = Z_0(-I_2)$ or $(-I_2) = Y_0 V_2$.

Then $\qquad V_1 = A V_2 - B I_2 = (A + B Y_0) V_2$

and $\qquad I_1 = C V_2 - D I_2 = (C + D Y_0) V_2$

so that $\qquad a_1 = \dfrac{V_1 + Z_0 I_1}{2 \sqrt{Z_0}} = \dfrac{V_2}{2 \sqrt{Z_0}} (A + B Y_0 + C Z_0 + D)$

$\qquad\qquad b_1 = \dfrac{V_1 - Z_0 I_1}{2 \sqrt{Z_0}} = \dfrac{V_2}{2 \sqrt{Z_0}} (A + B Y_0 - C Z_0 - D)$

$$b_2 = \frac{V_2 - Z_0 I_2}{2\sqrt{Z_0}} = \frac{V_2}{\sqrt{Z_0}} \qquad (2.32)$$

and, of course, $a_2 = 0$.

Thus
$$s_i = \frac{b_1}{a_1}\bigg|_{a_2=0} = \frac{(A - D) + (BY_0 - CZ_0)}{(A + D) + (BY_0 + CZ_0)} \qquad (2.33)$$

and
$$s_f = \frac{b_2}{a_1}\bigg|_{a_2=0} = \frac{2}{(A + D) + (BY_0 + CZ_0)} \qquad (2.34)$$

s_0 and s_r may be derived from the expressions for s_i and s_f. All that is required is to treat port 2 as the input for test purposes and invert (2.26), changing the signs of I_1 and I_2 to give

$$\begin{bmatrix} V_2 \\ I_2 \end{bmatrix} = \frac{1}{\Delta} \begin{bmatrix} D & B \\ C & A \end{bmatrix} \begin{bmatrix} V_1 \\ -I_1 \end{bmatrix} \qquad (2.35)$$

where Δ is the determinant of the ABCD-matrix. Substituting the new transmission parameters for the appropriate ones in the previously derived expressions for s_i and s_f gives

$$s_0 = \frac{(D - A) + (BY_0 - CZ_0)}{(A + D) + (BY_0 + CZ_0)} \qquad (2.36)$$

and
$$s_r = \frac{2\Delta}{(A + D) + (BY_0 + CZ_0)} \qquad (2.37)$$

Example 2.4

Derive the Π- and T-equivalent circuits of a lossless transmission line.

Solution:

A symmetrical Π-equivalent circuit is shown in Figure 2.13, with the impedance X_a in the series arm and admittances Y_b in the parallel arms. If the output is short-circuited, then clearly

$$X_a = \frac{V_1}{-I_2}\bigg|_{V_2=0}$$

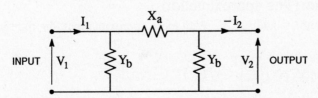

Figure 2.13 A symmetrical Π-equivalent circuit, drawn with an *impedance* jX_a and *admittances* jY_b

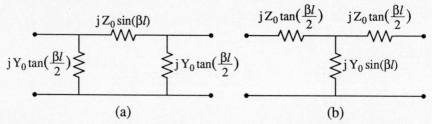

Figure 2.14 The Π- and T-equivalent circuits of a lossless transmission line (a) and (b) respectively. The values shown are *impedances* in the series arms and *admittances* in the parallel arms.

i.e. $X_a \; = \; B \; = \; j Z_0 \sin\beta l$

$$(2.38)$$

If the output is open-circuited, then the voltage division ratio

$$\frac{\left[\dfrac{1}{Y_b}\right]}{X_a + \left[\dfrac{1}{Y_b}\right]} \; = \; \left.\frac{V_2}{V_1}\right|_{I_2 = 0} \; = \; \frac{1}{A}$$

so that $1 + X_a Y_b \; = \; A$

and thus $Y_b \; = \; \dfrac{A - 1}{X_a}$

$$= \; \frac{\cos\beta l - 1}{j Z_0 \sin\beta l}$$

Since $1 - \cos\beta l = 2\sin^2(\frac{\beta l}{2})$ and $\sin\beta l = 2\sin(\frac{\beta l}{2})\cos(\frac{\beta l}{2})$

$$Y_b \; = \; j Y_0 \tan(\frac{\beta l}{2})$$

$$(2.39)$$

The T-equivalent circuit may be derived by proceeding in a similar manner. The Π- and T-equivalent circuits for the lossless transmission line are summarized in Figure 2.14.

2.2.2 The short line approximation

When $\beta l \ll 1$, $\sin\beta l \approx \tan\beta l \approx \beta l$. With these approximations, the right hand side of (2.38) becomes

$$j Z_0 \beta l \; = \; j\left\{\frac{L}{C}\right\}^{\frac{1}{2}}(\omega\sqrt{LC})\,l$$

$$= \; j\omega(L l)$$

$$(2.40)$$

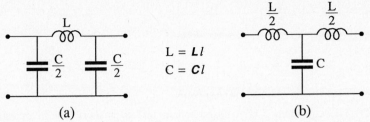

Figure 2.15 Equivalent circuits of a short length l of a lossless transmission line with (a) the Π form and (b) the T form

while similarly

$$j Y_0 \beta l = j\omega(Cl) \qquad (2.41)$$

so that the equivalent circuits in Figure 2.14 reduce to those in Figure 2.15, involving pure inductance L and pure capacitance C with

$$L = Ll \qquad (2.42)$$

and $$C = Cl \qquad (2.43)$$

L and C are, of course, the total series inductance and the total parallel capacitance in a line of length l.

The equivalent circuits of Figure 2.15 reduce further if $Z_0 \gg |Z_S|, |Z_L|$, as is the case for a high impedance line, or $Z_0 \ll |Z_S|, |Z_L|$ for a low impedance line. A high impedance line sees approximate short-circuit terminations, so that C can be ignored, yielding Figure 2.16a. Similarly, a low impedance line sees essentially open-circuit terminations, allowing L to be ignored and reducing the equivalent circuit to Figure 2.16b. Thus a short length of short-circuit terminated line behaves as an inductance, while a short length of open-circuit terminated line behaves as a capacitance.

If there are large standing waves on a line, then the approximations in Figures 2.16a and b are also valid at positions of zero or negligible total voltage and current respectively, for lines of any impedance. In circumstances when the approximations of Figure 2.16 are valid, they continue to be valid, even if a small load impedance Z_L loads the output in (a), or if a small load admittance Y_L loads the output in (b). That is, provided

$$|Z_L| \ll \omega Ll$$

or $$|Y_L| \ll \omega Cl \qquad (2.44)$$

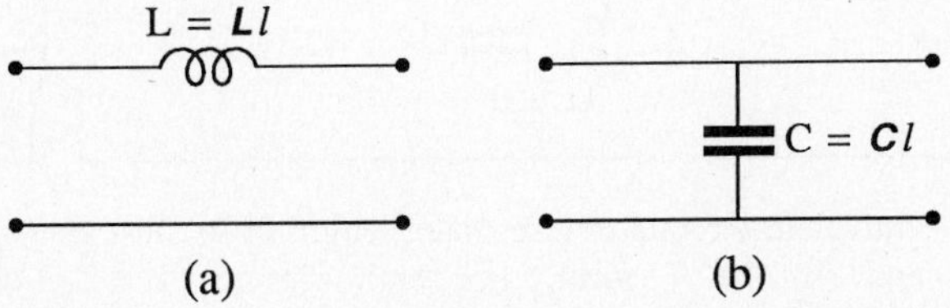

Figure 2.16 The equivalent circuit of a short length of (a) high impedance line and (b) low impedance line

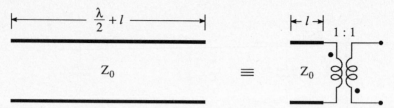

Figure 2.17 The equivalence of a near-resonant line to a short line in cascade with a phase-inverting transformer

then the immittance of the line plus load is

$$Z_{in} = j\omega L l + Z_L \qquad \text{in case (a)}$$

or $\qquad Y_{in} = j\omega C l + Y_L \qquad \text{in case (b)} \qquad (2.45)$

The short line approximations that have been developed in this section are important and will be used in numerous situations in subsequent chapters.

2.2.3 The resonant line approximation

When l is of the order of $\lambda/2$, one has a half-wave resonator. Unfortunately, the direct application of $l \approx \lambda/2$ to the circuits of Figure 2.14 does not yield a useful result. This is because the equivalent circuit has to produce a 180° phase shift from input to output when the line is $\lambda/2$ long. However, a useful result is obtained if the transformation of Figure 2.17 is used.

Figure 2.17 states that a line of length $\lambda/2 + l$ is equivalent to a line of length l in cascade with a $1:1$ phase-inverting ideal transformer, see Exercise 2.9. The equivalent circuit for a line with small l has already been derived and can now be used, with the proviso that l may be negative. When l goes negative, L becomes capacitive and C becomes inductive. Both these effects may be taken into account by

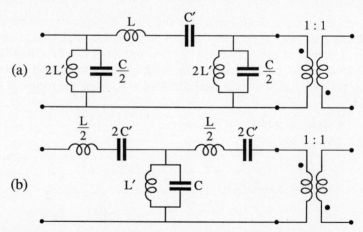

Figure 2.18 The equivalent circuits of a resonant length of lossless transmission line, with (a) the Π form and (b) the T form. For both circuits, $L = Z_0/(4f_0)$, $C = Y_0/(4f_0)$, $C' = (1/\omega_0^2 L)$ and $L' = (1/\omega_0^2 C)$, where ω_0 is the frequency for which the line is exactly $\lambda/2$ long.

replacing L with a series-tuned circuit and C by a parallel-tuned circuit, as in Figure 2.18. Equating the immittance slopes of $\omega(Ll)$ or $\omega(Cl)$, as the case may be, to the immittance slopes of the corresponding tuned circuits, gives the values quoted in Figure 2.18, after a little routine manipulation. It must also be remembered that l is now a function of frequency, namely $l \approx \pi v_{ph} \Delta\omega/\omega_0^2$ for small frequency excursion $\Delta\omega$ above the resonant frequency ω_0. Similarly, as for the short-length line, for a high characteristic impedance resonant line the series-tuned circuits may be ignored and for the low characteristic impedance resonant line the parallel-tuned circuits may be ignored.

2.2.4 Admittance (y) parameters

Referring again to Figure 2.11, the y-parameters are defined by

$$\begin{bmatrix} I_1 \\ I_2 \end{bmatrix} = \begin{bmatrix} y_i & y_r \\ y_f & y_o \end{bmatrix} \begin{bmatrix} V_1 \\ V_2 \end{bmatrix} \tag{2.46}$$

Any particular parameter may be determined from the defining equation with either V_1 or V_2 set to zero by short-circuiting the input or output respectively. Thus

$$y_i = \left. \frac{I_1}{V_1} \right|_{V_2=0} \qquad y_r = \left. \frac{I_1}{V_2} \right|_{V_1=0}$$

$$y_f = \left. \frac{I_2}{V_1} \right|_{V_2=0} \qquad y_o = \left. \frac{I_2}{V_2} \right|_{V_1=0} \tag{2.47}$$

The defining equations (2.46) may also be used to derive the equivalent circuit in Figure 2.19.

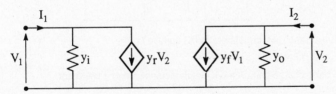

Figure 2.19 The y-parameter equivalent circuit of a two-port network

EXERCISES

2.1 Calculate the scattering parameters, all normalized to the characteristic impedance Z_0, for the following cases:

i) An ideal 10 dB attenuator that is matched to the characteristic impedance, Z_0.

ii) A length of transmission line with a characteristic impedance, Z_0

$$\longleftarrow \quad n\lambda \quad \longrightarrow$$

$$Z_0$$

and, in particular, when $n\lambda = \lambda/4$.

iii) An ideal n : 1 transformer

Check that a $\lambda/2$ line is equivalent to a 1 : 1 phase-inverting transformer.

iv) A series-connected impedance, Z, where
   a) $Z = Z_0$,
   b) $Z = j\omega L$. For this case, check that [S] is unitary.

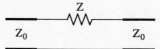

v) A length of transmission line with a different characteristic impedance for
   a) $n = \frac{1}{4}$,
   b) n in general.

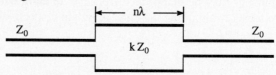

vi) A gyrator, where $V_1 = -\alpha I_2$ and $V_2 = +\alpha I_1$. Check that [S] is unitary.

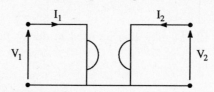

vii) An ideal three-port circulator. Is [S] unitary?

viii) An ideal isolator. Is [S] unitary?

ix) The basic charge-control model equivalent circuit of a transistor.

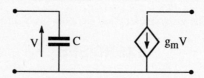

2.2 i)   Calculate the scattering parameters of the three three-port networks illustrated in (a), (b) and (c) below. The three-port networks consist of combinations of microstrip lines, of characteristic impedance Z_0 and length λ, and lumped resistors of specified magnitude. Normalize the scattering parameters to Z_0. Ignore the fringing field effects at the junctions.

    ii)   Compare these three circuits as power splitters on the basis of the results derived in part (i).

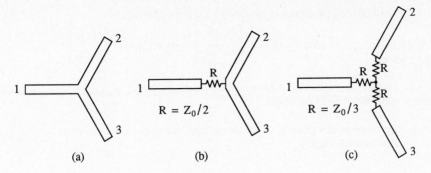

     (a)   (b)   (c)

2.3   Consider a 3 dB quadrature directional coupler that has two planes of symmetry: aa and bb.

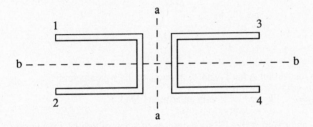

If ports 2, 3 and 4 are terminated in matched loads, an incident wave into port 1

a)   produces no reflection at port 1,

b)   delivers equal powers out of ports 2 and 3,

c)   port 3 output is in phase with port 1 input,

d)   port 2 output lags port 1 input by 90°,

e)   zero power is delivered out of port 4.

The circuit behavior when incident waves are applied to ports 2, 3 and 4 can be deduced from symmetry considerations.

i)   Calculate the scattering parameters of this directional coupler.

ii)   By permanently terminating port 4 in a matched load, the three-port network consisting of ports 1, 2 and 3 acts as a power divider/combiner. Calculate its scattering parameters.

iii)   The directional couplers described above are used to produce a balanced amplifier as illustrated below.

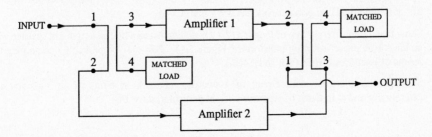

The scattering parameters of amplifiers 1 and 2 are related thus:

$$[S]_2 = [S]_1 + \begin{bmatrix} \delta s_i & \delta s_r \\ \delta s_f & \delta s_o \end{bmatrix} \quad \text{with} \quad [S]_1 = \begin{bmatrix} s_i & s_r \\ s_f & s_o \end{bmatrix}$$

Evaluate the scattering parameters of the balanced amplifier, i.e. between the terminals indicated as input and output.

2.4 i) Prove the results given in (2.24) for the input and output reflection coefficients of a two-port network, namely:

$$\Gamma_{in} = s_i - \frac{s_r s_f}{s_o - \dfrac{1}{\Gamma_L}}$$

and

$$\Gamma_{out} = s_o - \frac{s_r s_f}{s_i - \dfrac{1}{\Gamma_s}}$$

ii) Derive the formulae for T_f and T_r in terms of the s-parameters.

Note: $T_f = \dfrac{b_2}{a_1}$ with an arbitrary load, Γ_L.

iii) Use (i) and (ii) to derive [S] for a cascade of a pair of two-port networks.

iv) Use (i) to check the properties of a quarter-wave length of transmission line.

v) Use (i) to show that two gyrators in cascade are equivalent to an ideal transformer.

2.5 Prove that [S] is a unitary matrix for a lossless network, i.e. $[S^*]^T[S] = [\,I\,]$, where T stands for the transpose and $[\,I\,]$ is the unit matrix.

2.6 Prove Equation (2.14).

2.7 Consider a four-port network with the scattering parameters s_{mn}; m,n = 1,2,3,4 and having the following properties:

a) $s_{mn} = 0$ for m = n, i.e. the network is matched in every port.

b) $s_{mn} = s_{nm}$, i.e. the network is reciprocal.

c) [S] is unitary, i.e. the network is lossless.

Prove that at least one of the following is true

$$s_{14} = s_{23} = 0, \; s_{12} = s_{34} = 0 \text{ and } s_{13} = s_{24} = 0$$

2.8 Prove that in terms of the ABCD-parameters, with the determinant of the matrix, Δ

i) a reciprocal two-port network has $\Delta = 1$,

ii) a unilateral two-port network has $\Delta = 0$, and

iii) a symmetrical two-port network has $\Delta = 1$ *and* A = D.

2.9 Show that a transmission line of length $\lambda/2 + l$ is equivalent to a line of length l in cascade with an ideal 1 : 1 phase-inverting transformer, Figure 2.17. This may be proved by combining the results of Exercises 2.1(ii), (iii), and 2.4(iii).

2.10 Starting with the equivalent circuit of a two-port network in terms of its y-parameters, determine s_f and s_r in terms of y-parameters. In particular, show that

$$\frac{s_f}{s_r} = \frac{y_f}{y_r}$$

2.11 Consider the circuit shown below.
- i) First take $\theta = 0$ and use the result in (2.19) to derive the expressions for V_1, V_2 in terms of V_S, Γ_s and Γ_L.
- ii) Now take $\theta \neq 0$ and prove that

$$V_1 = \frac{V_S}{2}\left[1 - \frac{\Gamma_s - \Gamma_L e^{-j2\theta}}{1 - \Gamma_s \Gamma_L e^{-j2\theta}}\right]$$

and
$$V_2 = \frac{V_S}{2}\left[\frac{(1 - \Gamma_s)(1 + \Gamma_L)e^{-j\theta}}{1 - \Gamma_s \Gamma_L e^{-j2\theta}}\right]$$

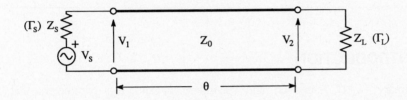

REFERENCES

The following is a useful bibliography for this chapter.

[2.1] Carson, R. S., *High-frequency Amplifiers*, Wiley, New York, 2nd edn, 1982.

[2.2] Gonzalez, G., *Microwave Transistor Amplifiers*, Prentice-Hall, Englewood Cliffs, NJ, 1984.

[2.3] Ha, T. T., *Solid-state Microwave Amplifier Design*, Wiley, New York, 1981.

[2.4] Hewlett-Packard, *S-parameters, Circuit Analysis and Design*, Application Note 95, September 1968.

[2.5] Vendelin, G. D., *Design of Amplifiers and Oscillators by the S-parameter Method*, Wiley, New York, 1982.

3 Microstrip transmission lines — basic theory

3.1 INTRODUCTION

The primary (L, C, R and G) and secondary (γ and Z_0) parameters that are associated with any two-conductor transmission line in a homogeneous dielectric medium have been considered in Chapter 1. A microstrip transmission line may be seen as a logical transformation in stages from the familiar coaxial line as is seen in Figure 3.1.

The coaxial line is normally analyzed in cylindrical coordinates with its conducting surfaces lying on circles with constant radius. The current flow in the coaxial line is longitudinal, i.e. in the direction of propagation of the wave. A longitudinal slot in the outer conductor has a minimal effect on the current and fields. A probe may be inserted through the slot to measure the electric field strength in the standing wave pattern. This form of construction required close tolerances to be maintained and was unsuitable for accurate standing wave detectors. The slab line was developed [3.1] specifically for its superior mechanical properties in the design of a standing wave detector that was compatible with a coaxial line system. In the air-filled slab line, the fields are confined mainly near the center conductor and decay rapidly out towards infinity, allowing the field strength to be sampled with a moving probe that traverses longitudinally between the parallel plates. The balanced stripline has the circular center conductor replaced by a thin strip conductor and provides a transmission line structure that is easy to construct. With a thin center conductor, the air gap will be minimal and the transmission line is essentially one with a uniform dielectric medium. As practical transmission lines, both the slab line and the

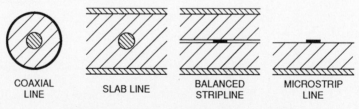

COAXIAL LINE   SLAB LINE   BALANCED STRIPLINE   MICROSTRIP LINE

Figure 3.1 Transmission line configurations

balanced stripline will also support transverse electromagnetic (TEM) waves that may travel in directions other than along the center conductor. An asymmetry in the line structure may set up these unwanted parallel plate modes.

At first sight the microstrip line appears to be one half of a balanced stripline configuration without the complication of the parallel plate modes. However there are two further complications that arise and must be considered. The strip is separated in practice from the ground plane by a supporting dielectric substrate and the complete transmission line structure is no longer in a homogeneous dielectric medium. While the fields are tightly bound in the vicinity of the strip, any discontinuity in the strip geometry may set up either radiation modes or surface wave modes across the substrate surface. The implications of these modes of propagation will be discussed in §4.5.

3.1.1 The coaxial line

The primary parameters may be rigorously derived for a coaxial line that is uniformly filled with dielectric material, Figure 3.2. The dielectric material has a relative permittivity ε_r with respect to the free space value, $\varepsilon_0 = 8.854 \times 10^{-12}$ F.m^{-1}. The coaxial line supports radial electric fields and circumferential magnetic fields, with a longitudinal current flow in the conductors. The capacitance is derived from the application of Gauss's Law around the charge on the center conductor, q per unit length, leading to the electric flux density, D_r, and the radial electric field strength, E_r. The potential difference between the conductors due to the charge leads to the capacitance per unit length

$$C = \frac{q}{V_1 - V_2} \tag{3.1}$$

where

$$V_1 - V_2 = \int_a^b E_r(r)\,dr = \int_a^b \frac{q}{2\pi\varepsilon_r\varepsilon_0} \cdot \frac{dr}{r} \tag{3.2}$$

and a and b are the inner and outer radii. Thus it follows that

$$C = \frac{2\pi\varepsilon_r\varepsilon_0}{ln(b/a)} \quad \text{F.m}^{-1} \tag{3.3}$$

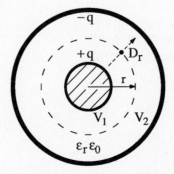

Figure 3.2 Capacitance evaluation for a coaxial line

In circuit and component design, the characteristic impedance, Z_0, and the propagation coefficient, γ, are required. For a lossless line, with primary parameters L and C per unit length

$$Z_0 = \left[\frac{L}{C}\right]^{\frac{1}{2}}$$

(3.4)

and $\qquad \gamma = j\beta = j\omega \sqrt{LC}$ (3.5)

Let C_d and C_0 be capacitances of the transmission line configuration with and without the dielectric filling respectively. For a homogeneously filled transmission line

$$C_d = \varepsilon_r C_0$$

(3.6)

The inductance per unit length for any transmission line is independent of the dielectric properties of the line. From (1.27), it may therefore be expressed in terms of the capacitance of the line as

$$L = \frac{1}{c^2 C_0}$$

(3.7)

where the velocity of electromagnetic wave propagation in free space is given by $c = 2.997925 \times 10^8$ m.s^{-1}. Thus the characteristic impedance, Z_0, is given by

$$Z_0 = \left[\frac{L}{C_d}\right]^{\frac{1}{2}} = \frac{1}{c\sqrt{\varepsilon_r} C_0}$$

(3.8)

with $\qquad v_{\text{phase}} = \frac{c}{\sqrt{\varepsilon_r}}$ (3.9)

For a coaxial line, substituting from (3.3) gives the characteristic impedance as

$$Z_0 = \frac{59.96}{\sqrt{\varepsilon_r}} \ln(b/a) \quad \Omega$$

(3.10)

Example 3.1

A $50\,\Omega$ characteristic impedance coaxial transmission line (0.141 *inch semi-rigid coaxial cable*) has inner and outer diameters of 0.914 mm and 3.00 mm respectively. Calculate

i)   the relative permittivity of the solid dielectric material between the conductors,
ii) the capacitance per meter for the line,
iii) the wavelength in the line at 100 MHz.

Solution:

i)   From (3.10), it is seen that

$$\varepsilon_r = \left[\frac{59.96}{Z_0} \ln(b/a)\right]^2$$

giving $\qquad \varepsilon_r = 2.03$

ii) From (3.3)

$$C = \frac{2\pi\varepsilon_r\varepsilon_0}{ln\,(b/a)} = \frac{2\pi\times2.03\times8.854\times10^{-12}}{ln\,(3.28)} \quad F.m^{-1}$$

i.e. $C = 95\ pF.m^{-1}$

iii) For a transverse electromagnetic wave (TEM), propagating in a uniformly filled dielectric line

$$v_{phase} = \frac{c}{\sqrt{\varepsilon_r}} \quad \text{and} \quad \lambda = \frac{\lambda_0}{\sqrt{\varepsilon_r}}$$

where λ_0 is the free space wavelength. In a coaxial line at 100 MHz

$$\lambda = \frac{0.3}{\sqrt{\varepsilon_r}} = 0.211\ m$$

3.1.2 The symmetrical strip transmission line

For the balanced or symmetrical strip transmission line, uniformly filled with dielectric material, the strip is equidistant between the two ground planes and supported by a homogeneous dielectric material as illustrated in Figure 3.3. The capacitance may be rigorously derived by conformal transformations that eventually equate a quarter of the line geometry to an ideal parallel-plate capacitor [3.2].

   If the strip has a width w and zero thickness and the separation between the two ground planes, that are assumed of infinite extent, is 2h, then

$$C = 4\varepsilon_r\varepsilon_0\frac{K(k')}{K(k)} \quad F.m^{-1} \tag{3.11}$$

where $k = sech\left[\dfrac{\pi}{2}\cdot\dfrac{w}{2h}\right]$ and $(k)^2 + (k')^2 = 1$ (3.12)

and the function K(k), the complete elliptic integral of the first kind, is given by

$$K(k) = \int_0^1 \frac{d\zeta}{\left[(1-\zeta^2)(1-k^2\zeta^2)\right]^{\frac{1}{2}}} \tag{3.13}$$

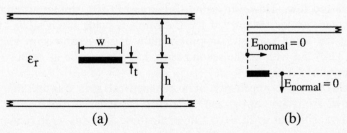

(a)   (b)

Figure 3.3 The symmetrical strip transmission line showing (a) the cross-sectional geometry and (b) the symmetry planes where the normal component of electric field is zero

While any degree of accuracy for a theoretical problem may be obtained using numerical integration of the integrals, such integrals are to be avoided for computational ease in practical situations.

For large w/2h ratio, from (3.12), $k \to 0$. Now, from [3.3]

$$\exp\left\{-\pi\frac{K(k')}{K(k)}\right\} = \frac{k^2}{16} + 8\left[\frac{k^2}{16}\right]^2 + \cdots \tag{3.14}$$

Taking only the first term of this expansion gives

$$\frac{K(k')}{K(k)} = \frac{4}{\pi}\ln 2 - \frac{2}{\pi}\ln k \tag{3.15}$$

and in turn, using (3.12)

$$-\frac{2}{\pi}\ln k \approx -\frac{2}{\pi}\ln\left\{2 \div \exp\left[\frac{\pi}{2} \times \frac{w}{2h}\right]\right\} \tag{3.16}$$

$$= \frac{w}{2h} - \frac{2}{\pi}\ln 2 \tag{3.17}$$

Hence $$\frac{K(k')}{K(k)} \to \frac{w}{2h} + \frac{2}{\pi}\ln 2 \tag{3.18}$$

giving

$$C = 4\varepsilon_r\varepsilon_0\left\{\frac{w}{2h} + \frac{2}{\pi}\ln 2\right\} \quad \text{F.m}^{-1} \tag{3.19}$$

The first term represents the parallel plate capacitance while the second term is a first approximation to the capacitance contribution of the fringing electric fields. The errors involved in deriving (3.15) and (3.16) tend to cancel, so that the final result in (3.19) is much more accurate than in each of the individual approximations.

For a small w/2h ratio, an approximation for the elliptic integrals [3.4] gives

$$C = 2\pi\varepsilon_r\varepsilon_0\left[\ln\left[\frac{8}{\pi} \cdot \frac{2h}{w}\right] + \frac{\pi^2}{48}\left[\frac{w}{2h}\right]^2\right]^{-1} \quad \text{F.m}^{-1} \tag{3.20}$$

The distinction made by Wheeler [3.4] between wide and narrow strips occurs when $K(k')=K(k)$, i.e. $k = k'$ and thus $k^2 = 0.25$, giving $w/2h = 0.561$. At this transition point the wide-strip approximation, giving a capacitance value that is always more than the true value, has an accuracy better than 0.2% while the narrow-strip value is almost 0.5% too low. However, using (3.19) and (3.20), similar accuracies for the two approximations are obtained if the transition is made when $w/2h \approx 0.5$. Cohn [3.5], using the same wide-strip approximation and a narrow-strip approximation similar to (3.20) but without the second term, has a transition at $w/2h \approx 0.35$ with a maximum error of 1.2%.

The characteristic impedance of a symmetrical strip transmission line given below follows from the narrow and wide-strip capacitance formulae in a modified form with a transition at $w/2h = 0.6$. In an air-filled line, this w/2h ratio corresponds to a characteristic impedance of 90.6Ω. The impedance errors are now less than 0.1% for all w/h.

$$Z_0 = \frac{59.96}{\sqrt{\varepsilon_r}} \left\{ ln\left[\frac{8}{\pi} \cdot \frac{2h}{w}\right] + 0.185 \left[\frac{w}{2h}\right]^2 \right\} \ \Omega, \quad \frac{w}{2h} \leq 0.6 \qquad (3.21)$$

and

$$Z_0 = \frac{94.18}{\sqrt{\varepsilon_r}} \left[\frac{w}{2h} + 0.44\right]^{-1} \ \Omega, \qquad\qquad \frac{w}{2h} \geq 0.6 \qquad (3.22)$$

Example 3.2

A symmetrical strip transmission line has a 6.0 mm wide center strip that has negligible thickness. The strip is supported by dielectric sheets, $\varepsilon_r = 2.32$, between two ground planes that are spaced 10.0 mm apart. Calculate the characteristic impedance for the line.

Solution:

For this transmission line with $w = 6$ mm and $h = 10$ mm, the ratio $w/2h = 0.6$. As this represents the transition point between the narrow and wide strip formulae, (3.21) and (3.22), the characteristic impedance using both equations will be evaluated. From (3.21), the narrow-strip approximation gives

$$Z_0 = \frac{59.96}{\sqrt{\varepsilon_r}} \left\{ ln\left[\frac{8}{\pi \times 0.6}\right] + 0.185 \times (0.6)^2 \right\} = 59.53 \ \Omega$$

From (3.22), for the wide-strip approximation

$$Z_0 = \frac{94.18}{\sqrt{\varepsilon_r}} \left\{0.6 + 0.44\right\}^{-1} = 59.45 \ \Omega$$

Thus taking an average value, the characteristic impedance of the symmetric strip transmission line is 59.5Ω.

3.1.3 The microstrip transmission line

For the microstrip transmission line, Figure 3.4, the transmission line is only partially filled with dielectric material, with the material being between the strip and an infinite ground plane. Thus the solution for the line properties becomes a two-dielectric problem. The concept of capacitance per unit length for a mixed dielectric transmission line involves a low frequency approximation that may not be valid when

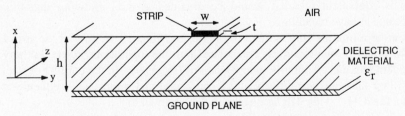

Figure 3.4 The structure of a microstrip transmission line

the cross-section dimensions of the strip become a significant fraction of the wavelength of the transmitted signal.

Consider a substrate material with $\varepsilon = \varepsilon_r \varepsilon_0$ and $\mu = \mu_0$. Using Maxwell's Equation, $\text{curl}\,\mathbf{E} = -\dot{\mathbf{B}}$, and the condition on the air/dielectric boundary that the tangential magnetic field component, H_y, is continuous across the boundary, then

$$\left[\frac{\partial E_z}{\partial x} - \frac{\partial E_x}{\partial z}\right]_{diel} = -j\omega\mu H_y = \left[\frac{\partial E_z}{\partial x} - \frac{\partial E_x}{\partial z}\right]_{air} \qquad (3.23)$$

Since D_x is continuous across the boundary, then

$$\varepsilon_r \left.\frac{\partial E_x}{\partial z}\right|_{diel} = \left.\frac{\partial E_x}{\partial z}\right|_{air} \qquad (3.24)$$

and

$$\left.\frac{\partial E_z}{\partial x}\right|_{diel} - \left.\frac{\partial E_z}{\partial x}\right|_{air} = (1 - \varepsilon_r)\left.\frac{\partial E_x}{\partial z}\right|_{diel} \qquad (3.25)$$

Since E_x is non-zero along the boundary and varies as $e^{-\gamma z}$, then the left hand side of (3.25) is also non-zero, which can only be the case if there is a longitudinal electric field component. Likewise, it may be shown that a longitudinal magnetic field component must also exist. As it may be shown from Maxwell's Equations that each transverse field component can be expressed in terms of E_z and H_z, the general field solution for the lowest order mode in a microstrip line requires all six field components to be present.

At low frequencies with wavelengths that are long compared with the width and height dimensions of the microstrip line, the right hand side of (3.25) tends to zero. Consequently the longitudinal field components diminish in importance. Thus, a low frequency or quasi-static condition exists where the field components are predominantly in the transverse plane. This quasi-static condition is really a particular example of a more general result [3.6]. From Maxwell's Equations, it can be shown that

$$\frac{\partial^2 \mathbf{E}}{\partial x^2} + \frac{\partial^2 \mathbf{E}}{\partial y^2} + \frac{\partial^2 \mathbf{E}}{\partial z^2} + \left[\frac{2\pi}{\lambda}\right]^2 \mathbf{E} = 0 \qquad (3.26)$$

where λ is the wavelength.

When λ is very much greater than the dimensions over which a substantial change of geometry occurs, the last term on the left hand side of (3.26) tends to zero and the equation reduces to Laplace's Equation. That is, the field distribution that results is equivalent to what would be obtained at d.c., allowing quantities like capacitance to have meaning.

Furthermore, the confinement of the fields within the vicinity of the strip is only slightly affected by changing frequency and the operation $(\partial/\partial x)$ on any field component remains essentially independent of frequency. In the context of microstrip lines with a substrate permittivity of 2.5 and height of 1.5 mm, low frequency may be below a few GHz. This will be discussed further in §4.4.

3.2 MICROSTRIP CAPACITANCE EVALUATION

3.2.1 Using conformal transformations

Consider two complex planes, $w = (u + jv)$ and $z = (x + jy)$. To calculate the capacitance of a transmission line in the z-plane, the stored energy in the electrostatic field is evaluated as

$$\frac{1}{2}\varepsilon_r \varepsilon_0 \int_{\text{plane}} \left\{ \left[\frac{\partial V}{\partial x}\right]^2 + \left[\frac{\partial V}{\partial y}\right]^2 \right\} dx\,dy = \frac{1}{2}C(V_2 - V_1)^2 \tag{3.27}$$

where V_1 and V_2 are the potentials of the two conductors. The left hand side of (3.27) may be evaluated once Laplace's Equation in two dimensions

$$\frac{\partial^2 V}{\partial x^2} + \frac{\partial^2 V}{\partial y^2} = 0 \tag{3.28}$$

has been solved with the appropriate boundary conditions in the z-plane. Now the complex function transformation

$$F(x + jy) = u + jv \tag{3.29}$$

where $F(z)$ is an analytic function of z, will transform (3.28) to an identical equation in the w-plane, namely

$$\frac{\partial^2 V}{\partial u^2} + \frac{\partial^2 V}{\partial v^2} = 0 \tag{3.30}$$

leading to a similar equation to (3.27) with an identical value for the stored energy in the w-plane. Thus the capacitance of the new transmission line configuration in the w-plane is identical to the original capacitance in the z-plane. This, then, in conjuction with the ability to make one or more transformations that will lead to a conductor configuration that is readily solved, is the essential ingredient of the conformal transformation method. The method is described in greater depth in [3.2].

A parallel plate capacitor with a plate width w, separation 2h and in a homogeneous dielectric medium, has an equipotential surface along the plane of symmetry between and parallel to the two plates, Palmer [3.7]. The capacitance is one half that of the corresponding microstrip transmission line at a distance h from an infinite ground plane that acts as the equipotential surface. Black and Higgins [3.8] used this approach to develop the theory for a general finite width ground plane. Their method, while capable of giving an accurate value for the capacitance, has two limitations: namely, the line is completely immersed in a homogeneous dielectric medium and the method is a doubly iterative one, where a capacitance value is assumed and elliptic integrals of a complex argument are evaluated, leading eventually to an improved capacitance value.

Judicious approximations combined with conformal mapping were introduced by Wheeler [3.4, 3.9] to overcome these earlier limitations, leading to relations in a simplified form that are suitable for either analysis or synthesis, as they no longer require the solution of elliptic integrals.

Of great importance was the recognition of the need not only for analysis, where the electrical properties are derived from the physical structure of the transmission line, but also for synthesis, obtaining the line dimensions for the desired electrical requirements, in particular for a desired characteristic impedance.

An empirical formula for synthesis using thin strips [3.10], that has better than 2% accuracy and the correct asymptotic behavior for wide strips for all substrate permittivities and for narrow strips for the high and low permittivity extremes [3.4, 3.9], is

$$\frac{w}{h} = \frac{8\left\{\dfrac{7\varepsilon_r + 4}{11\varepsilon_r} A + \dfrac{\varepsilon_r + 1}{0.81\varepsilon_r}\right\}^{\frac{1}{2}}}{A}$$

(3.31)

where

$$A = \exp\left\{\frac{Z_0}{42.4}\sqrt{\varepsilon_r + 1}\right\} - 1$$

(3.32)

For analysis, these two equations are reversed as

$$Z_0 = \frac{42.4}{\sqrt{\varepsilon_r + 1}} \, ln(1 + A)$$

(3.33)

where A is now the positive root of

$$A^2 - \frac{7\varepsilon_r + 4}{11\varepsilon_r} \cdot \left[\frac{8h}{w}\right]^2 A - \frac{\varepsilon_r + 1}{0.81\varepsilon_r} \cdot \left[\frac{8h}{w}\right]^2 = 0$$

(3.34)

3.2.2 Using the finite difference method

The capacitance of any two-conductor transmission line may be found from a knowledge of the charge on the conductors and the potential difference between them. Let V(x,y) be the potential function throughout the cross-section of the transmission line, with the strip at a potential of one volt above that of the ground plane. The continuous function V(x,y) must be a solution to Laplace's Equation in two dimensions

$$\frac{\partial^2 V}{\partial x^2} + \frac{\partial^2 V}{\partial y^2} = 0$$

(3.35)

subject to the appropriate boundary conditions. For the finite difference method [3.11, 3.12], a fine mesh is superimposed on the cross-section of the transmission line, Figure 3.5, and the continuous potential function V(x,y) is replaced by a function ϕ_n that has discrete values at the nodes of the mesh.

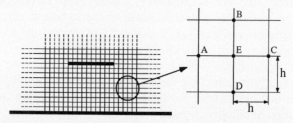

Figure 3.5 A typical mesh for the finite difference method

The details of setting up the equations from which the node potentials, ϕ_n, may be calculated when there are unity and zero potentials on the strip and ground planes respectively are given in Appendix 1. The total charge on either the strip or the ground plane is required with its magnitude being equal to the line capacitance for a 1 V potential difference between the strip and ground plane. To avoid any problems with the edge discontinuities of the strip itself, the normal component of the electric flux density, D_n, is calculated across the ground plane from the node potentials adjacent to the plane. Integrating D_n across the plane gives the total charge and hence the line capacitance.

Compared with conformal transformations, a significant disadvantage of the finite difference method is that the method only provides points for a curve that gives the line capacitance as a function of the width to substrate height ratio for the selected substrate material, rather than as an equation from which either accurate or asymptotic behavior of capacitance may be determined.

An accurate solution requires the use of a very fine mesh with nodes going out to infinity. The effect of a finite mesh size may be allowed for by observing the behavior of capacitance as the mesh size is reduced and extrapolating to an infinitesimal mesh size for improved accuracy. Several approaches that avoid the infinite cross-section dimensions are useful. A conducting boundary may be placed at a large distance where the potentials are assumed to be zero. This form of approximation represents a shielding enclosure around the line, as described in §4.2.3, and is an ideal approach for its solution.

It is desirable to have the greatest density of nodes with their potentials ϕ_n in the region where $V(x,y)$ has the greatest variation and, in particular, near the edges of the strip. This has been achieved both with a graded mesh size [3.13], which allows the shielding box to be placed further away from the strip, and with a coordinate transformation [3.14], which modifies Laplace's Equation for the new coordinate system, transforms infinity for both transverse directions to a finite distance and, by using a uniform mesh in the new coordinate system, gives the greatest density of nodes in the vicinity of the strip.

Example 3.3

From [3.14], when $w/h = 1.0$ and $\varepsilon_r = 1.0$, the extrapolated value for an infinitesimal mesh size gives a microstrip line capacitance of 26.419pF.m^{-1}. Compare this value with that obtained using Wheeler's empirical formulae.

Solution:

This is an analysis problem where the formulae (3.33) and (3.34) may be used in conjunction with

$$C = \frac{\sqrt{\varepsilon_r}}{c\,Z_0} \quad \text{F.m}^{-1}$$

From (3.34)

$$A^2 - 64.0\,A - 158.02 = 0$$

Solving this equation for the positive root gives

$$A = 66.38$$

From (3.33)

$$Z_0 = \frac{42.4}{\sqrt{\varepsilon_r}} \, ln\,(66.38 + 1) = 126.23 \, \Omega$$

Hence the microstrip transmission line capacitance from the empirical formulae

$$C = \frac{1}{c\,Z_0} = 26.425 \text{ pF.m}^{-1}$$

A more accurate value using data from [3.15] gives a capacitance of $26.385\,\text{pF.m}^{-1}$. The three values are all in close agreement.

3.2.3   Using the method of sub-areas

Consider an ideal uniform two-conductor transmission line such as the parallel plate line of width, w, and separation, 2h, that is infinitely long and in free space. It has been seen in §3.2.1 that the parallel plate line, now illustrated as a two-dimensional problem in Figure 3.6, is related to the microstrip transmission line. The surface of the conductors may be subdivided into a number of areas, $i = 1 \cdots n$, that are not necessarily equal in size. For a unit length of line, each area may be represented by a length, ds_i, with a geometric center at (x_i, y_i). It is assumed that ds_i is small, such that the charge density, $\rho(x,y)$, along its length is constant and may be represented by a charge at the center of the element

$$q_i = \rho.ds_i \qquad (3.36)$$

The potential $V(x_j, y_j)$ at any point (x_j, y_j) due to each surface charge q_i will be evaluated on the assumption that the distance from the surface is measured from (x_i, y_i). Thus

$$V(x_j, y_j) = -\int_\infty^r E_r \cdot dr = -\frac{q_i}{2\pi\varepsilon_0} \int_\infty^r \frac{dr}{r} \qquad (3.37)$$

$$= -\frac{q_i}{2\pi\varepsilon_0} \, ln\,(r) \qquad (3.38)$$

The potential at any field point due to a line-source charge q per unit length is thus evaluated by determining the electric field strength as a function of the radial

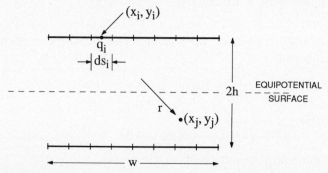

Figure 3.6   The line geometry for the method of sub-areas

distance from the source and integrating in (3.37) from infinity to the field point at a radial distance r. The term, $ln(\infty)$, for the lower limit of integration is constant and present for all the calculations. Since the potential difference between two conductors is the final requirement, the term will cancel out in the final capacitance equations. For simplicity, the term is omitted throughout the intermediate stages of calculation.

Expressing the integral in summation form over the n point charges

$$V(x_j, y_j) = -\frac{1}{2\pi\varepsilon_0} \sum_{i=1}^{n} ln\left[\sqrt{(x_i - x_j)^2 + (y_i - y_j)^2}\right] \cdot q_i$$
(3.39)

A system of n-equations that represent the discrete form of the integral will give the potential at each point *on the conductors*, (x_i, y_i), in terms of the charges on the n-areas and the distances from them. The matrix equation

$$V = [p] q$$
(3.40)

is formed from (3.39), where V is a column matrix of the voltages at the points (x_i, y_i) and q is a column matrix of the charges at (x_i, y_i). In this case as it is V that is known, (3.40) is inverted to give

$$q = [p]^{-1} V$$
(3.41)

With the V elements set in this case as either ± 1, summing q_i for all i over one conductor gives the total charge on that conductor. Hence the capacitance of the line is determined. The method is explained with examples in greater detail in Appendix 2.

3.2.4 Accurate capacitance results

The method of sub-areas has been presented in the previous section and is developed more completely in terms of Green's Functions in Appendix 2. In particular, the two-dielectric problem that is important when there is a dielectric substrate is developed. The important and accurate capacitance results for a microstrip transmission line in free space as evaluated by Kobayashi [3.15] using this method, are presented in Table 3.1.

Table 3.1 The normalized capacitance [3.15] and derived characteristic impedance for a microstrip line in free space, $\varepsilon_r = 1.0$

$\dfrac{w}{h}$	$\dfrac{C}{\varepsilon_0}$	Z_0, Ω
0.04	1.18587	317.683
0.10	1.43375	262.759
0.20	1.70270	221.255
0.40	2.09393	179.915
0.70	2.56365	146.951
1.00	2.97991	126.423
2.00	4.23158	89.028
4.00	6.52698	57.719
7.00	9.79686	38.454
10.00	12.9814	29.021
20.00	23.3628	16.125

3.3 THE CHARACTERISTIC IMPEDANCE — ANALYSIS

The equations that are presented here, and in §3.4 for synthesis, each cover the complete range of microstrip line parameters that may be met in practice. Discontinuities that occur in the equations, typically found in the literature, on changing from narrow- to wide-line formulae are thus avoided. Such discontinuities can produce problems in the computer simulation of circuits. These problems will therefore be eliminated by using only the appropriate wide-ranging equation.

Table 3.2 Analysis formulae

Given w/h and with $\varepsilon_r = 1$, the characteristic impedance is given by

$$Z_0 \equiv Z_{fs} = e^x \tag{3.42}$$

where the exponent

$$x = \sum_{i=0}^{6} X_i \left\{ ln(w/h) \right\}^i$$

i	X_i
0	4.8394(0)
1	−4.5016(−1)
2	−7.7456(−2)
3	−6.5863(−3)
4	1.6510(−3)
5	2.3168(−4)
6	−3.7508(−5)

The X_i coefficients with power of 10 for Equation 3.42.

Given w/h and ε_r, the effective filling fraction, q, is given by

$$q = \sum_{i=0}^{6} \sum_{j=0}^{3} Q_{ij} x^i y^j \tag{3.43}$$

where $\quad x = ln \left[\dfrac{w}{h} + 0.125 \right] \quad$ and $\quad y = 1 - \dfrac{1}{\varepsilon_r}$

i	j = 0	1	2	3
0	6.51309(−1)	−2.25160(−2)	−3.32199(−3)	−3.85162(−3)
1	6.65212(−2)	−1.26976(−3)	−6.80530(−5)	−1.03524(−3)
2	1.64039(−2)	2.59784(−3)	2.74253(−4)	6.52501(−4)
3	3.95737(−4)	6.69075(−4)	6.84457(−5)	3.34213(−4)
4	−1.84365(−3)	−2.11987(−4)	2.13720(−5)	−9.56514(−5)
5	−1.54945(−5)	−9.69139(−5)	−2.71033(−5)	−3.65429(−5)
6	6.53285(−5)	2.47067(−5)	4.06141(−6)	1.01064(−5)

The Q_{ij} coefficients with power of 10 for Equation 3.43.

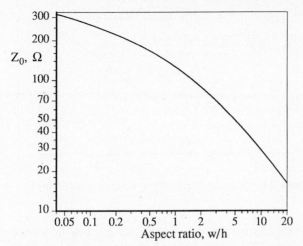

Figure 3.7 The variation of characteristic impedance with w/h for a microstrip line in free space

3.3.1 The microstrip line in free space

The analysis of a microstrip transmission line involves the determination of the characteristic impedance of the line from the dimension ratio, w/h. At first, the case of a line in a homogeneous dielectric material (free space) is considered. Using the values for C/ε_0 given in Table 3.1, the variation of Z_0 with w/h can be derived from (3.8). The results are also included in Table 3.1 and illustrated with the curve in Figure 3.7.

A single equation that expresses Z_0 as a function of ln(w/h) for w/h in the range $0.04 \rightarrow 20$ has been derived for this book through numerical curve fitting techniques to fit the above results. The equation, covering most practical microstrip transmission lines that are likely to be encountered, is given in Table 3.2 as Equation 3.42 together with the appropriate coefficients. The accuracy of fit to the data points when the parameters are in the ranges

$$0.04 \ : \ w/h \ : \ 20$$
$$317.7 \ : \ Z_0 \ : \ 16.13$$

is $\Delta Z_0 = 0.04\%$ r.m.s., with a peak percentage error $<0.07\%$ at any one of the eleven data points.

3.3.2 The effective relative permittivity

In the previous section, the microstrip line was considered to be entirely in free space. Now the effects of other relative permittivity substrates are considered in terms of an effective filling fraction, q, [3.4].

If the whole region has a uniform dielectric material with relative permittivity, ε_r, then

$$Z_0 = \frac{Z_{fs}}{\sqrt{\varepsilon_r}}$$

(3.44)

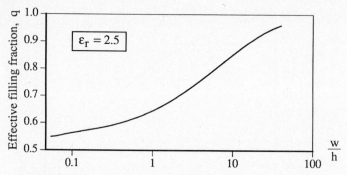

Figure 3.8   The microstrip effective filling fraction for $\varepsilon_r = 2.5$

where Z_{fs} is the characteristic impedance of a line that has the same dimensions but is entirely in free space. The wavelength is reduced by the same factor. However, the dielectric material does not completely fill the whole region but acts as a substrate between the strip and the ground plane. Thus, electric fields are present both in the air and the dielectric material and the stored electrostatic energy is divided between the regions as a function of both ε_r and w/h. The line capacitance, C_d, is related to that when there is no dielectric material present, C_0, by

$$C_d = \varepsilon_{eff} C_0$$

where $\qquad \varepsilon_{eff} = 1 + q(\varepsilon_r - 1) \qquad\qquad\qquad (3.45)$

The microstrip effective filling fraction, q, is thus

$$q = \frac{(\varepsilon_{eff} - 1)}{(\varepsilon_r - 1)}$$

$\qquad\qquad\qquad\qquad\qquad\qquad\qquad\qquad\qquad\qquad\qquad (3.46)$

With a value $0.5 < q < 1.0$, it is related to the portion of the electrostatic energy stored in the substrate region and is both a function of w/h and weakly dependent on ε_r. It is plotted as a function of w/h for $\varepsilon_r = 2.5$ in Figure 3.8, where it is apparent that $q \to 1$ as $w/h \to \infty$ and $q \to 0.5$ as $w/h \to 0$. While this is true for all permittivity substrates, the transition between the two asymptotic values is a function of ε_r, reflecting the permittivity dependence of the fringing fields at the edges of the strip. Typical small but significant variations of the effective filling fraction as a function of ε_r are illustrated in Figure 3.9.

The effective relative permittivity is required, not only for the evaluation of the characteristic impedance, but also for the phase coefficient, β, and the transmission line wavelengths. A two-dimensional polynomial fit to the results of Kobayashi [3.15] for the effective filling fraction as a function of both w/h over the range of 0.04 to 20 and all ε_r is also presented in Table 3.2.

A number of examples and exercises in later chapters are based on a substrate material with $\varepsilon_r = 2.5$. Simplified formulae and data based on the results of Tables 3.2 and 3.5 are presented for convenience in Appendix 3.

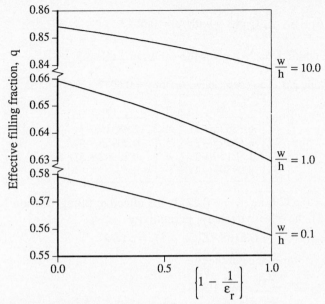

Figure 3.9 The influence of permittivity on the effective filling fraction, with w/h as a parameter, adapted from Kobayashi [3.15] (© 1978, IEEE).

Example 3.4

Two microstrip transmission lines that form part of a circuit are 1.0 and 2.0 mm wide respectively. The substrate has a relative permittivity of 2.53 and is 1.58 mm thick. Assuming that the strip thickness is negligible, calculate the characteristic impedance and low frequency phase velocity for each line.

Solution:

Consider the 1.0 mm wide line where w/h = 0.6329. From (3.42), for this line aspect ratio in free space

$$Z_{fs} = \exp\left\{\sum_{i=0}^{6} X_i x^i\right\} \qquad \text{with } x = ln(w/h) = -0.4574$$

with the X_i coefficients as given in Table 3.2, giving

$$Z_{fs} = 152.9\,\Omega \qquad \text{for } w = 1.0 \text{ mm}$$

For any one ε_r value of the substrate, an equation for the effective filling fraction as a function of w/h may be derived. Here, $\varepsilon_r = 2.53$. Thus

$$y = 1 - \frac{1}{\varepsilon_r} = 0.6047$$

The double summation (3.43) may be written in the form

$$q = \sum_{i=0}^{6} A_i x^i \qquad \text{with } x = ln(w/h + 0.125)$$

and with $\quad A_i = \sum_{j=0}^{3} Q_{ij}\, y^j \quad$ with $y = 0.6047$

Evaluating the coefficients, A_i, gives the values in Table 3.3.

Table 3.3   The coefficients, A_i, for $y = 0.6047$, i.e. $\varepsilon_r = 2.53$

i	A_i	i	A_i
0	6.35626(−1)	4	−1.98518(−3)
1	6.54995(−2)	5	−9.20965(−5)
2	1.82195(−2)	6	8.39902(−5)
3	8.99303(−4)		

Now for $w/h = 0.6329$, i.e. $x = -0.2772$, the effective filling fraction is 0.6188 and, from (3.45), the effective relative permittivity

$$\varepsilon_{eff} = 1 + 0.6188 \times (2.53 - 1) = 1.947$$

The characteristic impedance of the line

$$Z_0 = \frac{Z_{fs}}{\sqrt{\varepsilon_{eff}}} = 109.6\,\Omega$$

The phase velocity

$$v_{ph} = \frac{c}{\sqrt{\varepsilon_{eff}}} = 2.149 \times 10^8 \ \text{m.s}^{-1}$$

The results that have been obtained for $w = 1.0$ mm line are presented in Table 3.4, together with the values that may be similarly derived for the 2.0 mm case.

Table 3.4 A summary of results

w, mm	$\dfrac{w}{h}$	q	ε_{eff}	Z_{fs}, Ω	Z_0, Ω	v_{ph}, m.s^{-1}
1.0	0.6329	0.6188	1.947	152.9	109.6	2.149×10^8
2.0	1.2658	0.6592	2.009	113.2	79.85	2.115×10^8

3.4 THE CHARACTERISTIC IMPEDANCE — SYNTHESIS

As described in the previous sections, it is possible to derive the characteristic impedance and propagation coefficient for a microstrip transmission line, given values for the substrate permittivity and line dimensions. However, in the design of transmission line components and matching networks, it is necessary to derive w/h for a specific value of the characteristic impedance, assuming that the substrate permittivity is known.

As a guide to a typical range of characteristic impedances that may be obtained, Figure 3.10 illustrates the relationship between w/h and ε_r with $\sqrt{\varepsilon_r}\, Z_0$ plotted as a parameter for the curves. Here, the characteristic impedance Z_0 is the actual characteristic impedance of the transmission line for the ε_r substrate, while Z_{fs} is the characteristic impedance of a line that has the same dimensions but is entirely in free space.

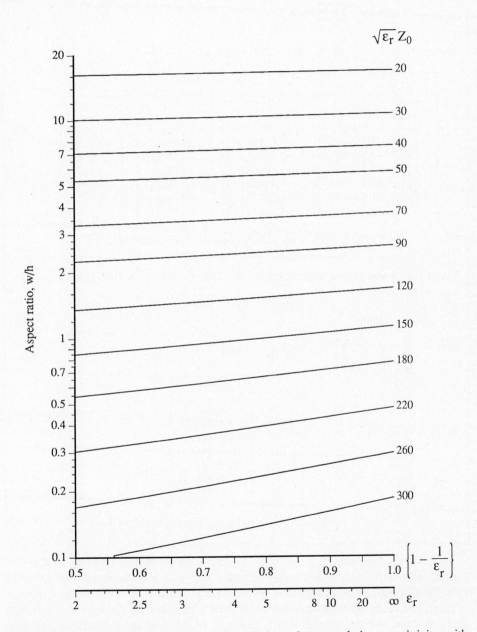

Figure 3.10 The relationship between w/h and the substrate relative permittivity with $\sqrt{\varepsilon_r}\, Z_0$ as the parameter

Table 3.5 Synthesis formulae

Given the desired characteristic impedance, Z_0, with a substrate permittivity, ε_r, the effective filling fraction is given by

$$q = \sum_{i=0}^{6}\sum_{j=0}^{3} R_{ij} x^i y^j \tag{3.47}$$

where $\quad x = \{ ln(\sqrt{\varepsilon_r}\, Z_0) - 4.0 \}$ and $y = 1 - \dfrac{1}{\varepsilon_r}$

i	j			
	0	1	2	3
0	7.80057(−1)	−1.34226(−3)	−3.01251(−3)	−1.06181(−3)
1	−1.35581(−1)	4.04927(−3)	3.48945(−3)	−3.92521(−4)
2	−2.36989(−2)	−2.73642(−3)	3.98886(−3)	2.89673(−3)
3	1.30331(−2)	−1.38608(−3)	−2.61336(−3)	−3.16976(−3)
4	6.31324(−3)	2.14458(−3)	−1.17603(−3)	−1.59925(−3)
5	−1.54467(−3)	−4.56127(−4)	1.17440(−3)	1.94180(−3)
6	−2.64802(−4)	−1.69567(−4)	−2.15831(−4)	−3.35670(−4)

The R_{ij} coefficients with power of 10 for Equation 3.47.

Given the characteristic impedance of the line geometry in free space, Z_{fs}, the ratio w/h is given by

$$\frac{w}{h} = e^y \tag{3.48}$$

with $\quad y = \sum_{i=0}^{6} Y_i \{ ln(Z_{fs}) - 4.0 \}$

i	Y_i
0	1.4664(0)
1	−1.4386(0)
2	−2.2444(−1)
3	−9.7196(−2)
4	−6.8506(−2)
5	−2.4801(−2)
6	5.1597(−3)

The Y_i coefficients with power of 10 for Equation 3.48.

Synthesis formulae, e.g. [3.16, 3.17], often approach the problem in the two regions with w/h being either small or large, and may give formulae that are only valid for a limited range of substrate permittivities. Since w/h of the order of unity is required for many transmission lines, care must be taken in noting the limits that apply to them. An advantage, if it is required, of splitting the formulation into two parts is that the asymptotic values for very large or very small values of w/h will be maintained.

The approach presented here is valid for a wide range of aspect ratios, $0.04 \le w/h \le 20$, and for all isotropic substrate permittivities. The steps to be taken for synthesis are as follows:

i) The effective filling fraction, q, as a function of the required characteristic impedance and substrate permittivity is calculated using the equation that has been derived through numerical curve fitting techniques to fit the results for the effective filling factor as given by Kobayashi [3.15]. This equation is given in Table 3.5 as (3.47) together with the appropriate coefficients.

ii) Knowing q and ε_r, the effective relative permittivity, ε_{eff}, may be derived from (3.45).

iii) Use ε_{eff} to find the free space characteristic impedance Z_{fs} for the same, but still unknown, line dimensions from (3.44).

iv) w/h, as a function of the free space characteristic impedance that has just been determined, is given by (3.48) in Table 3.5.

The accuracy of fit to the data points, given in Table 3.1, is $\Delta Z_{fs} < 0.07\,\%$.

Example 3.5

Ignoring end-effect corrections, what are the dimensions of a $100\,\Omega$ characteristic impedance quarter-wavelength long line at 1.0GHz fabricated as a microstrip transmission line on a 1.58 mm thick substrate, $\varepsilon_r = 2.53$?

Solution:

From (3.47), with a substrate relative permittivity, $\varepsilon_r = 2.53$, the effective filling fraction

$$q = \sum_{i=0}^{6} B_i x^i$$

where $B_i = \sum_{j=0}^{3} R_{ij} \left(1 - \frac{1}{\varepsilon_r}\right)^j$ gives the values presented in Table 3.6.

Table 3.6 The coefficients, B_i, for $y = 0.6047$, i.e. $\varepsilon_r = 2.53$

i	B_i	i	B_i
0	7.77909(−1)	4	6.82637(−3)
1	−1.31943(−1)	5	−9.61560(−4)
2	−2.32543(−2)	6	−5.20517(−4)
3	1.05381(−2)		

These coefficients are valid for any line where $0.04 < w/h < 20.0$ on an $\varepsilon_r = 2.53$ substrate. For a 100Ω characteristic impedance microstrip transmission line where

$$x = ln\left(\sqrt{\varepsilon_r}\, Z_0\right) - 4 = 1.0693$$

the summation for the effective filling fraction gives $q = 0.6299$ and, from (3.45)

$$\varepsilon_{eff} = 1.964$$

From (3.19), for the same w/h ratio but without the dielectric substrate

$$Z_{fs} = \sqrt{\varepsilon_{eff}}\, Z_0 = 140.14\ \Omega$$

Hence (3.48) gives y = −0.2394 and w/h = 0.787.

At 1.0GHz, the free space wavelength, $\lambda_0 = 300$ mm, and the microstrip transmission line wavelength

$$\lambda = \frac{\lambda_0}{\sqrt{\varepsilon_{eff}}} = 214.1\ mm$$

Thus the dimensions of a $100\,\Omega$ characteristic impedance quarter-wavelength long line at 1.0GHz are w/h = 0.787, i.e. w = 1.24 mm, with a length of 53.5 mm.

3.5 OTHER PUBLISHED MICROSTRIP LINE FORMULAE

Closed form expressions for microstrip line parameters have been developed from the results of conformal transformation methods of Wheeler [3.4], for a balanced strip configuration separated by a dielectric sheet, and by Schneider [3.18], who dealt directly with microstrip line geometries. These methods lead to pairs of expressions with one of the pair valid for narrow-strip and the other for wide-strip geometry. Owens [3.16] and Hammerstad [3.19] used the Green's function technique to test the closed form expressions against more accurate computational results for mid-range strip geometries. With this information, one can thus either determine the changeover point between narrow- and wide-strip formulae or modify the coefficients in the expressions for minimum errors over the range of interest.

Analysis — Characteristic impedance

From Owens [3.16, eqn.7], for narrow lines with w/h ≤ 2

$$Z_0 = \frac{119.9}{\sqrt{2(\varepsilon_r + 1)}} \left\{ H' - \frac{(\varepsilon_r - 1)}{2(\varepsilon_r + 1)} \left[0.4516 + \frac{0.2416}{\varepsilon_r} \right] \right\} \tag{3.49}$$

with

$$H' = ln\left[\frac{4h}{w} + \left\{ \left[\frac{4h}{w} \right]^2 + 2 \right\}^{\frac{1}{2}} \right] \tag{3.50}$$

From Owens [3.16, eqn.2] for wide lines with w/h ≥ 2, on simplifying the constant terms

$$Z_0 = \frac{376.7}{\sqrt{\varepsilon_r}} \left\{ \frac{w}{h} + 0.8825 + 0.1645 \left[\frac{\varepsilon_r - 1}{\varepsilon_r^2} \right] + \frac{\varepsilon_r + 1}{\pi \varepsilon_r} \left[1.4516 + ln\left(\frac{w}{2h} + 0.94 \right) \right] \right\}^{-1} \tag{3.51}$$

Analysis — Effective relative permittivity

From Owens [3.16, eqn.9], for narrow lines with w/h ≤ 1

$$\varepsilon_{eff} = \frac{\varepsilon_r + 1}{2} \left\{ 1 - \frac{(\varepsilon_r - 1)}{2 H'(\varepsilon_r + 1)} \left[0.4516 + \frac{0.2416}{\varepsilon_r} \right] \right\}^{-2} \tag{3.52}$$

with H′ as given by (3.50). For wide lines with w/h ≥ 1

$$\varepsilon_{eff} \;=\; \frac{\varepsilon_r + 1}{2} \;+\; \frac{\varepsilon_r - 1}{2} \times F \tag{3.53}$$

where Owens [3.16, eqn.12] gives

$$F \;=\; \left[1 + \frac{10\,h}{w}\right]^{-0.555} \tag{3.54}$$

and Hammerstad [3.19, eqn.4] gives

$$F \;=\; \left[1 + \frac{12\,h}{w}\right]^{-0.5} \tag{3.55}$$

Synthesis — Line geometry

For $8 \le \varepsilon_r \le 12$, the condition $Z_0 = (44 - 2\varepsilon_r)\Omega$ is used by Owens [3.16] to distinguish between narrow- and wide-line geometries. For narrow lines, Wheeler's formula [3.4] is rewritten as

$$\frac{w}{h} \;=\; \frac{8}{\exp(A) - 2\exp(-A)} \tag{3.56}$$

with

$$A \;=\; \frac{Z_0\sqrt{2(\varepsilon_r + 1)}}{119.9} \;+\; \frac{(\varepsilon_r - 1)}{2(\varepsilon_r + 1)}\left[0.4516 + \frac{0.2416}{\varepsilon_r}\right] \tag{3.57}$$

For wide strips

$$\frac{w}{h} \;=\; \frac{2}{\pi}\left\{(B - 1) - ln(2B - 1) + \frac{\varepsilon_r - 1}{2\,\varepsilon_r}\left[ln(B - 1) + 0.293 - \frac{0.517}{\varepsilon_r}\right]\right\} \tag{3.58}$$

with

$$B \;=\; \frac{59.96\,\pi^2}{Z_0\sqrt{\varepsilon_r}} \tag{3.59}$$

For $\varepsilon_r < 16$, Hammerstad [3.19] uses the same equations as above but with modified coefficients and a transition between the equations when w/h = 2. The modified equations in place of (3.57) and (3.58) are

$$A \;=\; \frac{Z_0\sqrt{2(\varepsilon_r + 1)}}{119.9} \;+\; \frac{(\varepsilon_r - 1)}{2(\varepsilon_r + 1)}\left[0.46 + \frac{0.22}{\varepsilon_r}\right] \tag{3.60}$$

and

$$\frac{w}{h} \;=\; \frac{2}{\pi}\left\{(B - 1) - ln(2B - 1) + \frac{\varepsilon_r - 1}{2\,\varepsilon_r}\left[ln(B - 1) + 0.39 - \frac{0.61}{\varepsilon_r}\right]\right\} \tag{3.61}$$

Synthesis — Effective relative permittivity

Given ε_r and with w/h known, ε_{eff} may be derived using the analysis equations, (3.52) and (3.53).

EXERCISES

3.1 Sketch the electric and magnetic field patterns for
    i)   a coaxial line,
    ii)   a balanced strip transmission line,
    iii) a microstrip transmission line.

3.2   A 50Ω characteristic impedance air-spaced coaxial cable has a solid cylindrical dielectric supporting bead ($\varepsilon_r = 2.25$) that is 5.0 mm long.

   i)   What is the additional shunt capacitance at low frequencies due to the dielectric material?

   ii)   Up to what frequency will this calculation be valid?

3.3   A balanced strip transmission line has a uniform dielectric material, $\varepsilon_r = 2.25$, and a 5.0 mm separation between the ground planes.

   i)   Calculate the strip width that is required for a 50Ω characteristic impedance line.

   ii)   At a plane where the dielectric material finishes so that a sliding short circuit may be used on the line, to what width must the strip be changed if the 50Ω characteristic impedance is to be maintained?

3.4   A 1.0 mm thick alumina substrate, $\varepsilon_r = 9.6$, is used for a microstrip circuit that requires a 50Ω characteristic impedance transmission line with the impedance to be maintained within 2% of the true value. What is the line width and the tolerance that must be maintained?

3.5   A 1.0 mm thick fused quartz substrate, $\varepsilon_r = 3.8$, is to be used for the construction of a microstrip circuit. If the line widths have to be within the limits of 0.2 to 6.0 mm, what is the range of characteristic impedances available to the circuit designer?

3.6   A microstrip circuit has been fabricated on a 1.58 mm thick substrate. The lines, which are connected to 50Ω characteristic impedance coaxial connectors, are themselves assumed to have a 50Ω characteristic impedance and are 4.7 mm wide. Estimate the relative permittivity for the substrate.

3.7   An open-circuit terminated stub microstrip transmission line on a 0.5 mm thick alumina substrate, $\varepsilon_r = 9.6$, is 1.0 mm wide and 6.0 mm long. What are the characteristic impedance of the line and its electrical length at 2.0GHz? Assume for this calculation that end-effects and dispersion may be ignored.

3.8   Consider the case of a microstrip transmission line in free space. The strip of width, w, and height, h, above the ground plane together with its image are each subdivided into four equal areas. Because of the double symmetry of the problem, the total of eight charged regions may be represented in terms of the magnitudes of two unknown charges only, say q_1 and q_2.

   i)   From the geometry of the problem, determine the elements of the matrix $[\,p\,]$ where

$$V = [\,p\,]\,q$$

   ii)   Solve the matrix equation for the unknown charges when the strip has unity potential with respect to the ground plane and calculate the line capacitances for w/h = 0.01, 0.1, 1.0 and 10.0. Results for the first three cases are given in Table A2.1, while for the fourth case of w/h = 10.0, a comparison with the value from Table 3.1 should be made.

3.9   i)   In Appendix 2, Point 2 highlighted the need to develop a special formula for the handling of self-potential. As the basic equation (A2.5) assumes that the source charge is concentrated at the center of the sub-area and is at a large distance from the field point where the potential, V, is evaluated, there may be a significant error for the term associated with two adjacent areas, ds_1 and ds_2, as illustrated below. Derive an equation for the potential at P_2 due to the distributed charge, q_1/ds_1 per unit width on the adjacent element.

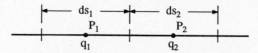

   ii)   Recalculate the two-point example illustrated above using the improved formulation for adjacent elements and with w/h = 0.1 and show that the error for the normalized capacitance is increased from 3.4% to 3.9%. What is the physical explanation for this apparent contradiction?

REFERENCES

[3.1]   Wholey, W. B. and Eldred, W. N., "A new type of slotted line section", *Proc. IRE*, Vol. 38, March 1950, pp. 244-8.

[3.2]   Collin, R. E., *Field Theory of Guided Waves*, McGraw-Hill, New York, 1960, Section 4.3.

[3.3]   Abramowitz, M. and Stegun, I. A., *Handbook of Mathematical Functions*, Dover Publications, Inc., New York, 1965. (Equation 17.3.21)

[3.4]   Wheeler, H. A., "Transmission-line properties of parallel strips separated by a dielectric sheet", *IEEE Trans. Microwave Theory Tech.*, Vol. MTT-13, No. 2, March 1965, pp. 172-85.

[3.5]   Cohn, S. B., "Characteristic impedance of the shielded-strip transmission line", *IRE Trans. Microwave Theory Tech.*, Vol. MTT-2, No. 2, July 1954, pp. 52-7.

[3.6]   Ramo, S., Whinnery, J. R. and van Duzer, T., *Fields and Waves in Communication Electronics*, Wiley, New York, 1965.

[3.7]   Palmer, H. B., "The capacitance of a parallel-plate capacitor by the Schwartz-Christoffel transformation", *Trans. AIEE*, Vol. 56, March 1937, pp. 363-6.

[3.8]   Black, K. G. and Higgins, T. J., "Rigorous determination of the parameters of microstrip transmission lines", *IRE Trans. Microwave Theory Tech.*, Vol. MTT-3, No. 2, March 1955, pp. 93-113.

[3.9]   Wheeler, H. A., "Transmission-line properties of parallel wide strips by a conformal-mapping approximation", *IEEE Trans. Microwave Theory Tech.*, Vol. MTT-12, No. 5, May 1964, pp. 280-9.

[3.10] Wheeler, H. A., "Transmission-line properties of a strip on a dielectric sheet on a plane", *IEEE Trans. Microwave Theory and Techniques*, Vol. MTT-25, No. 8, August 1977, pp. 631-47.

[3.11] Green, H. E., "The numerical solution of some important transmission-line problems", *IEEE Trans. Microwave Theory Tech.*, Vol. MTT-13, September 1965, pp. 676-92.

[3.12] Wexler, A., "Computation of electromagnetic fields", *IEEE Trans. Microwave Theory and Techniques*, Vol. MTT-17, No. 8, August 1969, pp. 416-39.

[3.13] Corr, D. G. and Davies, J. B., "Computer analysis of the fundamental and higher order modes in single and coupled microstrip", *IEEE Trans. Microwave Theory and Techniques*, Vol. MTT-20, No. 10, October 1972, pp. 669-78.

[3.14] Fooks, E. H. and Ladbrooke, P. H., "A co-ordinate transformation for the numerical analysis of microstrip transmission lines", *Proc. IREE Aust.*, Vol. 41, No. 2, June 1980, pp. 74-8.

[3.15] Kobayashi, M., "Analysis of the microstrip and the electrooptic light modulator", *IEEE Trans. Microwave Theory Tech.*, Vol. MTT-26, No. 2, February 1978, pp. 119-26.

[3.16] Owens, R. P., "Accurate analytical determination of quasi-static microstrip line parameters", *The Radio and Electronic Engineer*, Vol. 46, No. 7, July 1976, pp. 360-4.

[3.17] Edwards, T. C., *Foundations for Microstrip Circuit Design*, Wiley, Chichester, 1981.

[3.18] Schneider, M. V., "Microstrip lines for microwave integrated circuits", *Bell Syst. Tech. J.*, Vol. 48, No. 3, May-June 1969, pp. 1421-44.

[3.19] Hammerstad, E. O., "Equations for microstrip circuit design", *Proc. 5th European Microwave Conference*, Hamburg, September 1975, pp. 268-72.

4 Microstrip transmission lines — further considerations

4.1 INTRODUCTION

There are many reasons why a practical microstrip line does not behave exactly as described by the low frequency models for the ideal line in the previous chapter. In this chapter, the practical line is considered wherever possible in terms of perturbation effects on the ideal line. A finite thickness strip is taken care of through an effective increase in the line width of a zero-thickness line. The finite conductivity of metals and lossy dielectric substrates do not cause any significant changes to the characteristic impedance or wavelength, but are responsible for introducing an attenuation to the transmitted signal.

Quasi-static approximations are based on the assumption of TEM fields in microstrip lines and are assumed in the design of most components and circuits. It is, however, important to realize that this TEM assumption may be a severe limitation at higher frequencies unless one knows how to circumvent it. An appreciation of the dispersion problem appears in §4.4, where limits to the low frequency approximations are given together with a description of the variation of effective relative permittivity and characteristic impedance with frequency.

A microstrip line is capable of radiating from any line discontinuity as well as setting up other modes that are guided by the air-substrate interface without requiring the presence of a metal strip. A microstrip circuit is normally enclosed in a shielded environment to minimize these effects and to protect the circuit from external influences. However, the shield itself will influence the line parameters and will also allow other resonant modes to be excited in the resulting cavities. These effects are among the further considerations that must be taken care of, before completing the design of microstrip components and circuits.

4.2 PRACTICAL MICROSTRIP LINES

4.2.1 Finite strip thickness

The conductors of practical microstrip lines have a finite thickness, t, that must be accounted for in accurate calculations of characteristic impedance and propagation coefficient. Formulae that were derived in Chapter 3 for zero-thickness lines may be

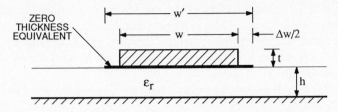

Figure 4.1 The equivalence between finite- and zero-thickness microstrip lines

used when an equivalence is established between a practical line with parameters (w,h,t) and a zero-thickness line (w′,h,t=0), where

$$w' = w + \Delta w \tag{4.1}$$

These parameters are illustrated in Figure 4.1. For the majority of microstrip lines, t ≪ h and, except for the very narrow high impedance lines, t ≪ w.

The secondary parameters of any TEM line, namely Z_0 and γ, are derived in §1.2 in terms of **L, C, R** and **G** and, in particular for a lossless line, in terms of **L** and **C**. Thus it is necessary to know the influence of finite line thickness on **L** and **C** and hence deduce its effect on both the characteristic impedance and, through ε_{eff}, on the propagation coefficient.

The influence of line thickness with $\varepsilon_r = 1$ was studied by Wheeler [4.1] and formulae for the line width corrections for narrow and wide strips were obtained. One of the wide-strip terms was later corrected [4.2], as a part of the original derivation was based on the assumption of an unlimited width strip. Bahl and Garg [4.3] modified the original Wheeler formulae by increasing all the correction terms to w/h by 25%, when used in characteristic impedance calculations, and thus found a close fit to existing data at that time. From [4.3]

$$\frac{w'}{h} = \frac{w}{h} + \frac{1.25\,t}{\pi h}\left\{1 + ln\left[\frac{4\pi w}{t}\right]\right\}, \qquad \frac{w}{h} \le \frac{1}{2\pi} \tag{4.2}$$

$$\frac{w'}{h} = \frac{w}{h} + \frac{1.25\,t}{\pi h}\left\{1 + ln\left[\frac{2h}{t}\right]\right\}, \qquad \frac{w}{h} \ge \frac{1}{2\pi} \tag{4.3}$$

Furthermore, the effective relative permittivity was also reduced for a finite line thickness to give

$$\varepsilon'_{eff} = \varepsilon_{eff} - \frac{\varepsilon_r - 1}{4.6} \times \frac{t/h}{\sqrt{w/h}} \tag{4.4}$$

Equations (4.2) to (4.4) apply for t/h ≤ 0.2, 0.1 ≤ w/h ≤ 20 and $\varepsilon_r \le 16$. Wheeler [4.2] introduced an additional term that extended the width formulae for greater thickness lines and combined both narrow- and wide-line formulae into a unified one, giving w′/h for characteristic impedance calculations. The correction term, suitable for both analysis and synthesis, is

$$\Delta w = \frac{t}{\pi}\left[(1 + ln4) - \frac{1}{2}ln\left[\left\{\frac{t}{h}\right\}^2 + \left\{\frac{1}{\pi\xi}\right\}^2\right]\right], \quad \text{with } \varepsilon_r = 1 \tag{4.5}$$

with either $\xi = w/h + 1.10$ or $\xi = w'/h - 0.26$, depending on whether the actual width or the equivalent zero-thickness width is already known. Equation (4.5) is valid for thicknesses up to a square cross-section for narrow strips and up to moderate thicknesses, i.e. $t < h$, for wide strips. This equation may be used for all inductance calculations since for any TEM-wave

$$L = \frac{1}{c^2 \, C_{(\varepsilon_r = 1)}} \quad \text{H.m}^{-1}$$

(4.6)

The width adjustment for inductance calculations is independent of the substrate permittivity and is most easily given through the thickness influence on $C_{\varepsilon_r = 1}$ in (4.6). However, the effect of strip thickness on the capacitance will be dependent on the substrate permittivity. Δw will be smaller when $\varepsilon_r > 1$ than when $\varepsilon_r = 1$, since the edge of the strip is in a region where the electric field strength for a given voltage is reduced as the substrate permittivity increases. From [4.2], the modified correction term to be used for capacitance calculations is $\Delta w/\varepsilon_r$, so that for characteristic impedance calculations

$$\Delta w' = \frac{1 + (1/\varepsilon_r)}{2} \, \Delta w$$

(4.7)

The effective relative permittivity is used primarily in the evaluation of the wavelength along the line at the operating frequency. Using the definition that $\varepsilon'_{\text{eff}} = (\lambda_0/\lambda)^2$ and substituting for each wavelength from a relationship of the form $\beta = 2\pi/\lambda = \omega\sqrt{LC}$, gives

$$\varepsilon'_{\text{eff}} = \left\{ \frac{C(\text{with substrate})}{C(\text{air})} \right\}$$

(4.8)

since the inductance is unaffected by the presence of the substrate. Thus

$$\varepsilon'_{\text{eff}} = \left\{ \frac{Z_0(\text{air})}{Z_0(\text{with substrate})} \right\}^2$$

(4.9)

Allowing for the finite thickness of the strip, (4.9) becomes

$$\varepsilon'_{\text{eff}} = \left\{ \frac{Z_0(w' = w + \Delta w, \ \varepsilon_r = 1)}{Z_0(w'' = w + \Delta w', \ \varepsilon_r)} \right\}^2$$

(4.10)

In this chapter, variations of characteristic impedance are required as functions of several parameters. For clarity, the impedance is specified as a function of the parameters (w, h, t, ε_r) as appropriate.

Example 4.1

Design a $50\,\Omega$ characteristic impedance microstrip line on a 1.58 mm thick, $\varepsilon_r = 2.5$ substrate with a conductor thickness of 0.1 mm. Calculate the effective relative permittivity for the line and compare the results with those obtained using the Bahl and Garg expressions.

Solution:

From Appendix 3, for a zero-thickness $50\,\Omega$ line, $\varepsilon_{\text{eff}} = 2.090$ and $w'/h = 2.837$, giving $w' = 4.482$ mm. In (4.5), with $\xi = 2.577$

$$\Delta w \;=\; 0.139 \text{ mm}$$

while
$$\Delta w' \;=\; \frac{1+0.4}{2}\times\Delta w \;=\; 0.097 \text{ mm}$$

giving the actual line width of $4.482 - 0.097 = 4.385$ mm.

This change in line width represents a correction term for the 1.5% error that would otherwise have occurred in the characteristic impedance. The effective relative permittivity is found from (4.10) as

$$\varepsilon'_{\text{eff}} \;=\; \left\{ \frac{Z_0(w' = 4.385 + 0.139, \ \varepsilon_r = 1)}{Z_0(w'' = 4.385 + 0.097, \ \varepsilon_r = 2.5)} \right\}^2$$

$$=\; \left\{ \frac{71.87}{50.01} \right\}^2 \;=\; 2.065$$

For comparison, with $w' = 4.482$, $h = 1.58$, and $t = 0.1$ mm, the Bahl and Garg expressions from (4.3) give an actual line width of 4.370 mm, irrespective of the substrate permittivity, while from (4.4) $\varepsilon'_{\text{eff}}$ is 2.078 as compared with the negligible line thickness value of $\varepsilon_{\text{eff}} = 2.090$ and the 2.065 above.

4.2.2 Losses

Dielectric losses

The dielectric substrate has a complex relative permittivity given by $\varepsilon = \varepsilon_r \varepsilon_0 (1 - j \tan\delta)$ where $\tan\delta = (G/\omega C)$ is the loss tangent for the substrate. The loss tangent represents the ratio of conduction to displacement currents that flow in the dielectric region. For a TEM-wave propagating along a low-loss uniformly-filled transmission line, from (1.22), the attenuation due to the losses in the dielectric material is

$$\alpha_d \;=\; \frac{G Z_0}{2} \;=\; \frac{\omega C Z_0}{2} \times \tan\delta \quad \text{neper.m}^{-1} \tag{4.11}$$

A dielectric-loss effective filling factor, q_d, is introduced now to allow for the fact that the lossy dielectric material does not completely fill the whole microstrip cross-section. It is clear that q_d will differ from a similar effective filling factor for capacitance calculations since, on the one hand, for evaluating q_d there is no conductance component associated with the air region above the substrate, even though there is energy transfer through this region, while on the other hand, for ε_{eff} calculations there is a capacitive component for the air region. In (4.11), C is the total capacitance per unit length of line and, from (1.29) with

$$C Z_0 \;=\; \frac{\sqrt{\varepsilon_{\text{eff}}}}{c} \tag{4.12}$$

it follows that

$$\alpha_d \;=\; \frac{\pi f \sqrt{\varepsilon_{\text{eff}}}}{c} \, q_d \tan\delta \tag{4.13}$$

As derived by Schneider [4.4], the effective filling factor for the dielectric loss tangent

$$q_d = \frac{\text{(electric energy stored per meter in the substrate)}}{\text{(electric energy stored per meter in the complete line)}} \equiv \frac{W_s}{W} \quad (4.14)$$

Furthermore, in [4.4, Appendix 1], it is shown that

$$\frac{\partial W}{\partial \varepsilon_r} = \frac{W_s}{\varepsilon_r} \quad \text{and} \quad \frac{\partial W}{\partial \varepsilon_{eff}} = \frac{W}{\varepsilon_{eff}} \quad (4.15)$$

giving

$$\frac{\partial \varepsilon_{eff}}{\partial \varepsilon_r} = \frac{\varepsilon_{eff}}{\varepsilon_r} \times \frac{W_s}{W} = \frac{\varepsilon_{eff}}{\varepsilon_r} \times q_d \quad (4.16)$$

Thus

$$q_d = \frac{\varepsilon_r}{\varepsilon_{eff}} \times \frac{\partial \varepsilon_{eff}}{\partial \varepsilon_r} \quad (4.17)$$

The variation of ε_{eff} with ε_r for a given line geometry is found by using (3.45), i.e. $\varepsilon_{eff} = 1 + q(\varepsilon_r - 1)$, from which

$$q_d = \frac{\varepsilon_r}{\varepsilon_{eff}} \left\{ q + (\varepsilon_r - 1)\frac{\partial q}{\partial \varepsilon_r} \right\} \quad (4.18)$$

From Figure 3.9, it is observed that for all substrate permittivities, $0.5 < q < 1.0$, and for $w/h = 1$, for which the slope is about the largest

$$q \approx 0.66 - 0.03\left\{1 - \frac{1}{\varepsilon_r}\right\}, \quad \text{i.e.} \quad \frac{\partial q}{\partial \varepsilon_r} = -\frac{0.03}{\varepsilon_r^2} \quad (4.19)$$

Thus the expression

$$q_d = \frac{\varepsilon_r}{\varepsilon_{eff}} \times q \quad (4.20)$$

will slightly overestimate the dielectric loss, but not by more than about 1% if $\varepsilon_r \geq 2.5$. Hence (4.13) becomes

$$\alpha_d = \frac{\pi f \sqrt{\varepsilon_{eff}}}{c} \frac{\varepsilon_r}{\varepsilon_{eff}} \left\{\frac{\varepsilon_{eff} - 1}{\varepsilon_r - 1}\right\} \tan\delta \quad (4.21)$$

or on rearranging

$$\alpha_d = \frac{\pi \sqrt{\varepsilon_{eff}}}{\lambda_0} \left\{\frac{1 - (\varepsilon_{eff})^{-1}}{1 - (\varepsilon_r)^{-1}}\right\} \tan\delta \quad \text{neper.m}^{-1} \quad (4.22)$$

where λ_0 is the free space wavelength.

Example 4.2

Calculate the attenuation due to dielectric losses at 1.0GHz for a 50Ω line on a 1.58 mm thick, $\varepsilon_r = 2.5$, substrate. The dielectric loss tangent for the substrate is 0.001.

Solution:

From Appendix 3, the effective relative permittivity for the line is 2.09.

Substituting into (4.22), gives

$$\alpha_d = \frac{\pi\sqrt{2.09}}{0.30}\left\{\frac{1-0.478}{1-0.4}\right\}\times 0.001$$

$$= 0.0132 \quad \text{neper.m}^{-1}$$

or $\qquad \alpha_d = 0.114 \quad \text{dB.m}^{-1}$

Conductor losses

At microwave frequencies, the current flows through a thin layer on the outside surface of the microstrip conductors. With the transverse dimensions much greater than the measure of this layer thickness, called the skin depth, the skin effects of an actual conductor surface may be analyzed in terms of a plane wave propagating along the normal into the conductor. For a plane wave propagating in a good conductor, the transmission line model of Chapter 1 can be used with **L** and **G** now evaluated for a unit cross-sectional area, as well as per unit length. From (1.5) with no magnetic losses and with the conductivity $\sigma \gg \omega\varepsilon$, the propagation coefficient

$$\gamma = \sqrt{j\omega LG} \;\Rightarrow\; \alpha = \beta = \left[\frac{\omega LG}{2}\right]^{\frac{1}{2}}$$

$$\Rightarrow \quad \alpha = \beta = \left[\frac{\omega\mu\sigma}{2}\right]^{\frac{1}{2}} \tag{4.23}$$

where μ is the permeability of the conductor that, for non-magnetic materials, is taken as μ_0, the permeability of free space. Thus the fields and currents decay exponentially into the conductor and, at one skin depth δ, have decayed to e^{-1} of their surface values, i.e. $\delta = \alpha^{-1}$. Hence the skin depth

$$\delta = \left[\frac{2}{\omega\mu\sigma}\right]^{\frac{1}{2}} \quad \text{m} \tag{4.24}$$

This definition of skin depth is useful since the total current for a uniform current density J_0 within this depth, as illustrated in Figure 4.2, is identical to that of the actual current with the same current density J_0 at the surface and an exponential variation into the conductor. It may be assumed that if there are at least three skin depths of conductor thickness from each surface, i.e. to the 5% maximum current density level, then the assumption of the uniform current distribution will produce errors that are negligible in comparison with other possible sources of error, such as the effects of surface roughness.

Example 4.3

Calculate the skin depth for a copper conductor at 1.0GHz and compare it with the thickness of metal for a substrate with a 1 oz. copper cladding (i.e. $t = 35\ \mu m$).

Solution:

From (4.24), with $\sigma = 5.8\times 10^7\ \text{S.m}^{-1}$ and $\mu = \mu_0 = 4\pi\times 10^{-7}\ \text{H.m}^{-1}$

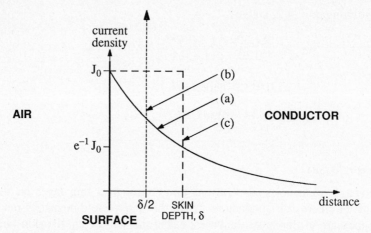

Figure 4.2 Current density variations with depth from the surface of a conductor, showing (a) the exponential decay of the current density, (b) the plane for the current sheet for inductance calculations and (c) the equivalent current distribution for surface resistivity calculations

$$\delta = \left\{ \frac{2}{2\pi 10^9 \times 4\pi 10^{-7} \times 5.8 \times 10^7} \right\}^{\frac{1}{2}} \, \text{m}$$

i.e. $\delta = 2.1 \, \mu\text{m}$

A copper conductor that is 35 μm thick represents a thickness of some 17 skin depths and there will not be any significant interference between the currents on opposite faces of the conductor.

Example 4.3 shows that the current flows within a very thin layer on the surface of the conductor. A conductor that is deposited onto a substrate will have a surface roughness similar to that of the substrate. The r.m.s. surface roughness may easily be 1 μm or greater and will cause the current to flow along a longer path across the surface of the conductor, giving increased conductor losses along the microstrip line. If α_c is the conductor loss with perfectly smooth conductors, then the practical loss α_c' accounting for surface roughness, as given by Hammerstad and Bekkadel [4.5], is

$$\alpha_c' = \alpha_c \left[1 + \frac{2}{\pi} \arctan(1.4\Delta/\delta) \right] \tag{4.25}$$

where Δ is the r.m.s. surface roughness. This equation fits the trends of the theoretical excess conductor loss when $\Delta < 2\delta$, as evaluated by Morgan [4.6] for three surfaces with known surface roughness.

The surface impedance Z_S of a conductor is required for attenuation calculations. It is determined by the voltage/current ratio at the surface and is equal to the input wave impedance seen by a wave traveling from the surface into an infinite thickness of the conductor. Thus, from (1.12), with no magnetic losses and with $\sigma \gg \omega\varepsilon$

$$Z_S = R_S + jX_S = \left\{ \frac{j\omega\mu}{\sigma} \right\}^{\frac{1}{2}} = (1+j)\left\{ \frac{\omega\mu}{2\sigma} \right\}^{\frac{1}{2}}$$

i.e.
$$Z_S = \frac{1+j}{\sigma\delta} \tag{4.26}$$

This surface impedance may also be written in the form

$$Z_S = \frac{1}{\sigma\delta} + j\omega\mu\left\{ \frac{\delta}{2} \right\} \tag{4.27}$$

from which it is now clear that the surface resistance is given by a uniform current flow over a depth δ and the surface reactance is inductive. The surface inductance $L_{surface}$ will be given by moving the current sheet from the surface, where it would be in the perfect conductor case, to a depth $\delta/2$. The true current density and the equivalent profiles for R and L calculations are illustrated in Figure 4.2. For a transmission line that has a surface with length l and width w

$$R = X = \frac{R_S l}{w} \tag{4.28}$$

Calculating X and deducing R from $R = X$ of (4.28) is better than calculating R directly using (4.26). This is because in practice there are problems in the direct estimation of R, due to the difficulties in finding the spatial dependence of the current that flows in the conductors. The inductance per unit length of a line at high frequencies normally assumes that all the current flows on the surface of the conductor, i.e. inductance external to the conductor L_{ext} is calculated, which ignores the magnetic flux that is actually within the conductive medium. However, as seen earlier, it is more precise to assume that the current sheet, rather than flowing on the surface, flows at a depth $\delta/2$ as shown in Figure 4.2. This calculation will give a greater value for the total inductance where the difference is attributed to the surface inductance, $L_{surface}$. Thus

$$L_{surface} = L_{total} - L_{ext} \tag{4.29}$$

It is $L_{surface}$ that is responsible for the reactance X in (4.28), so that $R = \omega L_{surface}$ may now be evaluated if the change in line inductance is calculated as perfect conducting walls recede by $\delta/2$ as shown in Figure 4.3. This procedure, known as the "incremental-inductance rule" of Wheeler [4.7], gives the line resistance due to the skin effect, but is based entirely on inductance calculations.

For inductance calculations that lead to the resistance, it is only necessary to

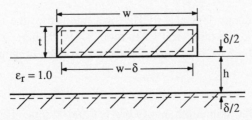

Figure 4.3 Receding surfaces for the calculation of internal inductance

have the conductors in air by taking $\varepsilon_r = 1.0$ for the substrate region. Using (4.5) for the equivalent width of a conductor may make negligible difference to the characteristic impedance of a line with thin conductors. However, as discussed by Pucel et al. [4.8], the thickness must be included in incremental inductance calculations, since the change in thickness of the line when taking the receding surfaces is significant in the difference calculations. The geometry of the problem is illustrated in Figure 4.3. From (1.17) and (1.27), the inductance may be given in terms of the characteristic impedance as

$$L = \frac{Z_0(air)}{c} \tag{4.30}$$

Thus, $R = \omega L_{surface} = \frac{\omega}{c} \left\{ Z_0(w-\delta, h+\delta, t-\delta, \varepsilon_r = 1) - Z_0(w, h, t, \varepsilon_r = 1) \right\}$

$$\tag{4.31}$$

The variation of power transmitted along a lossy line is

$$P(z) = P_0 e^{-2\alpha z} \tag{4.32}$$

On differentiating this expression with respect to z, the attenuation coefficient is found as

$$\alpha = \frac{(\text{Power loss per unit length})}{2 \times (\text{Power transmitted})} \tag{4.33}$$

In terms of total voltages and currents of a single traveling wave and the parameters of the line

$$\alpha = \frac{I^2 R}{2VI} \quad \left[= \frac{R}{2Z_0} \right] \tag{4.34}$$

This equation is of course equivalent to (1.22). However, at higher frequencies when the quasi-static model of the line is no longer accurate, (4.33) is still valid as the basis for attenuation calculations, with the transverse surface currents included in the power loss calculations.

Thus

$$\alpha_c = \frac{R}{2Z_0(\varepsilon_r)}$$

$$= \frac{\pi}{\lambda_0} \frac{Z_0(w-\delta, h+\delta, t-\delta, \varepsilon_r = 1) - Z_0(w, h, t, \varepsilon_r = 1)}{Z_0(w, h, t, \varepsilon_r)} \quad \text{neper.m}^{-1}$$

$$\tag{4.35}$$

Example 4.4

Calculate the conductor loss at 4.0GHz for a microstrip line that has the following parameters: $w = 0.508$ mm, $h = 1.270$ mm, $t = 0.009$ mm, $\varepsilon_r = 9.6$, and   copper conductors with $\sigma = 5.8 \times 10^7$ S.m^{-1}. The r.m.s. surface roughness of the conductors is 0.001 mm.

Solution:

From (4.24), the skin depth at 4.0GHz is 0.00104 mm. The thickness correction terms to the width for the strip conductor from (4.5) are:

i) For t = 0.009 mm and $\xi = 1.50$, then
$$\Delta w_1 = 0.01128 \text{ mm.}$$

ii) For t = 0.009 − δ = 0.00796 mm and $\xi = 1.50$, then
$$\Delta w_2 = 0.00997 \text{ mm.}$$

From Table 3.2, the following characteristic impedance calculations are made:

i) Using L_{ext} with $w' = w + \Delta w_1$ gives
$$Z_0(w' = 0.51928, h = 1.270, t' = 0, \varepsilon_r = 1.0) = 178.604 \ \Omega$$

ii) Using L_{total} with $w' = w - \delta + \Delta w_2$ and $h' = h + \delta$ gives
$$Z_0(w' = 0.51797, h = 1.27104, t' = 0, \varepsilon_r = 1.0) = 178.802 \ \Omega$$

iii) Using L_{ext}, but with the substrate present, from (4.7) gives $\Delta w' = 0.006$, and
$$Z_0(w' = 0.514, h = 1.27, t' = 0, \varepsilon_r = 9.6) = 72.63 \ \Omega$$

At 4.0 GHz, $\lambda_0 = 7.5$ cm and, from (4.35)
$$\alpha_c = \frac{8.686\,\pi}{7.5} \times \frac{178.802 - 178.604}{72.63} \ \text{dB.cm}^{-1}$$

i.e.
$$\alpha_c = 0.0099 \ \text{dB.cm}^{-1}$$

The surface roughness will increase the attenuation along the line. From (4.25)
$$\alpha_c' = \alpha_c \left\{ 1 + \frac{2}{\pi} \arctan \left[\frac{1.4 \times 0.001}{0.00104} \right] \right\}$$

giving an overall conductor loss of 0.016 dB.cm^{-1}.

The characteristic impedances that have been evaluated in this example using formulae in Table 3.2 are not accurate to the quoted number of significant figures. However, for small variations in the line parameters, any systematic errors in impedance will cancel when a difference value is found.

4.2.3 Shielding enclosures

In Chapter 3, the characteristic impedance and relative effective permittivity were derived for microstrip lines on the assumptions that only uniform lines were considered, that no additional lines were nearby and that both the ground plane and the region above the substrate were infinite in extent. In later chapters, when component designs are considered, the infinite nature of the ground plane and air region will be retained. However, in a practical system it will be necessary to have finite dimensions for the overall circuit. A modular system of enclosed housings, either at a component or subsystem level, not only provides easier handling and strength, but also may be required for internal or external electromagnetic shielding and hermetic sealing to reduce circuit deterioration from external elements.

The presence of sidewalls and a top plate modify the geometry of the ground plane. These additional conducting planes increase the line capacitance per unit length for a given w/h ratio and, in particular, increase the percentage of the energy that is transmitted through the air region. Thus both the characteristic impedance and effective relative permittivity of any line are decreased. A compromise for the

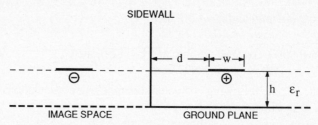

Figure 4.4 The presence of a sidewall, showing the odd-mode equivalent structure

housing between minimal circuit interference and overall compactness generally means that circuit effects should be small and may be considered as separate perturbations from one of the sidewalls and from the top plate.

A single sidewall

Figure 4.4 illustrates the geometry of a single microstrip line in the vicinity of a sidewall. The boundary conditions at the sidewall are maintained with the use of an image line. The combination of the line and its image is identical to the odd-mode configuration that will be described for parallel-coupled lines in Chapter 8.

The equations in Chapter 8 for the odd-mode impedance, where two lines separated by a distance $s = 2d$ are driven in anti-phase, are used directly to give the impedance of a single line in the proximity of a sidewall. The results, giving the line impedance normalized to Z_{0o}, the value of Z_0 as $d \to \infty$, and as a function of d/h, for

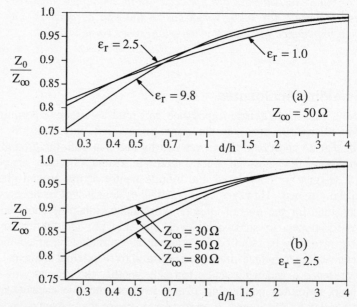

Figure 4.5 The reduction of characteristic impedance with the proximity to a sidewall showing (a) variations for 50Ω lines on three different substrates and (b) variations for three line-impedances on a single substrate

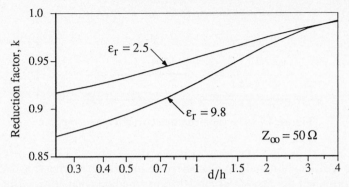

Figure 4.6 The reduction of ε_{eff} by a factor, k, with the proximity to a sidewall

three substrate relative permittivities (1.0, 2.5 and 9.8), are plotted for lines that have $Z_0 = 50\,\Omega$ in Figure 4.5a. The plots are fairly insensitive to ε_r and may be used to estimate the effects on the impedance over the permittivity range $1.0 < \varepsilon_r < 10.0$. The procedure for designing a $50\,\Omega$ line that has to pass close to a sidewall is illustrated in Example 4.5. Further plots that show the reduction of line impedance for 30, 50 and $80\,\Omega$ lines on an $\varepsilon_r = 2.5$ substrate are illustrated in Figure 4.5b.

Not only is the characteristic impedance reduced by the presence of the wall, but so also is the effective relative permittivity, now that there is a higher percentage of energy transfer above the substrate. ε_{eff} will be required if any wavelength dependent lines appear as parts of the circuit near the sidewall. In Figure 4.6, the reduction factor, k, for ε_{eff} is shown as a function of the distance of a nominal $50\,\Omega$ line from the sidewall for $\varepsilon_r = 2.5$ and 9.8. More precise calculations may be made for other line combinations by evaluating, from Appendix 4, the odd-mode effective permittivity for a coupled line.

Example 4.5

A $50\,\Omega$ line on a 1.58 mm thick, $\varepsilon_r = 2.5$ substrate is situated adjacent to a sidewall of an enclosure with 2.0 mm clear substrate between the line and the wall. What is the required width for the line?

Solution:

From the dimensions that are given, with d/h = 1.27, a $50\,\Omega$ impedance line with w/h = 2.837 is reduced by a factor of 0.95 to $47.5\,\Omega$. Assuming that this factor is insensitive to small impedance changes, the geometry of a $50/0.95 = 52.6\,\Omega$ line is required. From Appendix 3, for this line w/h = 2.625, giving w = 4.15 mm. In this case, a line width reduction of 0.34 mm is required for a $50\,\Omega$ line due to the proximity of the wall. The edge of the line remains at 2.0 mm from the wall.

The top plate

Consider a microstrip line with a metallic shield placed as a top or cover plate at a distance H above the substrate and strip as illustrated in Figure 4.7. The line

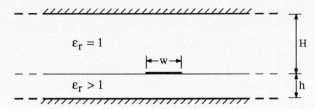

Figure 4.7 The geometry of a covered microstrip line

capacitance will increase and the characteristic impedance decrease as H is reduced. Bahl [4.9] provided the first set of closed-form equations that allowed for the effects of the top plate. These were corrected and extended by March [4.10], on whose expressions the following equations are based.

With reference to Figure 4.7 and with $T = (1 + H/h)$ and $V = w/h$, the characteristic impedance of the unshielded microstrip line *with air dielectric for the substrate region* is modified to give

$$Z_0(\text{air, shielded}) = Z_0(\text{air, unshielded}) - \Delta Z_0 \tag{4.36}$$

where

$$\Delta Z_0 = 270\left\{1 - \tanh\left[1.192 + 0.706\sqrt{T} - \frac{1.389}{T}\right]\right\} \times$$

$$\times \left\{1.0109 - \tanh^{-1}\left[\frac{0.012\,V + 0.177\,V^2 - 0.027\,V^3}{T^2}\right]\right\} \tag{4.37}$$

The effective relative permittivity is reduced by the presence of the cover plate. It is still expressed in terms of an effective filling factor and the substrate permittivity with

$$\varepsilon_{\text{eff}} = 1 + q(\text{shielded}) \times (\varepsilon_r - 1) \tag{4.38}$$

where

$$q(\text{shielded}) = 0.5 + q' \times \big(q(\text{unshielded}) - 0.5\big)$$

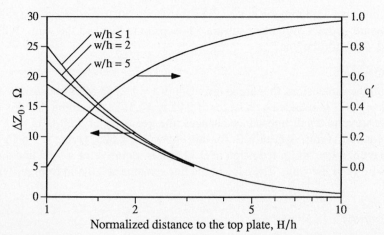

Figure 4.8 Correction factors for the presence of a top plate in the case of a shielded microstrip line

and
$$q' = \tanh\left[0.922 + 0.121\,T - \frac{1.164}{(T-1)}\right], \quad H \geq h \tag{4.39}$$

When $H = h$, giving $T = 2$ and $q' = 0.5$, the electric flux density is symmetrical about the plane of the strip and $\varepsilon_{eff} = (\varepsilon_r + 1)/2$ for all values of w/h. Finally, the characteristic impedance of the shielded line is given using (4.36) and (4.38) as

$$Z_0(\varepsilon_r,\text{ shielded}) = \frac{Z_0(\text{air, shielded})}{\sqrt{\varepsilon_{eff}}} \tag{4.40}$$

ΔZ_0 from (4.36) for an air line is plotted as a function of H/h with w/h as a parameter in Figure 4.8. The correction factor q' is also shown in the figure.

4.3 SUBSTRATE MATERIALS

It will have been observed that, up to this stage, the majority of examples have been concerned with a substrate with $\varepsilon_r = 2.5$. While this value is typical of a range of materials and has been used in Appendix 3 for calculations, there are also many other suitable substrate materials. In Table 4.1, a brief description of the properties of some of the substrate materials is given, highlighting some of the advantages and disadvantages of them. A more thorough discussion of this topic may be found in Hoffmann [4.11].

4.4 DISPERSION

In Chapter 3, a microstrip line was considered with low frequency approximations that allowed the line to be described in terms of its capacitance per unit length both with and without the presence of a dielectric substrate material. The majority of topics in this book will be presented in terms of the low frequency or quasi-static assumptions to allow the underlying principles of components and systems to be studied. However, an attractive feature of the microstrip line is that it may be used at high frequencies with compact circuit configurations, but where the cross-section dimensions of the line are no longer very small compared with the operating wavelength. The quasi-static assumptions that require predominantly transverse electromagnetic fields are inadequate at high frequencies and a full-wave solution allowing for the complete matching of the fields at the dielectric-air interface is needed.

4.4.1 The effective relative permittivity

Among the early work on the dispersive effects in microstrip lines, Chudobiak et al. [4.12] derived empirical expressions to fit both their experimental ε_{eff} results and those of Troughton [4.13] and Arnold [4.14]. Their expressions were straight line approximations for the frequency dependence of ε_{eff} for frequencies greater than f_0, the frequency above which dispersion no longer may be neglected. This frequency limit is given [4.12] as

$$f_0 = 0.95 \times \left\{\frac{Z_0}{h\sqrt{\varepsilon_r - 1}}\right\}^{\frac{1}{2}} \quad \text{GHz, with } h \text{ in mm and } Z_0 \text{ in } \Omega \tag{4.41}$$

Table 4.1 The properties of substrate materials

Material	ε_r	$\tan\delta$ @ 10GHz	Advantages	Disadvantages	†
Air	1.0006	≈ 0	Pure TEM wave. Assumed as free space.	No support for strip. Large physical size.	
PTFE	2.1	0.0003		High thermal expansion. Poor mechanical properties.	
Reinforced plastics	2.3 - 2.6	< 0.001	Improved mechanical properties. Large sizes — low cost.	High thermal expansion. Variability between batches.	1
Quartz	3.78	0.0001	Reproducible substrates. Useful in the mm range. Low thermal expansion.		
Ceramic loaded plastics	2.5 - 10	< 0.002	Good mechanical properties.	Variability between batches.	
Alumina 99.5% pure	9.8	0.0001	Reproducible substrates — improving with purity.	Brittle. Slightly anisotropic.	1
Sapphire Al_2O_3	9.4 11.6	0.0001	Reproducible substrates. Very smooth surfaces.	Anisotropic crystal. Small size — high cost.	2
GaAs	12.9	0.002	Integration with high frequency active devices.	Thin substrates (~ 0.15 mm).	3
Rutile TiO_2	85	0.004	Reduced size components.	Rough surfaces. Temperature sensitive ε_r.	4

† *Notes*

1.   Readily available and popular substrate materials.
2.   The cut with the C-axis normal to the ground plane gives electrical properties that are independent of the direction of propagation across the substrate. For electric fields along the C-axis, $\varepsilon_r = 11.6$.
3. $\tan\delta < 0.001$ for the high resistivity material.
4.   A negative thermal coefficient of permittivity.

The model upon which (4.41) is based is a linear piecewise fit to the low frequency variations of ε_{eff}. A significant advance was made by Getsinger [4.15] with an analysis of a waveguide model that possessed similar field characteristics to the microstrip line along the dielectric-air interface near the strip. Although it is stated that the analysis only holds for thin substrates, $h < \lambda_s/4$, and lines with $w < \lambda_s/3$ where λ_s is the wavelength of a plane wave in the substrate material, the analysis does show the distinctive features of the variation of effective permittivity with frequency. If $\varepsilon_{eff}(f)$ is the effective relative permittivity as a function of frequency, then it should possess the following properties:

i) $\varepsilon_{eff}(f)$ has a zero derivative with respect to frequency at $f = 0$ and increases smoothly with frequency.

ii) At high frequencies, $\varepsilon_{eff}(f)$ approaches an asymptotic value that is equal to ε_r, when all the energy is transmitted through the substrate region.

iii) An inflection frequency exists at which the second derivative of $\varepsilon_{eff}(f)$ with frequency is zero.

Atwater [4.16] compares seven published formulae for $\varepsilon_{eff}(f)$ with 120 data values that have been selected from a wide range of publications and concludes that the closed-form design equation of Kirschning and Jansen [4.17] shows the lowest average deviation from the measured results. Further measurements by Deibele and Beyer [4.18] support the use of this expression for circuits that have a high (frequency) × (substrate height) product, expressed in GHz.mm, up to the quoted limit of validity of the expressions, namely 39 GHz.mm. This limit corresponds to a maximum value for the substrate thickness of 0.13 × (free space wavelength). The Kirschning and Jansen expression is based on extensive numerical data, that were derived using the full-wave theoretical analysis of Jansen [4.19]. Other more recent variations in the approach to the full-wave solution are given by Kobayashi and Ando [4.20] and Shih et al. [4.21]. In these papers, emphasis is placed on the two-dimensional representation of the currents in the strip, as the incorrect treatment of the singularities associated with the currents may otherwise be a major source of error.

As the emphasis throughout this book is towards the lower frequency analysis of microstrip circuits, a simpler expression by Pramanick and Bhartia [4.22] is presented here. This expression, shown by Atwater to be one of the more accurate formulations, will give satisfactory results unless the high range of $(f \times h)$ products are to be used. As derived from the equations in [4.22]

$$\varepsilon_{eff}(f) = \varepsilon_r - \frac{\varepsilon_r - \varepsilon_{eff}(0)}{1 + (f/f_T)^2} \tag{4.42}$$

with
$$f_T = \left\{ \frac{\varepsilon_r}{\varepsilon_{eff}(0)} \right\}^{\frac{1}{2}} \frac{Z_0}{2\mu_0 h}, \quad \text{h in meters} \tag{4.43}$$

An accurate, but more complicated, dispersion formula by Kobayashi [4.23] has been developed from analytical results [4.20] for use at higher frequencies in microwave computer aided design.

Example 4.6

With quasi-static approximations, a 50Ω microstrip line on $\varepsilon_r = 2.5$ substrate has the parameters $w/h = 2.837$ and $\varepsilon_{eff}(0) = 2.09$. If the substrate thickness is 1.58 mm, plot the variation of the effective relative permittivity for frequencies up to 10.0 GHz. Indicate the frequency range over which the quasi-static approximations may be used.

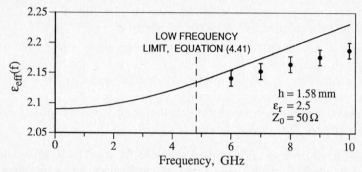

Figure 4.9 The frequency dependence of the effective relative permittivity from Equation (4.42). Specific points with bars for their quoted accuracy are derived from the equations of Kirschning and Jansen [4.17].

Solution:

In this solution, dispersion results for $(f \times h)$ up to $15.8\,\text{GHz.mm}$ are given. From (4.43)

$$f_T = \left[\frac{2.5}{2.09}\right]^{\frac{1}{2}} \times \frac{50}{2 \times 4\pi\,10^{-7} \times 0.00158}$$

$$= 13.8\,\text{GHz}.$$

Now with the frequency in GHz, (4.42) becomes

$$\varepsilon_{\text{eff}}(f) = 2.5 - \frac{0.41}{1 + (f/13.8)^2}$$

This expression, together with the quasi-static limit of (4.41) and specific values calculated from the more comprehensive equations of [4.17], are plotted in Figure 4.9.

4.4.2 The characteristic impedance

The concept of characteristic impedance may only be applied in a rigorous manner to a pure TEM transmission line as the ratio of voltage to current for a propagating wave along the line. The hybrid nature of the fundamental mode of a microstrip line requires that careful consideration be given to the meaning of characteristic impedance before the frequency dependence of the value can be determined. Associated with the propagating wave, there are three quantities — the average power flow, P_{av}, the magnitude of the total longitudinal current in each conductor, I_z, and the potential difference, V_x, between the conductors along the line of maximum electric field strength between the centers of the conductors. Definitions of the characteristic impedance may be made in terms of any two of the quantities, noting that V_x and I_z will be the peak quantities. Thus

$$Z_0(\text{P,I}) = \frac{2\,P_{av}}{I_z^2} \tag{4.44}$$

$$Z_0(P,V) \;=\; \frac{V_x^2}{2\,P_{av}} \tag{4.45}$$

and $\qquad Z_0(V,I) \;=\; \frac{V_x}{I_z} \;=\; \sqrt{Z_0(P,I)\times Z_0(P,V)} \tag{4.46}$

Once the quasi-static approximations are no longer valid, each definition leads to a different value for the characteristic impedance with $Z_0(P,I)$ being the geometric mean of the other two values. Jansen and Kirschning [4.24] discuss how the appropriate choice of expression is made in such a way that the theoretical full-wave solution and the use of characteristic impedance in circuit analysis give the best agreement for the majority of line configurations. The power in the propagating wave must be one of the terms in the impedance definition, if the power properties of associated network matrices in Chapter 2 are to be described correctly. Further, in [4.24], it is argued that since I_z is much less dependent than V_x with the increase in frequency for a constant power, then $I_z(f)$ is more suitable for use in the TEM-equivalent concept of characteristic impedance.

The use of (4.44) with its insensitive dependence on frequency has been supported by measurement, particularly when broadband and low reflection transitions have been considered between 50Ω coaxial and microstrip lines (England [4.25] and Majewski et al. [4.26]). As with the dispersion effects on the effective relative permittivity, so Jansen and Kirschning [4.24] have published a detailed expression for this $Z_0(f)$, based on data generated by the full-wave solution. Within the context of this chapter, it suffices to present typical variations in Figure 4.10 for lines on $\varepsilon_r = 2.5$ and 9.8 substrates. These curves show typical trends for $Z_0(f)$ with

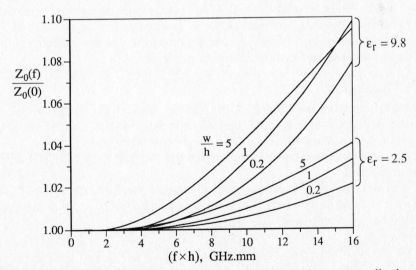

Figure 4.10 The frequency dependence of characteristic impedance, normalized to the quasi-static characteristic impedance and based on the equations of Jansen and Kirschning [4.24]

$(f \times h)$ up to a maximum of about 15 GHz.mm, corresponding to 10 GHz for a 1.5 mm thick substrate.

Example 4.7

Estimate the line width for a $50\,\Omega$ microstrip line on a 1.58 mm thick, $\varepsilon_r = 2.5$ substrate at 6.3 GHz.

Solution:

> For this substrate at 6.3 GHz, $(f \times h) = 10.0$ GHz.mm. From Figure 4.10, with $w/h \approx 2.8$ for a $50\,\Omega$ line (from Appendix 3)
>
> $$Z_0(f) \ = \ 50\,\Omega \ = \ 1.025\,Z_0(0)$$
>
> from which it is seen that a $48.8\,\Omega$ quasi-static characteristic impedance will provide the required $50\,\Omega$ impedance at 6.3 GHz. From Appendix 3, for a $48.8\,\Omega$ line
>
> $$w/h = 2.943, \quad \text{i.e. } w = 4.65 \text{ mm}$$
>
> The effect of dispersion may be judged, if this value is compared with $w = 4.48$ mm for a quasi-static $50\,\Omega$ line.

4.5 OTHER MODES OF PROPAGATION

A microstrip transmission line is an unbounded structure in the transverse plane and any discontinuity in the uniform nature of the line may generate other modes of propagation. A surface wave mode that is guided in the plane of the substrate, with the substrate on the ground plane providing the guiding structure, is discussed in §4.5.2 but is a less significant effect than that of radiation into the surroundings of the line. Nevertheless, it is important that all unwanted modes are reduced to a minimum to prevent cross-coupling and interference in other parts of a circuit. The choice of substrate parameters, line dimensions and the dimensions of any shielding structures all play an important part in this minimization process.

4.5.1 Radiation

Radiation from a microstrip line is a desirable effect in the design of microstrip antennas in Chapter 10 but what is desired for the design of non-radiating circuits is narrow high-impedance lines on thin high-permittivity substrates, in shielding structures that possess a cut-off frequency above the frequency of operation of the microstrip circuit.

Radiation power calculations generally follow the approach of Lewin [4.27], who has evaluated the power radiated from a discontinuity, P_{rad}. From the results of [4.27], for an incident power P_{in}

$$\frac{P_{rad}}{P_{in}} \ = \ 2\pi \frac{\eta_0}{Z_0} \left[\frac{h}{\lambda_0} \right]^2 F(\varepsilon_r)$$

(4.47)

where $\eta_0 \approx 376.7\,\Omega$ is the intrinsic impedance of free space and $F(\varepsilon_r)$ is a factor that is to be determined for each type of discontinuity.

The small percentage of power radiated from the open-circuit termination of a microstrip line may be modeled in terms of a shunt conductance, G_{rad}, across the open circuit. Ideally, $G_{rad} \ll 1/Z_0$ and the voltage across it is essentially that across a perfect open circuit, namely

$$V_{oc} \approx 2V_{in} = 2\sqrt{P_{in}Z_0} \tag{4.48}$$

where V_{in} is the voltage of the wave incident upon the open circuit. Thus

$$G_{rad} = \frac{P_{rad}}{V_{oc}^2} = \frac{1}{4Z_0} \times \frac{2\pi\eta_0}{Z_0}\left[\frac{h}{\lambda_0}\right]^2 F(\varepsilon_r) \tag{4.49}$$

$$= \frac{60\pi^2}{Z_0^2}\left[\frac{h}{\lambda_0}\right]^2 F(\varepsilon_r) \tag{4.50}$$

Van der Pauw [4.28] derived more exact expressions than those of Lewin but, following the work of Abouzahra and Lewin [4.29], better agreement between the two approaches was obtained, once the effective permittivity derived by Lewin was replaced in the various parts of the expressions for $F(\varepsilon_r)$ by ε_r or ε_{eff} as appropriate. Abouzahra [4.30] derived a more general formula for $F(\varepsilon_r, \varepsilon_{eff})$, in terms of the current reflection coefficient of the termination on the microstrip line, and applied it to both matched and short-circuit terminations, as well as the open-ended line. It should be noted that in all these derived expressions, the termination is at a plane coincident with the edge of the substrate and ground plane.

For an open circuit, with a current reflection coefficient of -1, $F(\varepsilon_r, \varepsilon_{eff})$ is plotted, Figure 4.11, as a function of the characteristic impedance of the line for a substrate relative permittivity of 2.5. Increasing the permittivity of the substrate reduces the value of $F(\varepsilon_r, \varepsilon_{eff})$, as is seen from the asymptotic expression for large ε_r, given by Abouzahra and Lewin [4.29], namely

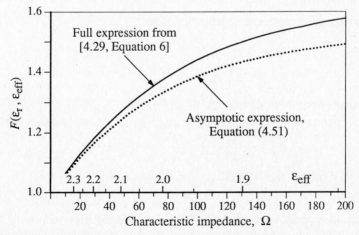

Figure 4.11 The radiation factor, $F(\varepsilon_r, \varepsilon_{eff})$, for an open-circuit termination on a substrate with $\varepsilon_r = 2.5$ as a function of the line impedance, i.e. as w/h is varied

$$F(\varepsilon_r, \varepsilon_{eff}) \rightarrow \frac{8}{3\varepsilon_{eff}}\left\{1 + \frac{4}{5\varepsilon_{eff}} - \frac{1}{\varepsilon_r} + \cdots\right\} \tag{4.51}$$

The accuracy of (4.51) improves for lower impedance lines with their higher effective permittivities on any given substrate, as seen in Figure 4.11, where a comparison is made with the full expression for $\varepsilon_r = 2.5$. It also improves with the use of higher relative permittivity substrates.

Example 4.8

Consider the open-circuit termination on a 50Ω characteristic impedance microstrip line that is used as a matching element at 2.0GHz. The substrate is 1.58 mm thick with $\varepsilon_r = 2.5$ and with no shielding around it. Calculate

i)   the percentage power loss due to radiation from the open circuit,
ii) the equivalent radiation conductance, G_{rad},
iii) the V.S.W.R. on the microstrip line.

Solution:

i)   From Figure 4.11, the open-circuit termination radiation factor for a 50Ω characteristic impedance line on $\varepsilon_r = 2.5$ substrate is

$$F(\varepsilon_r, \varepsilon_{eff}) = 1.28$$

At 2.0GHz, the free space wavelength is 150 mm. Now using (4.47), the radiated power for one watt of incident power onto the termination is

$$P_{rad} = 2\pi \frac{376.7}{50}\left[\frac{1.58}{150}\right]^2 \times 1.28$$

$$= 6.7 \text{ mW}$$

Thus there is 0.67% power loss by radiation from the open circuit termination.

ii) The radiation conductance from (4.49) is

$$G_{rad} = \frac{1}{4 \times 50} \times 0.0067$$

$$= 33.6\,\mu S$$

iii) The 50Ω line is terminated by a radiation resistance of $29.7k\Omega$ instead of a true open circuit. This load resistance gives a voltage reflection coefficient of

$$|\Gamma| = \frac{29700 - 50}{29700 + 50} = 0.9966$$

and a V.S.W.R. of about 600.

Note
Practical transmission line losses on their own will probably reduce the V.S.W.R. from the ideal value of infinity to a value less than the one just calculated.

4.5.2 Surface waves

A dielectric sheet on a ground plane will act as a guiding structure for surface waves, see e.g. Collin [4.31]. With the ground plane in the y-z plane and with a propagating lossless wave in the z-direction given by $e^{j(\omega t - \beta z)}$, a solution for the complete set of TM and TE-modes may be derived. Now for situations where field components in the direction of propagation have to be included in the solution, it is useful to distinguish between the different phase coefficients, k_0 and β. In the direction of propagation, $\beta = 2\pi/\lambda$, while for a plane wave in free space, $k_0 = 2\pi/\lambda_0$. In the substrate, the field components are uniform in the y-direction and vary as $\frac{\sin}{\cos}(px)$ along the normal to the ground plane. Above the ground plane, the fields decay as e^{-ux}. The electric field lines for the lowest-order TM-mode are illustrated in Figure 4.12. The magnetic field lines are in the transverse x-y plane and are proportional and orthogonal to the transverse components of electric field. Now p and u, each controlling the field variations in the transverse plane, are related to β which must be equal in the two regions, giving

$$\beta^2 = k_0^2 + u^2 \quad \text{and} \quad \beta^2 = \varepsilon_r k_0^2 - p^2 \tag{4.52}$$

Thus, it follows that

$$u^2 + p^2 = (\varepsilon_r - 1) k_0^2 \tag{4.53}$$

The waves may propagate in any direction parallel to the surface and do not require the strip for guidance. The lowest order mode, which is the TM-mode, will propagate down to d.c., but is so weakly guided by the surface that the waveguide wavelength will be almost as long as the free space wavelength and vastly different from the microstrip-mode wavelength. Matching the H_y and E_z field components across the air-dielectric boundary gives [4.31]

$$\frac{ph}{\varepsilon_r} \tan(ph) = uh \tag{4.54}$$

The intersection of plots of (4.53) and (4.54) in the first quadrant of a diagram with uh and ph as the axes gives the solution for surface wave modes. Solving for (ph) for the lowest order mode, eliminate u from (4.53) and (4.54) and use

$$\left[\frac{u}{p}\right]^2 = \frac{u^2}{k_0^2} \times \frac{k_0^2}{p^2} = \frac{\beta^2 - k_0^2}{k_0^2} \times \frac{k_0^2}{\varepsilon_r k_0^2 - \beta^2} = (\varepsilon_{\text{eff}} - 1) \frac{1}{\varepsilon_r - \varepsilon_{\text{eff}}} \tag{4.55}$$

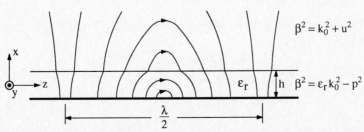

Figure 4.12 The electric fields of the lowest-order TM-mode surface wave, with no field variation in the y-direction of the transverse plane

where ε_{eff}, still defined as $(\lambda_0/\lambda)^2$, is the effective relative permittivity for the appropriate surface wave mode. Thus from (4.54)

$$ph = \arctan\left[\varepsilon_r \frac{u}{p}\right] = \arctan\left[\varepsilon_r \left\{\frac{\varepsilon_{eff}-1}{\varepsilon_r - \varepsilon_{eff}}\right\}^{\frac{1}{2}}\right] \tag{4.56}$$

Now, substituting for p and rearranging gives

$$\frac{h}{\lambda_0} = \frac{1}{2\pi\sqrt{\varepsilon_r - \varepsilon_{eff}}} \times \arctan\left[\varepsilon_r \left\{\frac{\varepsilon_{eff}-1}{\varepsilon_r - \varepsilon_{eff}}\right\}^{\frac{1}{2}}\right] \tag{4.57}$$

This equation gives the substrate thickness that is required if the lowest order TM-mode is to propagate with a guided wavelength such that the effective relative permittivity is ε_{eff}. It is the only mode that has no lower cut-off frequency and, at low frequencies, the majority of the energy is transmitted above the substrate so that $\varepsilon_{eff} \approx 1$. As the propagation coefficients for this surface wave mode and for the normal microstrip quasi-TEM mode are quite different, the coupling between the modes will be negligible at low frequencies. At high frequencies, the two phase coefficients can become comparable, since ε_{eff} for the surface wave will increase with frequency. This situation will most likely occur with very narrow microstrip lines since, when $w/h \to 0$, the microstrip $\varepsilon_{eff} \approx (\varepsilon_r + 1)/2$ and is its minimum value. Substituting this value into (4.57) gives the lowest frequency, f_s, at which substantial coupling to the surface wave may occur, i.e.

$$f_s = \frac{c \times \arctan(\varepsilon_r)}{\pi h \sqrt{2(\varepsilon_r - 1)}} \tag{4.58}$$

Coupling to the surface mode is avoided by always operating at frequencies below f_s.

Example 4.9

i)   Below what frequency should a 5.0 mm thick, $\varepsilon_r = 2.5$, substrate be used if surface wave coupling is to be avoided?

ii) If in fact the smallest line width to be used on the substrate is $w/h = 0.1$, at what frequency will synchronous coupling with the lowest-order TM-surface wave occur?

Solution:

i)   To avoid coupling with the surface wave, the maximum frequency to be used from (4.58) is

$$f_s = \frac{3 \times 10^8 \arctan(2.5)}{\pi \times 0.005 \times \sqrt{2(2.5 - 1)}}$$

$$= 13.1 \text{ GHz}.$$

ii) Using the data from Appendix 3, for $\varepsilon_r = 2.5$ and $w/h = 0.1$, the effective relative permittivity is 1.851. Now, from (4.57)

$$f_s = \frac{3 \times 10^8}{2\pi \times 0.005 \times \sqrt{2.5 - 1.851}} \times \arctan\left[2.5\left\{\frac{1.851 - 1}{2.5 - 1.851}\right\}^{\frac{1}{2}}\right]$$

$$= 14.6\,\text{GHz}.$$

4.5.3 Transverse microstrip resonance

A possible resonant structure results when two open-circuit planes are $\lambda/2$ apart. This situation occurs in the transverse plane of a microstrip line as illustrated in Figure 4.13. For the lowest resonant frequency in the transverse plane, the wave propagation is TEM in the y-direction, the fields are constant in the z-direction and (E_x, H_z) are the fields that are present. While the magnetic field in this idealized resonant mode is orthogonal to that in the usual propagating wave along a microstrip line, the presence of any form of line discontinuity may limit the extent of the resonant mode in the z-direction and, in closing the magnetic loops in the y-z plane, be sufficient to give coupling to the (E_x, H_y) fields of the usual propagating mode. The end-correction length, Δl, at the open circuit for a very wide line (*here, in the sense of a very large width along the z-axis*), is a function of ε_r and h and is considered in detail in §5.2. For dielectric substrates with $\varepsilon_r > 2$, it will be seen in Figure 5.2 that $\Delta l \approx 0.5\,\text{h}$. Thus, an estimate of resonance conditions in terms of the wavelength, λ_s, of a plane wave in the substrate is given by

$$w + 2\Delta l = \frac{\lambda_s}{2} = \frac{\lambda_0}{2\sqrt{\varepsilon_r}} \tag{4.59}$$

or

$$f_{res} = \frac{c}{2\sqrt{\varepsilon_r}(w + h)} \quad \text{with } 2\Delta l \approx h \tag{4.60}$$

The transverse microstrip resonance is only supported if transverse currents can flow in the strip. These currents may be suppressed with minimal effect on the propagating mode by having narrow longitudinal slots in the strip.

Example 4.10

Estimate the free space wavelength and the frequency of the lowest transverse microstrip resonance for a line with w/h = 3.0 on a 0.5 mm thick, $\varepsilon_r = 9.7$ substrate.

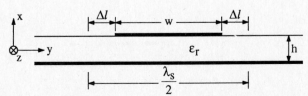

Figure 4.13 The line geometry for a transverse microstrip resonance

Solution:

From (4.59), the free space wavelength at resonance for the lowest order transverse mode is given by

$$\lambda_0 \approx 2\sqrt{\varepsilon_r}(w+h)$$

Substituting $\varepsilon_r = 9.7$, $h = 0.5$ mm and $w = 1.5$ mm, gives $\lambda_0 = 12.5$ mm and a resonant frequency of 24.0GHz.

4.5.4 Waveguide cavity resonance

The shielding of a microstrip circuit in an enclosed metallic box creates a waveguide cavity that is partially loaded with the dielectric material of the substrate as shown in cross-section in Figure 4.14.

Consider the cross-section of the waveguide without the substrate present. The lowest frequency for propagation is in the TE-mode that has a maximum electric field across the center of the waveguide between the centers of the broad faces. This is the dominant mode. From the theory of rectangular waveguides [4.32], the free space wavelength at the lowest frequency that may propagate, called the cut-off wavelength λ_c, is given for $A > B$ by

$$\lambda_c = 2A \tag{4.61}$$

The only field component not required in the complete field solution for the dominant mode when the substrate is present is H_x. Thus the magnetic fields lie in the y-z plane and the mode is described as being a Longitudinal Section Magnetic (LSM) mode. An approximate expression for the LSM-mode cut-off wavelength

$$\lambda_c = 2A \left\{ 1 + (\sqrt{\varepsilon_r} - 1)\frac{h}{B} \right\} \tag{4.62}$$

gives a small overestimate of λ_c for intermediate substrate heights. If the length (i.e. along the z-axis) of the shielding box is D, with $D > B$, the free space wavelength for the lowest order resonance when no dielectric is present is

$$\left[\frac{2}{\lambda_{res}}\right]^2 = \left[\frac{2}{\lambda_c}\right]^2 + \left[\frac{1}{D}\right]^2 \tag{4.63}$$

Note that with $\lambda_{res} < \lambda_c$, the resonant frequency of the shielding box is greater than the cut-off frequency of the transverse cross-section waveguide. The parasitic

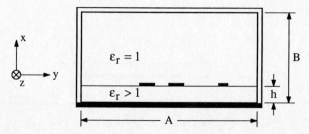

Figure 4.14 A typical cross-section of a shielded microstrip line circuit

waveguide resonator modes may be excited by the radiation from microstrip component discontinuities and coupled between any two such radiation points within the cavity. Since the microstrip conductors will perturb the cavity fields, it is not practical to calculate accurate resonant frequencies. Effective suppression of the resonant modes is required if a resonant frequency occurs near or below the frequencies used in the circuit. Suppression may involve moving the resonances to higher frequencies by using metallic posts in the x-direction and placed near the center line given by $y = A/2$. The posts must be kept clear of the microstrip lines. Further suppression of the modes may be achieved with lossy dielectric materials that are placed alongside the top or side plates, but outside the range of the wanted microstrip line fields.

EXERCISES

4.1 A 100Ω microstrip line on a 1.58 mm thick, $\varepsilon_r = 2.5$, substrate is constructed with a line width of 1.26 mm, assuming negligible thickness conductors. If the actual conductor thickness is 0.2 mm, what is the true characteristic impedance for the line?

4.2 What is the maximum normalized line thickness, t/h, that may be used if the impedance of a 50Ω line calculated for $t = 0$ on an $\varepsilon_r = 2.5$ substrate is not to be in error by more than 2%?

4.3 i) Calculate the attenuation per wavelength from dielectric losses for a 100Ω line at 1.0GHz, if the substrate parameters are $\varepsilon_r = 2.5$ and $\tan\delta = 0.002$.
   ii) Repeat part (i), but with $\varepsilon_r = 9.8$.

4.4 Calculate the skin depths at both 1.0GHz and 10.0GHz for the following materials:

Material	Conductivity
Silver	6.2×10^7 S.m^{-1}
Copper	5.8×10^7 S.m^{-1}
Gold	4.1×10^7 S.m^{-1}
Aluminium	3.8×10^7 S.m^{-1}

4.5 i) Plot the factor for increased attenuation due to the surface roughness of a conductor, as a function of r.m.s. surface roughness normalized to the skin depth, i.e. Δ/δ.
   ii) If the increased attenuation due to the surface roughness of the conductors is not to be greater than 20%, what is the maximum r.m.s. surface roughness allowable for each of the materials given in Exercise 4.4?

4.6 Calculate the conductor loss at 2.0GHz for a 50Ω microstrip line that has the following parameters: w = 4.385 mm, h = 1.58 mm, $\varepsilon_r = 2.5$ and t = 0.1 mm. The copper conductors have conductivity, $\sigma = 5.8 \times 10^7$ S.m^{-1}, and may be assumed to be perfectly smooth.

4.7 The characteristic impedance of an ideal 60Ω microstrip line on a 1.5 mm thick, $\varepsilon_r = 2.5$, substrate is influenced by its proximity to a side wall of a shielding box. Estimate the minimum distance between the strip and wall, if the characteristic impedance is not to be reduced by more than 2%.

4.8 A 50Ω microstrip line on 1.0 mm thick alumina substrate ($\varepsilon_r = 9.8$) has a width of 0.97 mm and $\varepsilon_{eff} = 6.56$ at low frequencies. Plot $\varepsilon_{eff}(f)$ for frequencies up to 20GHz and indicate over what frequency range quasi-static approximations may be used, if wavelength calculations are to be accurate to 5%.

4.9 The radiation factor for a right-angled corner in a microstrip line [4.27] is $F \approx 4/(3\varepsilon_r)$. Estimate the percentage power radiated at 10.0GHz from a corner in a 50Ω line on a 1.0 mm thick alumina substrate, $\varepsilon_r = 9.8$.

4.10 A microstrip circuit is constructed on a 40×80 mm piece of 1.58 mm thick, $\varepsilon_r = 2.5$, substrate. The circuit is completely shielded by a metallic box with the top plate at a height of 13 mm above the substrate. Calculate the expected frequency for the lowest-order waveguide resonance.

REFERENCES

[4.1] Wheeler, H. A., "Transmission-line properties of parallel strips separated by a dielectric sheet", *IEEE Trans. Microwave Theory and Techniques*, Vol. MTT-13, No. 2, March 1965, pp. 172-85.

[4.2] Wheeler, H. A., "Transmission-line properties of a strip on a dielectric sheet on a plane", *IEEE Trans. Microwave Theory and Techniques*, Vol. MTT-25, No. 8, August 1977, pp. 631-47.

[4.3] Bahl, I. J. and Garg, R., "Simple and accurate formulas for a microstrip with finite strip thickness", *Proc. IEEE*, Vol. 65, No. 11, November 1977, pp. 1611-12.

[4.4] Schneider, M. V., "Microstrip lines for microwave integrated circuits", *Bell Syst. Tech. J.*, Vol. 48, No. 3, May-June 1969, pp.1421-44.

[4.5] Hammerstad, E. O. (ed. Bekkadal, F.), *Microstrip Handbook*, ELAB Report STF44 A74169, University of Trondheim, Norway, February 1975.

[4.6] Morgan, S. P. Jr., "Effect of surface roughness on eddy current losses at microwave frequencies", *Journal of Applied Physics*, Vol. 20, No. 4, April 1949, pp. 352-62.

[4.7] Wheeler, H. A., "Formulas for the skin effect", *Proc. IRE*, Vol. 30, No. 9, September 1942, pp. 412-24.

[4.8] Pucel, R. A., Massé, D. J. and Hartwig, C. P., "Losses in microstrip", *IEEE Trans. Microwave Theory and Techniques*, Vol. MTT-16, No. 6, June 1968, pp. 342-50. Correction: *IEEE Trans. Microwave Theory and Techniques*, Vol. MTT-16, No. 12, December 1968, p. 1064.

[4.9] Bahl, I. J., "Use exact methods for microstrip design", *Microwaves*, Vol. 17, No. 12, December 1978, pp. 61-2.

[4.10] March, S., "Microstrip packaging: Watch the last step", *Microwaves*, Vol. 20, No. 12, December 1981, pp. 83-94.

[4.11] Hoffmann, R. K., *Handbook of Microwave Integrated Circuits*, Artech House, Norwood, MA, 1987.

[4.12] Chudobiak, W. J., Jain, O. P. and Makios, V., "Dispersion in microstrip", *IEEE Trans. Microwave Theory and Techniques*, Vol. MTT-19, No. 9, September 1971, pp. 783-4.

[4.13] Troughton, P., "Measurement techniques in microstrip", *Electronics Letters*, Vol. 5, No. 2, January 1969, pp. 25-6.

[4.14] Arnold, S., "Dispersive effects in microstrip on alumina substrates", *Electronics Letters*, Vol. 5, No. 26, December 1969, pp. 673-4.

[4.15] Getsinger, W. J., "Microstrip dispersion model", *IEEE Trans. Microwave Theory and Techniques*, Vol. MTT-21, No. 1, January 1973, pp. 34-9.

[4.16] Atwater, H. A., "Tests of microstrip dispersion formulas", *IEEE Trans. Microwave Theory and Techniques*, Vol. MTT-36, No. 3, March 1988, pp. 619-21.

[4.17] Kirschning, M. and Jansen, R. H., "Accurate model for effective dielectric constant of microstrip with validity up to millimetre-wave frequencies", *Electronics Letters*, Vol. 18, No. 6, March 1982, pp. 272-3.

[4.18] Deibele, S. and Beyer, J. B., "Measurements of microstrip effective relative permittivities", *IEEE Trans. Microwave Theory and Techniques*, Vol. MTT-35, No. 5, May 1987, pp. 535-8.

[4.19]   Jansen, R. H., "High-speed computation of single and coupled microstrip parameters including dispersion, high-order modes, loss and finite strip thickness", *IEEE Trans. Microwave Theory and Techniques*, Vol. MTT-26, No. 2, February 1978, pp. 75-82.

[4.20]   Kobayashi, M. and Ando, F., "Dispersion characteristics of open microstrip lines", *IEEE Trans. Microwave Theory and Techniques*, Vol. MTT-35, No. 2, February 1987, pp. 101-5.

[4.21]   Shih, C., Wu, R-B., Jeng, S-K. and Chen, C. H., "A full-wave analysis of microstrip lines by variational conformal mapping techniques", *IEEE Trans. Microwave Theory and Techniques*, Vol. MTT-36, No. 3, March 1988, pp. 576-81.

[4.22]   Pramanick, P. and Bhartia, P., "An accurate description of dispersion in microstrip", *Microwave Journal*, Vol. 26, No. 12, December 1983, pp. 89-92, 96.

[4.23]   Kobayashi, M., "A dispersion formula satisfying recent requirements in microstrip CAD", *IEEE Trans. Microwave Theory and Techniques*, Vol. MTT-36, No. 8, August 1988, pp. 1246-50.

[4.24]   Jansen, R. H. and Kirschning, M., "Arguments and an accurate model for the power-current formulation of microstrip characteristic impedance", *Arch. Elek. Übertragung (AEÜ)*, Vol. 37, No. 3, March 1983, pp. 108-12.

[4.25]   England, E. H., "A coaxial to microstrip transition", *IEEE Trans. Microwave Theory and Techniques*, Vol. MTT-24, No. 1, January 1976, pp. 47-8.

[4.26]   Majewski, M. L., Rose, R. W. and Scott, J. R., "Modeling and characterization of microstrip-to-coaxial transitions", *IEEE Trans. Microwave Theory and Techniques*, Vol. MTT-29, No. 8, August 1981, pp. 799-805.

[4.27]   Lewin, L., "Radiation from discontinuities in strip-line", *Proc. IEE*, Vol. 107, Part C, February 1960, pp. 163-70.

[4.28]   Van der Pauw, L. J., "The radiation of electromagnetic power by microstrip configurations", *IEEE Trans. Microwave Theory and Techniques*, Vol. MTT-25, No. 9, September 1977, pp. 719-25.

[4.29]   Abouzahra, M. D. and Lewin, L., "Radiation from microstrip discontinuities", *IEEE Trans. Microwave Theory and Techniques*, Vol. MTT-27, No. 8, August 1979, pp. 722-3.

[4.30]   Abouzahra, M. D., "On the radiation from microstrip discontinuities", *IEEE Trans. Microwave Theory and Techniques*, Vol. MTT-29, No. 7, July 1981, pp. 666-8.

[4.31]   Collin, R. E., *Field Theory of Guided Waves*, McGraw-Hill, New York, 1960.

[4.32]   Jordan, E. C. and Balmain, K. G., *Electromagnetic Waves and Radiating Systems*, Prentice-Hall, 2nd edn, Englewood Cliffs, NJ, 1968.

5 Discontinuities

5.1 INTRODUCTION

The accurate representation of the fields and currents of straight and uniform microstrip lines with constant characteristic impedance are well known and have been discussed in Chapter 3. As soon as such a line is used in a practical circuit, there will no longer be a continuous cross-section geometry, but there will be changes brought about by the joining together of lines, the changing of characteristic impedance or propagation direction and the connection to various loads. Three classes of discontinuity effects that may be considered are:

i) the presence of fringing quasi-static electric fields and the associated capacitance, for example when there are sudden changes in the width of the line and in particular at open-circuit terminations;

ii) the changes to the normal flow of conduction current and the associated series inductance;

iii) the launching of higher order modes and surface waves as well as unbounded radiation. These effects have been described in Chapter 4 and may be modeled by a shunt conductance to represent the loss of power from the line whenever their influence on the choice of substrate permittivity and thickness for microstrip systems is important.

When the discontinuity is produced by a change in the line transverse dimension, an effective way to represent the first two cases is to convert the discontinuity effects into an equivalent dimensional change in the line geometry of an idealized line, i.e. a line in which there are no fringing effects.

5.2 THE OPEN-CIRCUIT END CORRECTION

Open-circuit terminated microstrip transmission lines are commonly used in matching networks and filter structures because reasonable open circuits are easier to realize than short circuits. In practice, an approximate open circuit is constructed by using the open end of a transmission line. The ideal field patterns associated with a standing wave from the open-circuit terminated line are distorted by the abrupt termination with fringing electric fields from around the end of the line to the ground plane. At low frequencies the fringing fields, Figure 5.1a, and the increase in

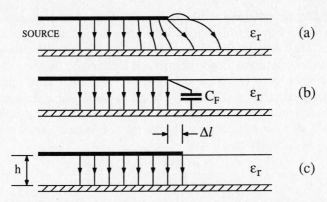

Figure 5.1 The electric fields at the open-end termination of a microstrip line, showing (a) the fringing electric fields, (b) the equivalent fringing capacitance, and (c) the equivalent line extension to an ideal open circuit

electrostatic energy as a result of the extra stored charge may be modeled by a capacitive termination, C_F, across a true open circuit at the terminating plane of the open-ended line as illustrated in Figure 5.1b. The fringing capacitance at the termination of the line is equivalent to extending the line by Δl in Figure 5.1c, i.e. the apparent line length is greater than the physical line length.

Using (1.54), the input impedance of an open-circuit terminated lossless line is

$$Z_{in} = -j Z_0 \cot(\beta l) \tag{5.1}$$

In (5.1), the correction length, $l \equiv \Delta l$, that gives the appropriate impedance for the fringing capacitance

$$Z_{in} = \frac{1}{j \omega C_F} \tag{5.2}$$

is therefore given by

$$\Delta l = \frac{1}{\beta} \arctan(\omega Z_0 C_F) \tag{5.3}$$

In practical situations, such as a single-stub matching network where the theoretical length of an open-circuit terminated line is known, it must be remembered that the calculated line length is shortened by the correction factor for the constructed circuit. As a correction term at low frequencies with $\Delta l \ll \lambda/16$, and using $\beta = \sqrt{\varepsilon_{eff}} (\omega/c)$, (5.3) may be written as

$$\frac{\Delta l}{h} \approx \frac{c Z_0}{\sqrt{\varepsilon_{eff}}} \left[\frac{w}{h} \right] \left[\frac{C_F}{w} \right] \tag{5.4}$$

This capacitive model is the first approximation to the correction term and neglects both the inductance to account for the redistribution of the current flow in the line and the conductance to model the loss of power by radiation. Equation (5.4) is consistent with the intuitive notion that $C_F = C \Delta l$, which is really an expression of the short line approximation of Figure 2.16b, valid at a current zero position. Silvester and Benedek [5.1] evaluate the excess capacitance associated with the fringing fields directly, avoiding the possibility of large errors that may occur when the difference is taken between the charge associated with the open-ended line and the same length of uniform line. They give empirical equations for six selected substrate permittivities, expressing the fringing capacitance normalized with respect to the width of the line, (C_F/w), as a function of (w/h) for $0.1 \leq (w/h) \leq 10$. Their error analysis suggests that the capacitance values will probably be on the low side of the true values. The six empirical equations for capacitance in Table 1 of [5.1] have been reduced by Hammerstad [5.2] to one equation such that the normalized line extension, $\Delta l/h$, is given in terms of ε_r and w/h as

$$\frac{\Delta l}{h} = 0.412 \times \frac{\varepsilon_{eff} + 0.3}{\varepsilon_{eff} - 0.258} \times \frac{w/h + 0.262}{w/h + 0.813} \tag{5.5}$$

The maximum error for Δl in (5.5) compared with the data that was used for fitting the expression is less than $0.05h$ for $\varepsilon_r = 1$ and less than $0.01h$ for $\varepsilon_r \geq 2.5$.

In a hybrid-mode analysis of the end effects, Jansen [5.3] formulated the problem in terms of the field and surface current distributions that occur in the plane of the microstrip. The complete termination is considered to be enclosed within a perfect metal wall structure, with the walls being sufficiently far from the microstrip line for them to have minimal influence on the solution. The method may also be used to predict the frequency dependence of the end-effect correction term.

An empirical formula that accurately models results derived from the method of Jansen [5.3] for the effect of an open end on a microstrip line has been given by Kirschning et al. [5.4]. From [5.4]

$$\Delta l/h = (\zeta_1 \zeta_2 \zeta_3 / \zeta_4) \tag{5.6}$$

with

$$\zeta_1 = 0.434907 \frac{\varepsilon_{eff}^{0.81} + 0.26}{\varepsilon_{eff}^{0.81} - 0.189} \times \frac{(w/h)^{0.8544} + 0.236}{(w/h)^{0.8544} + 0.87}$$

$$\zeta_2 = 1 + \frac{0.5274 \times \arctan(0.084 (w/h)^{1.9413/\chi})}{\varepsilon_{eff}^{0.9236}}$$

$$\text{with } \chi = 1 + \frac{(w/h)^{0.371}}{2.358 \, \varepsilon_r + 1}$$

$$\zeta_3 = 1 - 0.218 \exp(-7.5 w/h)$$

$$\zeta_4 = 1 + 0.0377 (6 - 5 \exp(0.036 (1 - \varepsilon_r))) \times \arctan(0.067 (w/h)^{1.456})$$

The accuracy of the fit of (5.6) to the theoretical results at 1.0 GHz is better than 2.5% for relative permittivities less than 50 and line geometries in the range $0.01 \leq w/h \leq 100$. While this error, when compared with the overall circuit dimensions, appears to be small for a total line length $l \gg h$, its effect may still have to be carefully considered in those cases where either the susceptance of a parallel-connected line that is approximately $\lambda/4$ long or the length of a $\lambda/2$ resonator is being computed.

Over the ranges $0.04 \leq w/h \leq 10$ and $2.2 \leq \varepsilon_r \leq 2.6$ that cover the majority of situations for microstrip lines using polystyrene-based substrate materials, a simplified expression to replace (5.6) that introduces less than 1% additional error at low frequencies has been derived for this book and is given by

$$\frac{\Delta l}{h} = \sum_{i=0}^{5} A_i x^i + (2.4 - \varepsilon_r)(0.022 + 0.008 \, w/h) \tag{5.7}$$

where $\quad x \equiv ln(w/h)$, and A_i is listed in Table 5.1.

Likewise, for $9.6 \le \varepsilon_r \le 10.0$

$$\frac{\Delta l}{h} = \sum_{i=0}^{5} B_i x^i \tag{5.8}$$

with B_i also given in Table 5.1.

Table 5.1 End-effect correction coefficients

i	A_i $2.2 \le \varepsilon_r \le 2.6$ Equation 5.7	B_i $9.6 \le \varepsilon_r \le 10.0$ Equation 5.8
0	0.3817	0.3173
1	0.1038	0.07592
2	0.00879	−0.00201
3	−0.00073	−0.00288
4	−0.00068	−0.00030
5	−0.00019	−0.00004

The $\Delta l / h$ of equation (5.6) is plotted in Figure 5.2 (the solid line) as a function

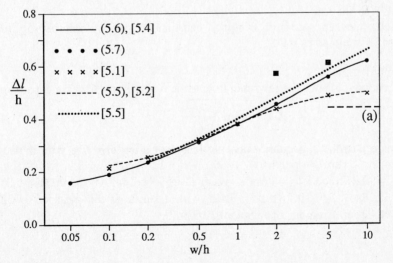

Figure 5.2 A comparison of theoretical results and empirical formulae for the low frequency open-circuit end correction factor with $\varepsilon_r = 2.5$. The asymptotic limit (a) from Wheeler [5.6] is for $\varepsilon_r \gg 1$. Experimental points [■ ■] are from [5.5] for $\varepsilon_r = 2.53$.

of w/h with $\varepsilon_r = 2.5$. Data points (• •) from (5.7) illustrate the fit of the simplified, but restricted permittivity range, equation to the more complete equation. The Silvester and Benedek results (× ×) and the derived equation by Hammerstad (the dashed line) tend to support the former authors' analysis that their results are expected to be low. The theoretical results of James and Tse [5.5] were extracted from a small published graph and plotted as the dotted curve. The experimental results, also from [5.5] (■ ■) but without any stated errors, were obtained for $\varepsilon_r = 2.53$. Figure 5.2 thus compares Silvester and Benedek's results with those based on Jansen's approach for the case of $\varepsilon_r = 2.5$. The agreement between the two methods improves for large ε_r, but substantially deteriorates as ε_r tends to unity.

Since the fringing fields at the end of a line are similar to the fringing fields at the edge of a line, asymptotic values for $\Delta l / h$ may be deduced for lines with large w/h from the expressions derived by Wheeler [5.6, 5.7]. These expressions allow for the fringing fields of a uniform line and give the increase of the effective line width over the actual line width. In one case, with $\varepsilon_r \gg 1$, from [5.6, eqn. 21], it is found that

$$\frac{\Delta l}{h} = \frac{ln\,4}{\pi} = 0.441 \tag{5.9}$$

giving an equivalent line extension of $0.441 \times h$ for open circuits on wide lines for high permittivity substrates. This asymptote is shown as (a) in Figure 5.2. Secondly, with $\varepsilon_r = 1.0$, from [5.7, eqn. 35]

$$\frac{\Delta l}{h} = \frac{1}{\pi} \left\{ ln \left[\frac{\pi w}{h} \right] + 1 \right\} \tag{5.10}$$

This equation gives $\Delta l / h = 1.097$ for w/h = 10 and shows a linear increase of $\Delta l / h$ with ln(w/h).

Example 5.1

Calculate the open-circuit fringing capacitance for a microstrip line with w = h = 2.0 mm and $\varepsilon_r = 2.5$.

Solution:

Equation (5.4) may be rewritten to give the open-circuit fringing capacitance

$$C_F = \Delta l\, C = \Delta l\, \frac{\sqrt{\varepsilon_{eff}}}{c\,Z_0}$$

where C is the capacitance per unit length of a uniform line with dimensions w and h. From Appendix 3, with w/h = 1.0 and $\varepsilon_r = 2.5$, it is found that $Z_0 = 90.2\,\Omega$ and $\varepsilon_{eff} = 1.966$, giving $C = 51.9\,\text{pF.m}^{-1}$. With $\Delta l / h = 0.379$ from (5.7), then $C_F = 19.7 \times h$ pF. Thus, with a substrate thickness of 0.002 m, the open-circuit fringing capacitance is 0.039 pF.

Example 5.2

A $\lambda/2$ 1.0 GHz resonator that has open-circuit terminations is constructed from a

100Ω microstrip line on a 5.0 mm thick, $\varepsilon_r = 2.53$, substrate. What physical correction factor is required to compensate for the fringing electric fields at each end of the resonator?

Solution:

For a 100Ω line on $\varepsilon_r = 2.53$ substrate

$$w/h = 0.787 \quad \text{and} \quad \varepsilon_{eff} = 1.964 \qquad \text{(See Example 3.5)}$$

giving $x \equiv ln(w/h) = -0.240$. Hence, substituting values in (5.7) gives

$$\frac{\Delta l}{h} = \sum_{i=0}^{5} A_i x^i - 0.004$$

$$= 0.353$$

leading to $\Delta l = 1.76$ mm as the length by which the resonator line must be shortened at each open-circuit termination in order to compensate for the fringing electric fields.

At 1.0GHz, the free space wavelength is 300 mm. Thus the half-wavelength resonator length, l, is given by

$$l = \frac{300}{2\sqrt{1.964}} - 2 \times 1.76$$

i.e. $\qquad l = 103.5$ mm

5.3 CORNERS

Corners are required, not only for the convenience of improving the usage of a given substrate area, but also for such components as directional couplers where it is necessary to bring two lines into close proximity for a known electrical length. If space is not a limiting factor in the design, Howe [5.8] shows that for a balanced stripline a rounded corner with a radius of curvature greater than 3w will give a corner that is hard to distinguish from a normal straight section of line. Nevertheless, it is the abrupt corner that will be used in the majority of situations, as it can be designed to create a minimal disturbance on the line.

Consider a microstrip corner with uniform lines leading up to the two reference planes as illustrated in Figure 5.3a. The corner, between planes R_1 and R_2, may be modeled by the T-network in Figure 5.3b. Here, L_c accounts for the current and stored magnetic energy and C_c for the charge and stored electric energy. The corner also affects the current and voltage distributions in the uniform connecting lines. These disturbances, however, can be considered as a part of the corner segment and thus lumped into the equivalent circuit in Figure 5.3b. In evaluating these disturbances to the uniform connecting lines in the theoretical analysis, auxiliary planes are introduced some distance away from the corner reference planes. At these

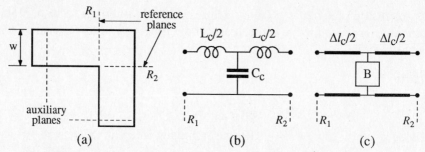

Figure 5.3 The geometry of a microstrip corner and its equivalent circuits

planes, it is assumed that the fields and currents are identical to those of the uniform transmission line and are not affected by the presence of the corner. L_c and C_c in Figure 5.3b have been evaluated in [5.9, 5.10] and [5.11] respectively.

A T-network is also an equivalent circuit for a short length of line that has a characteristic impedance of $\sqrt{L_c/C_c}$. However, because of the excess capacitance at a square corner, this characteristic impedance value will be lower than that of the uniform connecting lines. In Figure 5.3c, the inductance is accounted for by the line length, Δl_c, that has a characteristic impedance equal to that of the connecting lines. From Figure 2.15 and (2.42)

$$\Delta l_c \;=\; L_c/\boldsymbol{L} \tag{5.11}$$

The excess capacitance over that required for the Δl_c line is shown as the shunt susceptance, B. This latter model is to be preferred here as it is more useful in practical situations where the corner geometry is modified to reduce the excess capacitance and the equivalent line length Δl_c is given.

Inductance and capacitance values have been experimentally verified by Easter [5.12], using half- and full-wavelength L-shaped resonators with open-circuit terminations, and by Douville and James [5.13] using a square loop resonator that, when excited from a movable probe, could be arranged to have either an electric field maximum or a current maximum at each corner. L-shaped resonators are used in one of the experiments in Chapter 13. An electric field maximum and no current flow at a corner makes the capacitive effect dominant, while for a current maximum it is the inductance that is the dominant element. In both cases, a change in the resonator frequency is interpreted as due to appropriate line extensions, $\Delta x_{(C)}$ in the capacitive dominant case and $\Delta x_{(L)}$ in the other. Since the corner dimensions are very much less than the wavelength, the short line equivalent circuit is valid and the corner capacitance and inductance are obtained from (2.42) and (2.43) to give

$$C_c \;=\; \boldsymbol{C}\Delta x_{(C)} \;=\; \frac{\sqrt{\varepsilon_{\text{eff}}}}{c\,Z_0}\,\Delta x_{(C)} \tag{5.12}$$

and
$$L_c \;=\; \boldsymbol{L}\Delta x_{(L)} \;=\; \frac{\sqrt{\varepsilon_{\text{eff}}}\,Z_0}{c}\,\Delta x_{(L)} \tag{5.13}$$

Experimental details for the measurement of $\Delta x_{(C)}$ and $\Delta x_{(L)}$ are given in §13.4.

The reader will recall from (1.17) that the characteristic impedance of a line is given by

$$Z_0 = \left[\frac{L}{C}\right]^{\frac{1}{2}}$$

(5.14)

At a microstrip corner, the region between the two reference planes has inductance and capacitance values, L_c and C_c respectively. Thus, with the corner dimensions small compared with the microstrip line wavelength, the characteristic impedance of the corner section of line is

$$Z_{corner} = \left[\frac{L_c}{C_c}\right]^{\frac{1}{2}}$$

(5.15)

with an equivalent line length, not necessarily of a 50Ω line, of

$$\Delta l = \frac{c}{\sqrt{\varepsilon_{eff}}}\sqrt{L_c . C_c}$$

(5.16)

It turns out that Z_{corner} for the geometry of Figure 5.3a is significantly less than Z_0. Thus the corner may be considered to have either insufficient inductance or excessive capacitance. It is therefore necessary to either increase the corner inductance with a narrow slit in the line [5.14] or reduce the corner capacitance in a symmetrical manner, as illustrated in Figure 5.4, in order to obtain a low reflection from the corner. Reducing the corner capacitance has been the preferred technique [5.13], as it is both dimensionally less sensitive in practice, i.e. easier to construct, and more suitable for theoretical evaluation.

A comparison of theoretical results for a corner between equal width lines is given in Figure 5.5, with Neale and Gopinath results [5.15, from Figure 6] shown as a solid curve, those of Thomson and Gopinath [5.16, Figure 3] shown as a dashed curve, and the measurements of Easter [5.12] for the corner inductance as individual points. The results are plotted with the inductance, L_c, normalized to h and the inductance per unit length of the uniform transmission line, L. This latter inductance, independent of the substrate material, is given by

$$L = Z_{fs}/c$$

(5.17)

where Z_{fs} is the characteristic impedance of the line in free space. Negative incremental inductance values for small w/h ratios occur as a result of the redistribution of current flow in the lines leading up to the corner, giving a reduced inductance contribution to the overall effective inductance of the corner.

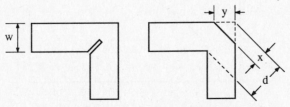

(a) increased inductance   (b) decreased capacitance

Figure 5.4 Compensation techniques for a microstrip corner

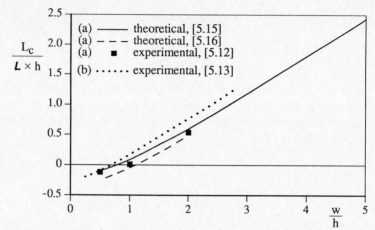

Figure 5.5 A comparison of theoretical and experimental results for the normalized inductance of a right-angle corner, showing results for (a) the basic corner and (b) the mitered corner

The results from Douville and James [5.13] for the inductance of a mitered corner are also included in Figure 5.5. It is seen that the miter increases the corner inductance. Since the mitered corner is matched for the characteristic impedance of the line, the equivalent line length of the corner may be derived as the normalized inductance value times the substrate height.

The corner capacitance for a 90° bend in a 50Ω line, deduced from the theoretical results of Silvester and Benedek [5.11], is presented in Figure 5.6. Care should be taken in using the closed form expressions for the capacitance given by Garg and Bahl [5.17] and quoted by others, as the expression for $w/h \leq 1$ does not take into account the major correction that was published by Silvester and Benedek to amend their earlier paper.

A detailed experimental analysis on the effects of reducing the corner capacitance by mitering the corner as in Figure 5.4b has been presented by Douville

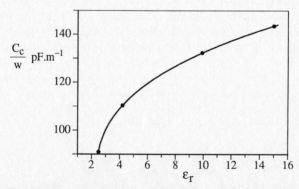

Figure 5.6 The capacitance of a right-angle corner in 50Ω line, deduced from results given by Silvester and Benedek [5.11]

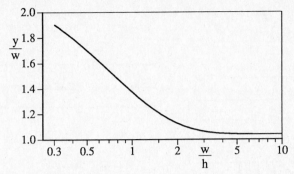

Figure 5.7 Data for cutting the mitered corner for a right-angle bend

and James [5.13]. The percentage miter of the corner is defined by

$$m = \frac{x}{d} \times 100\% \tag{5.18}$$

The m required for a good match depends on w/h, but has been found to be fairly insensitive to the substrate permittivity. Their single smooth empirical curve to fit the data is given as

$$m = 52 + 65 \exp(-1.35\, w/h) \% \tag{5.19}$$

where $w/h \geq 0.25$ and $\varepsilon_r \leq 25$. This implies that the miter is cut with a distance along each direction from the outer corner of y (see Figure 5.4b) given by

$$y = \frac{m}{50} w = \left(1.04 + 1.3 \exp(-1.35 w/h)\right) w \tag{5.20}$$

This distance is plotted as a function of w/h in Figure 5.7.

5.4 THE SYMMETRICAL STEP

The step discontinuity at the junction of two transmission lines with different characteristic impedances on a uniform substrate occurs frequently, for example in matching networks and in the design of filters. Consider the symmetrical step that is illustrated in Figure 5.8a, where a high impedance line with characteristic impedance, Z_{01}, joins one of lower impedance, Z_{02}. As shown, the step is symmetrical with respect to the center line of the strip conductor.

In the vicinity of the step there will be a transition region, assumed small compared with the wavelength, where the current flow from one line to the other is no longer that of either of the infinite uniform lines. This effect will be modeled by a series inductance for the step, L_S. The electric field will be distorted as the corners of the step are approached and, in particular, there will be fringing electric fields from the transition edge. The excess charge stored in this region will be modeled by the step capacitance, C_S. Thus, an equivalent circuit for the step is formed, Figure 5.8b, with the inductance component split into two equal parts.

Thomson and Gopinath [5.16] describe a method of calculating the microstrip

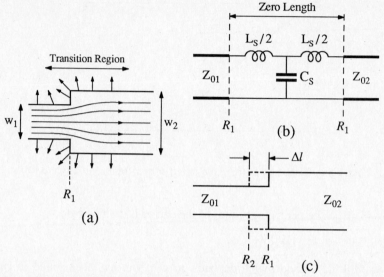

Figure 5.8 The symmetrical microstrip step discontinuity with (a) the fringing electric fields and current flow, (b) the quasi-static lumped equivalent circuit as seen from planes outside the transition region, and (c) the line shortened to retain the correct impedance at R_2

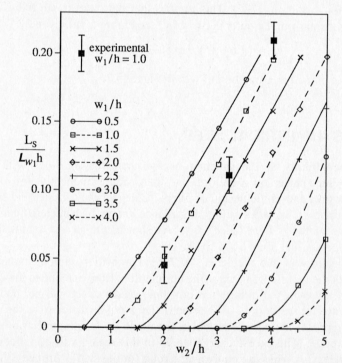

Figure 5.9 The theoretical values for the discontinuity inductance for a symmetrical step change of line width with experimental results for $w_1/h = 1.0$, from Gopinath et al. [5.18] (© 1976, IEEE)

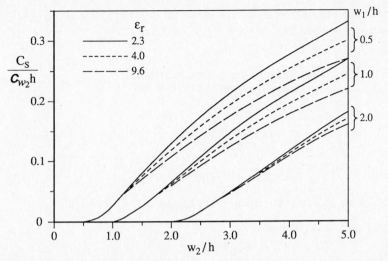

Figure 5.10 The discontinuity capacitance for a symmetrical step change of width, from Gupta and Gopinath [5.21] (© 1977, IEEE)

discontinuity inductance. The method was used to give detailed theoretical inductance values for a symmetrical step discontinuity by Gopinath et al. [5.18] and gave the total junction inductance normalized to both L_{w_1}, the inductance per unit length of the uniform microstrip line of width w_1, $(w_1 < w_2)$, and substrate height as plotted in Figure 5.9. Also shown for comparison in Figure 5.9 are experimental results [5.18] that were obtained for $w_1/h = 1.0$.

The excess capacitance associated with a microstrip symmetrical step change has been evaluated by Farrar and Adams [5.19] and Benedek and Silvester [5.20], with more detailed results given by Gupta and Gopinath [5.21]. The data from [5.21] is presented in Figure 5.10 for three substrate relative permittivities: 2.3, 4.0 and 9.6. The discontinuity capacitance is normalized with respect to both the substrate height and the capacitance per unit length for the uniform line of width w_2, C_{w_2} where $w_2 > w_1$.

Example 5.3

Derive an equivalent circuit that includes the discontinuity effects at the symmetrical junction between two microstrip lines, $w_1/h = 1.0$ and $w_2/h = 4.0$, where $h = 1.58$ mm and $\varepsilon_r = 2.3$ for the substrate.

Solution:

With $\varepsilon_r = 2.3$ and the line dimensions as given, the analysis formulae of Table 3.2 give characteristic impedances of $93.2\,\Omega$ and $40.9\,\Omega$ and effective relative permittivities of 1.838 and 1.989 for the narrow and wide lines respectively. Let the equivalent circuit take the form shown in Figure 5.8b, with L_S and C_S representing the total discontinuity inductance and capacitance.

From Figure 5.9, using the curve for $w_1/h = 1.0$ at $w_2/h = 4.0$, the step inductance is given by

$$L_S = 0.2 \times L_{w_1} \times h$$

The inductance per meter of the uniform narrow line

$$L_{w_1} = \frac{Z_{01}\sqrt{\varepsilon_{eff1}}}{c}$$

$$= \frac{93.2\sqrt{1.838}}{3 \times 10^8}$$

i.e. $\qquad L_{w_1} = 4.21 \times 10^{-7} \text{ H.m}^{-1}$

Substituting values in the equation for L_S, gives $L_S/2 = 67 \text{ pH}$.

From Figure 5.10, it is seen that

$$C_S = 0.22 \times C_{w_2} \times h$$

where, for the broad line

$$C_{w_2} = \frac{\sqrt{\varepsilon_{eff2}}}{Z_{02}\,c} = 1.15 \times 10^{-10} \text{ F.m}^{-1}$$

Thus, on substituting the relevant values

$$C_S = 40 \text{ fF}$$

The resultant equivalent circuit that includes the discontinuity effects is shown in Figure 5.11.

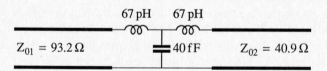

Figure 5.11   The equivalent circuit of the step discontinuity for Example 5.3

Compensation for the fringing capacitance may be achieved in several ways if its effect on the overall transmission line network is considered. However, the approach that will be taken here is one where the compensation takes place in the proximity of the junction. It is designed to give the same impedance at the reference plane R_2 in a practical network as is required for the theoretical design using idealized lines. The excess capacitance, as the dominant component in the equivalent circuit, has the effect of making the lower impedance (i.e. wider) line appear to be electrically longer by the length Δl, Figure 5.8c. This means that if the junction position, on the basis of ideal lines free from fringing effects, was calculated to be at the plane R_2, then in a practical circuit the junction would be moved by an amount Δl to R_1. The wider line would thus be shortened by Δl and the correct calculated impedance at R_2 would be retained.

On the assumption that the fringing capacitance is proportional to the step size, Hammerstad and Bekkadal [5.22] proposed that Δl just discussed is given by

$$\Delta l = \left[1 - \frac{w_1}{w_2} \right] \Delta l_{oc} , \qquad w_2 > w_1 \tag{5.21}$$

Here Δl_{oc} is the correction factor for an open-circuit termination on a strip of width, w_2. With information from Figures 5.10 and 5.2, it can be deduced that (5.21) overestimates the capacitive effect and length correction, even at an open-circuit plane, and represents a simplification of the situation that is now to be addressed.

It was seen in §5.3 for corners that either the discontinuity capacitance or inductance was dominant, if the discontinuity was at a plane of voltage maximum or minimum respectively. The same is true for the step discontinuity. However, the corner had equal width lines on each side of it and, by mitering the corner, it was possible to make it look like a constant length of line, irrespective of any standing wave that may be present. The same is not true for a step discontinuity which, by its very nature, includes a change in line impedance. Now, it is necessary to know the impedance at the step before accurate compensation can take place.

At a voltage maximum, the equivalent length to compensate for the capacitance, from (2.43) is

$$\Delta l_c = C_S / C = \frac{c \, Z_{02}}{\sqrt{\varepsilon_{eff2}}} \times C_S \tag{5.22}$$

This correction term will now give the same impedance at R_2 as is required at the transition between two ideal lines. As long as the approximations remain valid, Δl_c is independent of frequency.

> THE WIDE LINE IS SHORTENED BY THE LENGTH Δl_c

At a voltage minimum, the equivalent length to compensate for the inductance, from (2.42) is

$$\Delta l_L = L_S / L = \frac{c}{Z_{01} \sqrt{\varepsilon_{eff1}}} \times L_S \tag{5.23}$$

This is the length of the narrow line that compensates for the discontinuity inductance. Hence

> THE NARROW LINE MUST BE SHORTENED BY Δl_L

Equations (5.22) and (5.23) are approximations that are also valid for some of the cases where there is a high V.S.W.R. on the line. Consider two cases, (i) where there is a very high impedance, approximately an open circuit, at the discontinuity and (ii) a typical practical situation where there is still an impedance that is greater than the characteristic impedance of the line, but nowhere near an open-circuit value. The following analysis will show how the correction for case (i) is to be modified to take care of case (ii). Both cases will be expressed in graphical terms on the Smith Chart, shown as a part of an admittance chart, Figure 5.12, with the admittance values

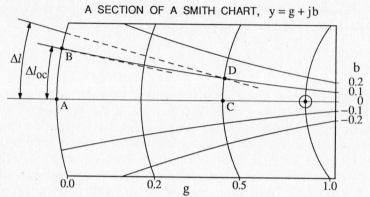

Figure 5.12 Illustrating the increase in line length that compensates for the step-discontinuity capacitance by returning the admittance to its original value

normalized to Y_{01}. For readers not yet familiar with the Smith Chart, its derivation and use are described in Chapter 6.

First consider the plane where $y_L = 0$ (an open circuit, A). A normalized capacitive susceptance $+j0.1$ in parallel with this load admittance gives B. Without having moved any physical distance, this is identical to a line extension towards the generator of Δl_{oc}, i.e. 0.0158λ. Cutting back the original line by this amount gives a combined admittance that is equivalent to an open circuit.

Reducing the length of the wider line at a step transition now follows on from the previous discussion, if a junction between the original line and an extremely high impedance line is considered. Now consider a load admittance that is real and less than unity, say $y_L = 0.5$ at point C on Figure 5.12, with the junction susceptance still $+j0.1$. The parallel combination of $0.5 + j0.1$ is plotted, D. In this region of the Smith Chart, the circles of constant conductance and of constant V.S.W.R. are almost identical and it is seen that the length correction to bring the admittance back to the real axis is now greater than the previous open-circuit case with the same shunt susceptance.

A multiplying correction factor, m, applied to the open-circuit approximation

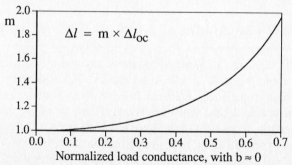

Figure 5.13 The correction factor, m, for the discontinuity capacitance

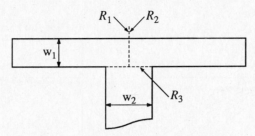

Figure 5.14 The geometry of a T-junction

will minimize the errors of discontinuity capacitance. This factor is plotted in Figure 5.13 as a function of the normalized real part of load admittance.

5.5 THE T-JUNCTION

At the junction where two transmission lines are joined in parallel to a common input line, the electric fields and current flow in the lines are distorted. This is a common situation in microstrip circuits, occurring, for example, in bias networks, stub matching networks, hybrid couplers and power dividers. A typical T-junction of three transmission lines is illustrated in Figure 5.14, with the shunt line having a different characteristic impedance to the main through transmission line. Reference planes, R_1, R_2 and R_3, are taken as shown.

The quasi-static equivalent circuit, Figure 5.15, assumes that line cross-sectional dimensions are very much smaller than the wavelength, and includes a transformer with turns ratio 1 : n. If there is no dispersion, n = 1.

The normalized discontinuity inductances, $L_A/(L_{w_1} \times h)$ and $L_B/(L_{w_2} \times h)$, are mostly negative and independent of the substrate material (assumed non-magnetic)

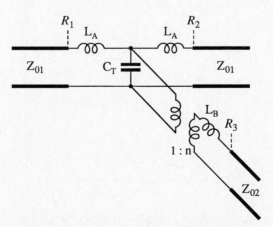

Figure 5.15 The T-junction equivalent circuit

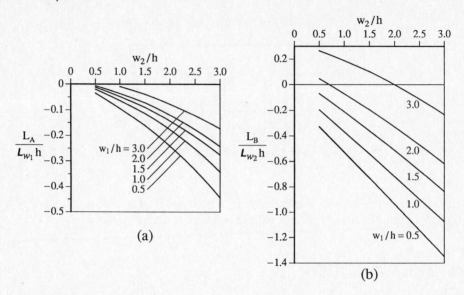

Figure 5.16 Normalized discontinuity inductance values for a T-junction, (a) for the straight-through arm and (b) for the stub arm, from Neale and Gopinath [5.15] (© 1978, IEEE)

and have been evaluated by Neale and Gopinath [5.15] for a number of cases. Their results are summarized in Figure 5.16.

The normalized discontinuity capacitance, C_T, is also expected to be negative when the reference planes are taken as in Figure 5.14, since some of the fringing electric fields, especially from the through line, will no longer be present. The normalized capacitance values from Gopinath and Gupta [5.23] for a T-junction that has equal width lines are presented in Figure 5.17. In [5.23], results are also given for

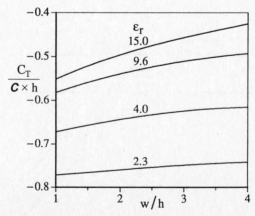

Figure 5.17 Normalized discontinuity capacitance values for a T-junction with equal width lines, from Gopinath and Gupta [5.23] (© 1978, IEEE)

unequal line widths, provided that w/h = 1 for the through line. These latter results may be of interest for designs on $\varepsilon_r = 9.6$ substrate material, where w/h = 1 gives a 50 Ω line.

Example 5.4

A single-stub matching circuit is used to match a load, $Z_L = 50 + j\,50\,\Omega$, to the input transmission line. 50 Ω characteristic impedance lines on a 1.0 mm thick, $\varepsilon_r = 9.6$ substrate are used throughout. The theoretical solution to the matching problem is illustrated in Figure 5.18. What are the line length corrections that have to be made to account for the T-junction discontinuity effects? Assume (i) low frequency approximations are valid and (ii) that the open-circuit end correction has been considered separately.

Solution:

With low frequency approximations, the dispersion effects will not be considered and the transformer turns ratio for the shunt arm will be assumed as 1 : 1. The three inductances in the equivalent circuit for the T-junction, Figure 5.15, are individually converted to their respective extensions in a 50 Ω line. The negative inductances between the planes R_1 and R_2 are equated to negative lengths of line, which represent in the practical line an extension of R_1 towards the source with a similar extension of R_2 towards the load. The negative lengths of line also have negative capacitance which has to be taken from C_T, i.e. the total junction capacitance C_J will become less negative and maybe even positive. A similar calculation for the stub line is also made. Now, the stub line in effect has some shunt capacitance in parallel with it at the junction, and its line length must be further adjusted to give the desired total shunt susceptance at this plane (i.e. +j0.02 S in this case).

For a 50 Ω characteristic impedance microstrip line with $\varepsilon_r = 9.6$, the line parameters are w/h = 0.991 and $\varepsilon_{eff} = 6.44$. A value of w/h = 1 will be assumed here. At the planes R_1 and R_2, the normalized discontinuity inductance

$$\frac{L_A}{L_{w_1}h} = -0.07$$

i.e. the inductance of a length of line, $\Delta l_1 = \Delta l_2 = -0.07$ h. For the stub arm

$$\frac{\Delta l_3}{h} = \frac{L_B}{L_{w_2}h} = -0.38 \qquad \text{(Here, } w_1 = w_2\text{)}$$

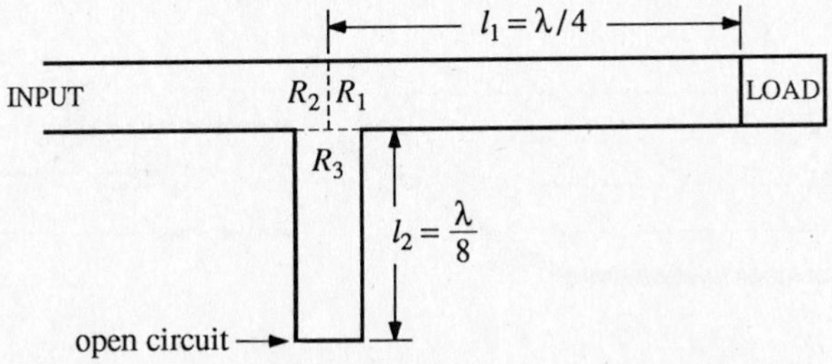

Figure 5.18 The matching circuit for $Z_L = 50 + j\,50\,\Omega$ to a 50 Ω line

Thus, with h = 1 mm, $\Delta l_1 = \Delta l_2 = -0.07$ mm and $\Delta l_3 = -0.38$ mm. The junction capacitance for the equivalent circuit from Figure 5.17 is $C_T = -0.58 (\boldsymbol{C} \times \text{h})$ pF. To this must be added the capacitance of the length $(\Delta l_1 + \Delta l_2 + \Delta l_3)$ of 50Ω transmission line. With $\varepsilon_{\text{eff}} = 6.44$, the total junction capacitance

$$C_J = \frac{\sqrt{\varepsilon_{\text{eff}}}}{c \times Z_0} \times (0.07 + 0.07 + 0.38 - 0.58)\text{h} = -0.017\,\text{pF}$$

The cancellation of C_J cannot be achieved by further reference plane adjustments at the junction and, from this point on, the solution will depend both on the frequency and the circuit environment. Taking the frequency as 2.0 GHz, the normalized shunt susceptance of C_J is $-j0.011$ at the plane R_3. Thus for single-stub matching, the stub admittance in this case must now be $+j1.011$, instead of being ideally $+j1.0$. This value can be achieved by increasing the stub length from 0.125λ to 0.1259λ. This increase in stub length is equivalent to moving R_3 out from the junction by 0.0009λ and keeping the stub length to R_3 unchanged at $\lambda/8$.

The microstrip wavelength at 2.0 GHz is $150/\sqrt{\varepsilon_{\text{eff}}} = 59.11$ mm. Hence for the final circuit, the distance from the load to the center line of the stub is

$$l_1 = 14.78 + 0.07 = 14.85\,\text{mm}$$

and the stub length

$$l_2 = 59.11 \times 0.1259 + 0.38 = 7.82\,\text{mm}$$

Corrections for the open-circuit termination of the stub line will further reduce l_2.

5.6 SERIES GAPS

The series gap in a microstrip line, Figure 5.19a, has the obvious property of a d.c. open circuit that may be useful in a bias-feeding network but, as illustrated, will probably also have a high impedance at microwave frequencies because of its small capacitance value. The gap may be used as a coupling element between resonators of a band-pass filter although, in Chapter 9, it will be seen that this is not the preferred coupling mechanism, especially if a large amount of coupling is required. By exciting a resonator through a series gap, measurements of discontinuity effects have

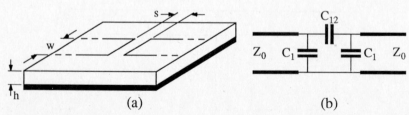

Figure 5.19 (a) A series gap in a microstrip line with (b) its equivalent circuit

been made with the discontinuity, e.g. a corner or a T-junction, built into a resonant length of line [5.12, 5.24].

The equivalent circuit for a gap may be represented as a π-network of capacitors, Figure 5.19b. At a voltage maximum along the line, the shunt capacitances may be equated to line extensions as was the case for an open-circuit terminated line, while the series capacitance component essentially remains unaltered. In this section, only gaps between two equal width lines are considered. For unequal line widths, the shunt capacitance components would be represented as different capacitance values.

The capacitance values for a very large gap separation approach that of the open-circuit terminated line capacitance for C_1 and zero for C_{12}. Conversely for very narrow gaps, C_1 tends to zero and C_{12} steadily increases as the separation is reduced.

The theoretical values of the equivalent circuit capacitances are found by considering the even- and odd-mode capacitances of the gap, excited as shown in Figure 5.20. The Benedek and Silvester method [5.20] is preferred to that of Farrar and Adams [5.19], since the former is concerned directly with the excess charge in the vicinity of the gaps, while the latter approach is one where the excess charge is determined as a difference of two much larger quantities.

From Figure 5.19, it is seen that the capacitance from the strips at +V to ground for the even mode is

$$C_{even} = 2C_1 \qquad (5.24)$$

while for the odd mode, between +V and ground

$$C_{odd} = C_1 + 2C_{12} \qquad (5.25)$$

Thus, if C_{even} and C_{odd} are known, then

$$C_1 = \frac{C_{even}}{2} \qquad (5.26)$$

and

$$C_{12} = \frac{C_{odd} - C_1}{2} \qquad (5.27)$$

The even- and odd-mode capacitances for gaps on an $\varepsilon_r = 2.5$ substrate are taken from Benedek and Silvester [5.20] and plotted as normalized capacitances C/w and as a function of normalized gap separation s/w for w/h = 0.5, 1 and 2 in Figure 5.21. Further capacitances curves are presented in [5.20] and Gopinath and Gupta [5.23], and the results have been presented as a closed form equation over a limited range of parameters for $\varepsilon_r = 9.9$ and with better than 7% accuracy by Garg and Bahl [5.17]. Measurements supporting the theory and including the frequency dependence of the capacitance values have been given by Özmehmet [5.25]. The following

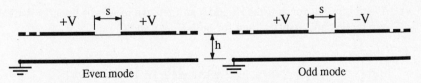

Figure 5.20 Even- and odd-mode excitation across a series gap

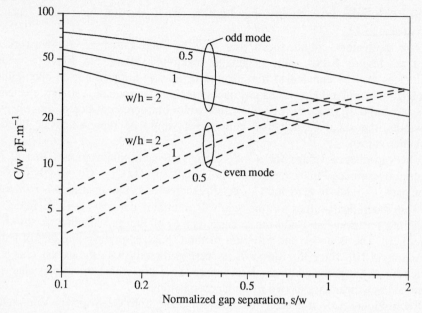

Figure 5.21 The normalized even- (dashed) and odd-mode (solid) capacitances of a series gap, separation s, in a microstrip line on an $\varepsilon_r = 2.5$ substrate, from Benedek and Silvester [5.20] (© 1972, IEEE)

example illustrates how the curves may be used to estimate the capacitances of a practical gap.

Example 5.1

Estimate the Π-network capacitances for a 1.5 mm gap in a 50Ω microstrip line that is constructed on a 1.58 mm thick, $\varepsilon_r = 2.5$ substrate.

Solution:

The estimate of the capacitances is made in the following manner with the results presented in the table:

i)   Calculate s/w for each of w/h = 1 and 2; these being the parameters for the curves in Figure 5.21.

ii) Read off the values for the normalized even- and odd-mode capacitances in pF.m^{-1} from Figure 5.21.

iii) Knowing w for each case, calculate the values for C_{even} and C_{odd}.

iv) From Appendix 3, for a 50Ω line it is found that w/h = 2.84.

v)   Assuming that for constant h and s an increase in w will linearly increase the two mode capacitances, then for both C_{even} and C_{odd}

$$C\Big|_{w/h=2.84} = C\Big|_{w/h=2.0} + \frac{(2.84-2.0)}{(2.0-1.0)}\left\{C\Big|_{w/h=2.0} - C\Big|_{w/h=1.0}\right\}$$

vi) Evaluate the Π-network capacitances, C_1 and C_{12}, from (5.26) and (5.27).

Step			w/h = 1.0	w/h = 2.0	w/h = 2.84
(i)	s/w		0.949	0.475	
(ii)	C_{even}/w	pF.m^{-1}	25.5	20.9	-
(ii)	C_{odd}/w	pF.m^{-1}	27.5	23.7	-
(iii), (v)	C_{even}	fF	40.3	66.1	87.7
(iii), (v)	C_{odd}	fF	43.5	74.7	100.8
(vi)	C_1	fF	-	-	43.8
(vi)	C_{12}	fF	-	-	28.5

Thus, an estimate of the series gap Π-network capacitances is $C_1 = 44$ fF and $C_{12} = 29$ fF.

EXERCISES

5.1 The conductor pattern for an open-circuit terminated half-wave resonator is fabricated on a 3.2 mm thick substrate with $\varepsilon_r = 2.3$. If the pattern has negligible thickness and is 6.0 mm wide by 110.0 mm long, calculate its resonant frequency. Ignore dispersion effects.

5.2 Estimate the equivalent line length and characteristic impedance for a right-angle corner in a 50 Ω line on an $\varepsilon_r = 2.5$ substrate
i)   with a square corner section,
ii) with a mitered corner section.

5.3 Calculate the impedances at 2.0 GHz that are associated with the discontinuity effects given by the equivalent circuit in Figure 5.11. If the 40.9 Ω line is terminated with a matched load, what is the input impedance, normalized to the 93.2 Ω line at the discontinuity
i)   ignoring discontinuity effects,
ii) including discontinuity effects?

5.4 The equivalent circuit for a typical step discontinuity is shown in Figure 5.8. If the discontinuity effects are small, i.e. $|Z_L| \ll Z_0$ and $|Y_c| \ll Y_0$, the T-equivalent circuit may be redrawn without significant additional error as one of the L-networks as shown below.

Consider the case of a discontinuity in an otherwise output matched transmission line ($Z_{02} = 50\,\Omega$) where Z_L and Y_c have discontinuity impedances of $+j1$ and $-j4000\,\Omega$ respectively.

Calculate the input impedance at plane A for each version of the equivalent circuit and comment on the importance of the order in which Z_L and Y_c appear as their values increase.

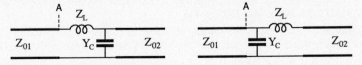

5.5 The theoretical position for an ideal 30 Ω quarter-wave transformer on a microstrip transmission line, $\varepsilon_r = 4.0$ and $h = 1.0$ mm, that matches a load at the end of a 50 Ω line to the 50 Ω

characteristic impedance input line is found to be between planes R_1 and R_2 as shown below. Calculate the inductance and capacitance for each discontinuity. What is the practical position of the transformer with respect to R_1 and R_2 when a correction is made for the step discontinuities?

Given for $\varepsilon_r = 4.0$		
Z_0	$\dfrac{w}{h}$	ε_{eff}
30Ω	4.35	3.29
50Ω	2.05	3.08

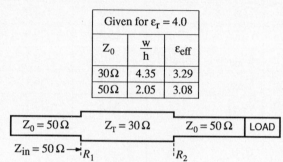

REFERENCES

[5.1]   Silvester, P. and Benedek, P., "Equivalent capacitances of microstrip open circuits", *IEEE Trans. Microwave Theory and Techniques*, Vol. MTT-20, No. 8, August 1972, pp. 511-16.

[5.2]   Hammerstad, E. O., "Equations for microstrip circuit design", *5th European Microwave Conference*, Hamburg, September 1975, pp. 268-72.

[5.3]   Jansen, R. H., "Hybrid mode analysis of end effects of planar microwave and millimetrewave transmission lines", *Proc. IEE*, Vol. 128, Pt. H, No. 2, April 1981, pp. 77-86.

[5.4]   Kirschning, M., Jansen, R. H. and Koster, N. H. L., "Accurate model for open end effect of microstrip lines", *Electronics Letters*, Vol. 17, No. 3, February 1981, pp. 123-5.

[5.5]   James, D. S. and Tse, S. H., "Microstrip end effects", *Electronics Letters*, Vol. 8, No. 2, January 1972, pp. 46-7.

[5.6]   Wheeler, H. A., "Transmission-line properties of parallel strips separated by a dielectric sheet", *IEEE Trans. Microwave Theory and Techniques*, Vol. MTT-13, No. 2, March 1965, pp. 172-85.

[5.7]   Wheeler, H. A., "Transmission-line properties of parallel wide strips by a conformal-mapping approximation", *IEEE Trans. Microwave Theory and Techniques*, Vol. MTT-12, No. 5, May 1964, pp. 280-9.

[5.8]   Howe, H. Jr., *Stripline Circuit Design*, Artech House, Dedham, MA, 1974.

[5.9]   Gopinath, A. and Silvester, P., "Calculation of inductance of finite-length strips and its variation with frequency", *IEEE Trans. Microwave Theory and Techniques*, Vol. MTT-21, No. 6, June 1973, pp. 380-6.

[5.10] Mehran, R., "Compensation of microstrip bends and Y-junctions with arbitrary angle", *IEEE Trans. Microwave Theory and Techniques*, Vol. MTT-26, No. 6, June 1978, pp. 400-5.

[5.11] Silvester, P. and Benedek, P., "Microstrip discontinuity capacitances for right-angle bends, T-junctions and crossings", *IEEE Trans. Microwave Theory and Techniques*, Vol. MTT-21, No. 5, May 1973, pp. 341-6. Correction: *IEEE Trans. Microwave Theory and Techniques*, Vol. MTT-23, No. 5, May 1975, p. 456.

[5.12] Easter, B., "The equivalent circuit of some microstrip discontinuities", *IEEE Trans. Microwave Theory and Techniques*, Vol. MTT-23, No. 8, August 1975, pp. 655-60.

[5.13] Douville, R. J. P. and James, D. S., "Experimental study of symmetric microstrip bends and their compensation", *IEEE Trans. Microwave Theory and Techniques*, Vol. MTT-26, No. 3, March 1978, pp. 175-82.

[5.14] Hoefer, W. J. R., "Equivalent series inductivity of a narrow transverse slit in microstrip", *IEEE Trans. Microwave Theory and Techniques*, Vol. MTT-25, No. 10, October 1977, pp. 822-4.

[5.15] Neale, B. M. and Gopinath, A., "Microstrip discontinuity inductances", *IEEE Trans. Microwave Theory and Techniques*, Vol. MTT-26, No. 10, October 1978, pp. 827-31.

[5.16] Thomson, A. F. and Gopinath, A., "Calculation of microstrip discontinuity inductance", *IEEE Trans. Microwave Theory and Techniques*, Vol. MTT-23, No. 8, August 1975, pp. 648-55.

[5.17] Garg, R. and Bahl, I. J., "Microstrip discontinuities", *International Journal of Electronics*, Vol. 45, July 1978.

[5.18] Gopinath, A., Thomson, A. F. and Stephenson, I. M., "Equivalent circuit parameters of microstrip step change in width and cross junctions", *IEEE Trans. Microwave Theory and Techniques*, Vol. MTT-24, No. 3, March 1976, pp. 142-4.

[5.19] Farrar, A. and Adams, A. T., "Matrix methods for microstrip three-dimensional problems", *IEEE Trans. Microwave Theory and Techniques*, Vol. MTT-20, No. 8, August 1972, pp. 497-504.

[5.20] Benedek, P. and Silvester, P., "Equivalent capacitances of microstrip gaps and steps", *IEEE Trans. Microwave Theory and Techniques*, Vol. MTT-20, No. 11, November 1972, pp. 729-33.

[5.21] Gupta, C. and Gopinath, A., "Equivalent circuit capacitance of microstrip step change in width", *IEEE Trans. Microwave Theory and Techniques*, Vol. MTT-25, No. 10, October 1977, pp. 819-22.

[5.22] Hammerstad, E. O. (ed. Bekkadal, F.), *Microstrip Handbook*, ELAB Report STF44 A74169, University of Trondheim, Norway, February 1975.

[5.23] Gopinath, A. and Gupta, C., "Capacitance parameters of discontinuities in microstriplines", *IEEE Trans. Microwave Theory and Techniques*, Vol. MTT-26, No. 10, October 1978, pp. 831-6.

[5.24] Kirschning, M., Jansen, R. H. and Koster, N. H. L., "Measurement and computer-aided modeling of microstrip discontinuities by an improved resonator method", 1983 *IEEE MTT-S Digest*, pp. 495-7.

[5.25] Özmehmet, K., "New frequency dependent equivalent circuit for gap discontinuities of microstriplines", *Proc. IEE*, Vol. 134, Pt. H, No. 3, June 1987, pp. 333-5.

6 The Smith Chart and its uses

6.1 INTRODUCTION

A load placed at the end of a transmission line with an impedance not equal to the characteristic impedance of the line will reflect some or all of the incident forward-traveling wave, to create a wave that now travels in the reverse direction. The combination of forward and reverse waves along the line sets up standing wave patterns. In this chapter the reflection coefficient at the load, as well as those that represent the ratio of reverse to forward waves at any plane along the line, will be taken as *voltage* reflection coefficients, i.e. the ratio of the voltages carried by the reflected and incident waves.

The presence of a standing wave along the line in many circuits may be detrimental to the overall operation of the system. For example:

i) With reflected power from the load, in most practical situations there will not be maximum power transfer to the load.

ii) The input impedance will depend critically on the length of the line from the input plane to the load.

iii) The interaction between two or more reflections will cause a frequency dependent response that is determined by the electrical lengths between the reflections. This may be ideal in the design of a band-pass filter but, unless specifically designed for, normally is detrimental to performance.

iv) Both the output power of an oscillator and its frequency may be perturbed by the uncertain load impedance that is attached to it.

v) For high power systems, although less likely to be found in microstrip circuits, the presence of peaks of voltage and current along the standing wave will reduce the power carrying capability.

These effects are reduced if one or both of the ports of a circuit are matched to the characteristic impedance of the connecting transmission line. The reader will recall from Chapter 2 that the word "matching" may refer either to matching for no reflections or to conjugate matching for optimum power transfer. In this chapter, while it is the matching for no reflections that is the main concern, conjugate matching is also considered.

The construction of the Smith Chart as a graphical aid for transmission line problems will be developed and important features, such as its relationship to the reflection coefficient and its use as either a normalized impedance or admittance chart, will be discussed. For clarity, most of the charts will be drawn in a simplfed form.

The load impedance is evaluated first from measurement data. Then a wide range of techniques will be introduced, that may be used to match a load to the characteristic impedance of the feeding line. There are advantages in being familiar with the majority of matching techniques for, while some of them may be more applicable for microstrip circuit design than others, an improved working knowledge of transmission line networks will be gained by a more thorough study of the topics. In general, the transformation between two complex impedances will require the determination of two real parameters associated with the matching network. These parameters will be either line lengths, as in single-stub matching, or a combination of a line length and a characteristic impedance, as in quarter-wave transformer matching. It is not usual practice for all the line lengths to be fixed and only two characteristic impedances to be evaluated as this approach is too restrictive on the range of load impedances that may be matched within the range of realizable characteristic impedances.

6.2 THE SMITH CHART

The Smith Chart is useful as a graphical aid in the design and understanding of transmission line problems. The chart is a plot of the reflection coefficient, Γ, in the complex plane and, as such, any value of Γ may be plotted with respect to the origin at the center of the chart. There is a one-to-one correspondence between Γ and the normalized impedance z, both of which are complex quantities, at any point along a transmission line. For easy conversion between Γ and the impedance that produces it, contours of constant resistance and reactance are overlaid on the complex plane. In a passive circuit with $|\Gamma| \leq 1$, the chart may be bounded by $|\Gamma| = 1$, still covering all possible impedances. Since the underlying basis of the chart is the reflection coefficient, moving along a lossless transmission line causes the reflection coefficient to move along a circle with its center at the center of the chart. A complete revolution is achieved in a half wavelength.

Consider the complex representation of Γ_L as illustrated in Figure 6.1 where

$$\Gamma_L = |\Gamma_L| \underline{/\phi} \tag{6.1}$$

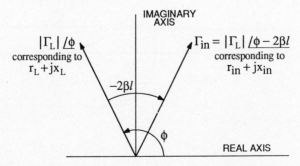

Figure 6.1 The reflection coefficient plotted on the complex plane

On this diagram of complex reflection coefficients, it is possible to draw the loci of constant normalized values of resistance and reactance, r and x. This will lead to the Smith Chart.

If the axes on Figure 6.1 are labeled u and jv respectively, then, in general

$$\Gamma = u + jv \tag{6.2}$$

However, from (1.44), the reflection coefficient and normalized impedance at any plane are related by

$$\Gamma = \frac{z-1}{z+1} = \frac{r+jx-1}{r+jx+1} \tag{6.3}$$

Equating (6.2) and (6.3) for Γ leads to two equations, one for the real parts and the other for the imaginary parts. From these equations, either one of the unknown variables (r or x) may be eliminated, giving the other as a function of u and v, namely

$$\left\{u - \frac{r}{r+1}\right\}^2 + v^2 = \frac{1}{(r+1)^2} \tag{6.4}$$

and

$$\left\{u - 1\right\}^2 + \left\{v - \frac{1}{x}\right\}^2 = \frac{1}{x^2} \tag{6.5}$$

If r and x are constant, (6.4) and (6.5) are equations of circles that form the basis of the Smith Chart. With $r = 0$, (6.4) becomes

$$u^2 + v^2 = 1 \tag{6.6}$$

giving the equation of a unit-radius circle with its center at (u=0,v=0). All impedance values with a positive real component lie within this circle, shown in Figure 6.2, with the loci of constant r and constant x being circles or portions of circles within this unit-radius circle. The position of the other major circles shown in

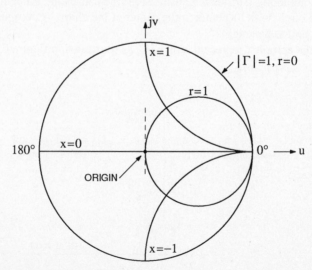

Figure 6.2   Major elements in the construction of the Smith Chart

Figure 6.2 may be verified from (6.4) and (6.5). To be useful as a graphical aid in transmission line work, the finer detail of the intermediate values of resistance and reactance values is desirable. This is provided on commercially available charts such as that illustrated in Figure 6.3. Design examples in this book will use skeleton charts for clarity.

Admittances

The normalized admittance, y = 1/z. However, with

$$z = \frac{1 + \Gamma}{1 - \Gamma} \tag{6.7}$$

where Γ is the voltage reflection coefficient, then

$$y = \frac{1 - \Gamma}{1 + \Gamma} \tag{6.8}$$

Thus the transformation to the normalized admittance point is obtained by taking the negative of the voltage reflection coefficient vector, this being equivalent to a 180° movement around the chart. The same chart may now be used as an admittance chart, with the resistance circles read as conductance circles, g, and the reactance circles becoming the susceptance circles, b.

6.3 MEASUREMENT OF A LOAD IMPEDANCE

6.3.1 Measurement based on the voltage reflection coefficient

When an unknown load impedance is attached to a transmission line that has a characteristic impedance Z_0, then, for every possible load impedance, there will be a unique complex voltage reflection coefficient at the plane of the load, given (1.44) by

$$\Gamma_L = \frac{Z_L - Z_0}{Z_L + Z_0} \tag{6.9}$$

In particular, it is seen that a short-circuit termination replacing the load impedance will give $\Gamma_L = -1$. The short circuit is used to provide a reference for both the magnitude and phase in reflection coefficient measurements, when for example a network analyzer is used to measure load impedances.

The normalized load impedance, z_L, may be derived directly from (6.9) as

$$z_L = \frac{\Gamma_L + 1}{\Gamma_L - 1} \tag{6.10}$$

or obtained directly from the Smith Chart once Γ_L has been plotted. As was seen in §6.2, *the radial distance on the chart is a linear measure of* $|\Gamma_L|$, varying from zero for a matched load at the center of the chart to unity for all pure reactances. *The angle of* Γ_L *is read in exactly the same way as angles in any complex plane*, by going counter-clockwise from the zero-angle position. An open circuit with $\Gamma_L = 1\underline{/0°}$ gives the zero-angle reference. These points are emphasized in Figure 6.4 for a normalized load impedance of 0.16 + j0.37. Since the phase angle of the reflection coefficient at some point becomes more negative as the distance l between that point

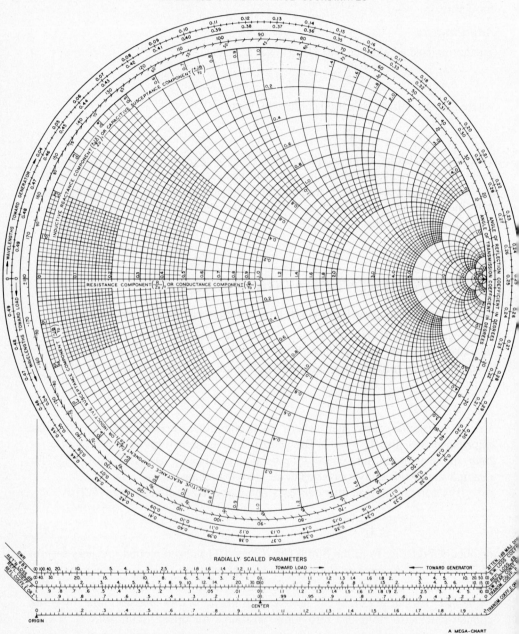

Figure 6.3   A commercially available Smith Chart *(This chart is made available courtesy of Analog Instruments Co., New Providence, NJ., USA.)*

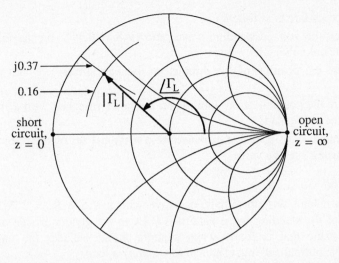

Figure 6.4 The determination of a load impedance from its voltage reflection coefficient. The Γ_L shown corresponds to $Z_L = 0.16 + j0.37$.

and the load is increased, see Figure 6.1, it follows that a *clockwise rotation* on the Smith Chart is required for movement along a transmission line towards the generator, and an *anti-clockwise rotation* for movement towards the load.

6.3.2 Measurement based on the standing wave pattern

A standing wave pattern is set up when a transmission line is terminated with an unmatched load. From measurements made on a typical voltage standing wave pattern, Figure 6.5, the normalized load impedance may either be deduced from the Smith Chart or calculated from first principles, using (6.10).

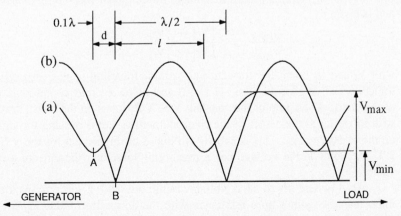

Figure 6.5 Standing wave measurements for the determination of the normalized load impedance (a) with the load termination and (b) with the load replaced by a short circuit

The procedure is as follows:

i) Measure the voltage standing wave ratio (V.S.W.R.), S, on the line terminated with the unknown load.

ii) Measure the position of the minima on the line terminated with the unknown load.

iii) Measure the position of the minima on the line terminated with a short circuit at the plane of the load.

The determination of a load impedance from a typical set of measurements is now illustrated using specific values.

Example 6.1

An unknown load sets up a standing wave with a V.S.W.R. = 3.0. The voltage minimum of the standing wave pattern is 0.1λ to the generator side of the voltage minimum plane that is found when a short circuit replaces the unknown load. Calculate the normalized impedance of the unknown load.

Solution:

The standing wave pattern that corresponds to this set of data is illustrated in Figure 6.5, with the separation between the two planes A and B being $d = 0.1\lambda$.

The standing wave pattern is repeated every half wavelength along the line, consistent with the fact that impedances on the line are equal at any two planes that are an integral number of half wavelengths apart on a lossless transmission line. A standing wave pattern that does not possess good symmetry and repetition indicates that serious errors may be caused by the presence of multiple frequencies, attenuation, or poor dimensional tolerances that cause uneven coupling of the detector to the standing wave pattern. The separation between the clearly defined voltage minima of the standing wave pattern for a short-circuit termination may be used to calculate the transmission line wavelength at the operating frequency. When the line is terminated with a resistive load with a magnitude SZ_0, it follows from (1.49) that the V.S.W.R. will be S. Here S is the ratio of the voltage maximum to voltage minimum of the standing wave pattern and is given by

$$S = \frac{V_{max}}{V_{min}} = \frac{|V_f| + |V_r|}{|V_f| - |V_r|}$$

with V_f and V_r being the forward and reverse traveling waves respectively. All impedances that give the same $|\Gamma|$ or S will lie on a circle, centered at the center of the chart. Thus, to plot the constant V.S.W.R. locus on the chart, draw a circle with its center at the center of the chart and with a radius passing through $z_L = S$ on the resistive axis. This is shown as Point 1 on Figure 6.6 for the V.S.W.R. of 3.0. At Point 1, the voltage has a maximum value and the current a minimum value.

At A where there is a voltage minimum and a current maximum, the impedance is again a pure resistance with the normalized value Z_0/S (Point 2). However, it is the impedance at the plane of the load or at any plane $n\lambda/2$ away that is required. When the load is replaced by a short circuit that gives a voltage

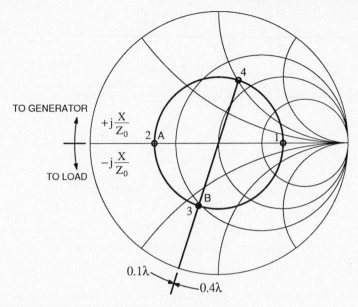

Figure 6.6 Plotting the standing wave measurements on the Smith Chart

minimum at the load plane, there will be voltage minima at all planes a half wavelength apart towards the generator. One such plane, B, will be considered to be equivalent to the load plane. The impedance at A, Point 2 on the Smith Chart, must therefore be moved through a distance of 0.1λ towards the load to bring it to B. The same load impedance is also obtained by moving 0.4λ towards the generator. Point 3 is therefore the load impedance, $z = 0.48 - j0.61$. If it is required, Point 4 gives the normalized load admittance, i.e. the numerical value of the normalized admittance is equal to the normalized impedance transformed through a distance of $\lambda/4$.

6.4 SINGLE-STUB MATCHING

In single-stub matching, the two variables are the electrical length of the stub l and its distance d from the load. This matching network is illustrated in Figure 6.7, with all the lines in this case having the same characteristic impedance.

A parallel-connected stub line with an open-circuit termination is easily fabricated as a microstrip line and is the only case to be considered in detail here. The sections of line, d and l, will generally have the same characteristic impedance as the input line (as in Figure 6.7) but, as will be seen in Example 6.3, this need not necessarily always be the case.

In carrying out the single-stub matching procedure, first of all plot the normalized load impedance, z_L, on the Smith Chart and convert it to a normalized admittance, y_L, in preparation for adding other admittances in parallel later. An

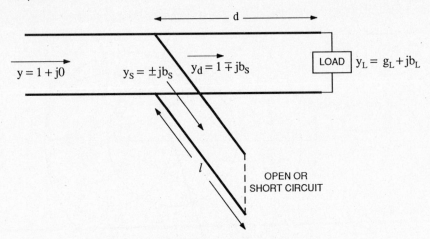

Figure 6.7 A single-stub matching network

open- or short-circuit terminated stub can present any possible value of normalized susceptance, $y_S = \pm j\,b_S$, at the junction plane. This is added to the load admittance after it has been transformed through the length d. The resultant must give $y = 1 + j\,0$ for the matched condition. Thus the load admittance must be transformed by moving towards the generator, until a plane where $y_d = 1 \mp j\,b_S$ is reached, i.e. move the load admittance towards the generator on a constant V.S.W.R. circle until an intersection with the unit conductance circle is made. The distance so moved gives d.

For the stub line, in general an open- or short-circuit termination may be chosen and used as the load admittance for the line but, as mentioned earlier, in microstrip circuits an open circuit is preferred in practice. Move towards the generator (there is no other direction to move if one is already at the load plane on the stub line) around the circumference of the chart, i.e. along the zero conductance circle, until the required admittance $\pm j\,b_S$ is achieved. The distance traversed gives l. Thus, the two electrical lengths d and l have been obtained.

Finally, the electrical lengths are converted to physical lengths, knowing the frequency of operation and the effective relative permittivity of the transmission lines. In practical microstrip circuits, corrections for discontinuity effects at both the open circuit and the T-junction will be required.

Example 6.2

Calculate the electrical lengths of a single-stub matching network with either a short-circuit or an open-circuit termination on the stub that will match a load impedance of $(30 + j\,70)\,\Omega$ to a $50\,\Omega$ input transmission line.

Solution:

The solution for this example is illustrated in Figure 6.8. The load impedance, normalized to the characteristic impedance of the line, is $z_L = 0.6 + j\,1.4$ and plotted as (1). The admittance of the load is therefore $y_L = 0.259 - j\,0.603$, (2).

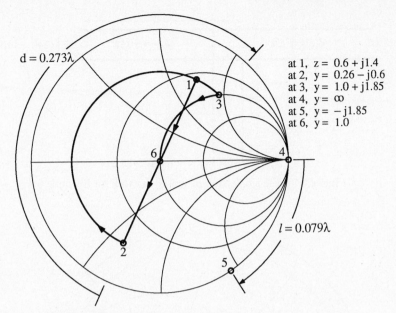

$d = 0.273\lambda$

at 1, $z = 0.6 + j1.4$
at 2, $y = 0.26 - j0.6$
at 3, $y = 1.0 + j1.85$
at 4, $y = \infty$
at 5, $y = -j1.85$
at 6, $y = 1.0$

$l = 0.079\lambda$

Figure 6.8 The Smith Chart used for single-stub matching

Now move a distance d wavelengths along the line towards the generator until the admittance on the line is of the form $1 \pm jb_S$, i.e. (3), where the admittance is $1 + j1.85$ and the distance d is 0.275λ. At this point on the line add a stub, in parallel, with a stub admittance of $-j1.85$.

A short-circuit terminated stub with an infinite admittance is considered here, (4). Now move around the Smith Chart on the $g = 0$ circle, i.e. the circumference of the chart with a constant unit magnitude for the reflection coefficient, away from the short circuit, which in this case is the load, until an admittance $0 - j1.85$ is reached, (5). The distance moved around the Smith Chart, 0.078λ, is the required length l of the short-circuit terminated stub at a distance 0.275λ from the load. The input admittance of the load and stub combined in parallel is now $1 + j0$, i.e. a perfect match at (6).

Using an open-circuit terminated stub, the stub length will differ from that for the short-circuit terminated stub. Thus, in this case, the stub length will be increased by 0.25λ to 0.328λ.

A further pair of solutions may be obtained if the length d is increased to a new point (3′) where the admittance is $1 - j1.85$. Evaluate d and l for this case.

Example 6.3

Consider the schematic of a single-stub matching network that is illustrated in Figure 6.9. The load, $Z_L = (140 - j70)\Omega$, is to be matched to the 50Ω input transmission line. The other transmission lines are no longer 50Ω, but have the characteristic impedances shown. Find the lengths d and l to achieve matching if a short-circuit terminated stub is to be used.

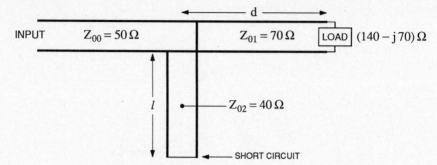

Figure 6.9 The single-stub matching network for Example 6.3

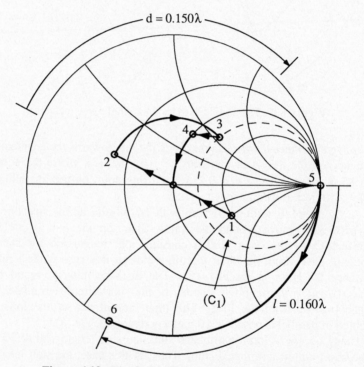

Figure 6.10 The Smith Chart solution for Example 6.3

Solution:

This example illustrates variations in single-stub matching brought about by having lines with characteristic impedances which differ from the input 50Ω line. For the three types of line

$$Z_{00} = 50\,\Omega \;\Rightarrow\; Y_{00} = 0.02 \quad \text{S,}$$
$$Z_{01} = 70\,\Omega \;\Rightarrow\; Y_{01} = 0.0143 \text{ S,}$$
$$Z_{02} = 40\,\Omega \;\Rightarrow\; Y_{02} = 0.025 \quad \text{S,}$$

$$Z_L = 140 - j70\,\Omega \;\Rightarrow\; Y_L = 0.00572 + j0.00286\text{ S}$$
$$\Rightarrow\; y_L = 0.4 + j0.2,\text{ normalized to }Y_{01}.$$

First consider the transformation of the load admittance through the distance d to the plane of the stub where an admittance $Y_d = 0.02 \pm jB_s$ is required. Normalized to the Y_{01} characteristic admittance line, this represents a normalized admittance of $1.4 \pm jb_s$.

Plot $z_L = 2.0 - j1.0$ at (1) in Figure 6.10 with $y_L = 0.4 + j0.2$, (2). Move towards the generator to intersect the $g = 1.4$ circle (C_1) at (3). Here $y_d = 1.4 + j1.114$, i.e. $Y_d = (0.02 + j0.0159)\text{ S}$, which when renormalized to the 0.02S line, is $(1.0 + j0.796)$.

Now consider the stub. With a short-circuit termination, the load admittance for the stub is infinite, (5). The stub line has a characteristic admittance of 0.025S and, for an actual input admittance of $-j0.0159$S, a normalized stub input admittance of $-j0.636$ is required, (6). Moving towards the generator from the short circuit (5) to (6) gives an electrical length of 0.160λ.

A matched network is given with $d = 0.150\lambda$ and $l = 0.160\lambda$.

6.5 DOUBLE-STUB MATCHING

In single-stub matching it is necessary to place the stub in the correct plane with respect to the load. In double-stub matching, the two stubs are normally fixed in relationship to each other and may be positioned at almost any plane with respect to the load. The adjustable quantities in a practical system are the lengths of the two stubs, l_1 and l_2. The two stubs are placed generally about $\lambda/8$ or $3\lambda/8$ apart, although this distance is not critical. However, as will be seen later, it may not be possible to match out every impedance and, for this reason, a third stub is often incorporated in stub-matching arrangements. If the stubs are $n \times \lambda/4$ apart and n is odd, the region representing unmatchable load impedances on the chart is a maximum, while with n even the stubs are effectively in parallel at the same load plane and the situation reduces to that of single-stub matching with a fixed position stub. Nevertheless, for stubs that are *close* to a multiple of $\lambda/2$ apart, most load impedances may be matched, but at the expense of larger standing waves on the connecting lines and a very narrow-band frequency response for the resultant matching circuit.

This matching technique is a good introduction to the idea of working a problem from two viewpoints: (i) from the load and (ii) from the generator so that, at some plane within the network, two plots are obtained. The first plot is of all possible impedances (or admittances) that may be derived at the plane from the known load

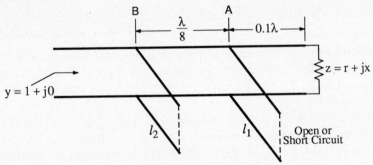

Figure 6.11 A double-stub matching network

with the variables of the network taken into account up to that plane. The second plot represents all desirable impedances, knowing that they may be matched out using the variables of the remainder of the network between the plane and the generator. The intersections of the two plots represent impedances that can be obtained from the load *and* matched out to the generator and are used to derive the final network parameters.

The procedure is shown by means of a particular case using stub-line positions that are preset at 0.1λ from the load and $\lambda/8$ apart. The double-stub matching network is illustrated in Figure 6.11 and the use of the Smith Chart to solve this case in Figure 6.12.

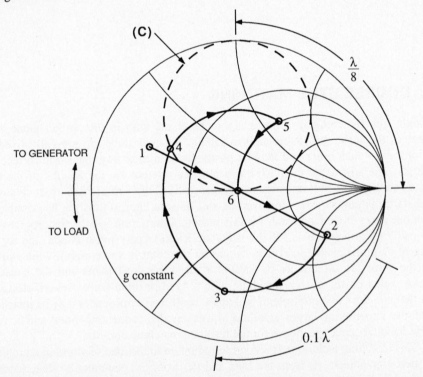

Figure 6.12 The Smith Chart used for double-stub matching. The solution, from the load to the generator, is obtained by going $1 \rightarrow 2 \rightarrow \cdots \rightarrow 6$.

Consider the parallel stub connected at B. This stub may add any required susceptance to provide the final matched input. Thus the admittance of the network to the right of B must lie on the unit conductance circle while the input admittance of the network at a plane just to the left of A must be such that, when it is transformed by $\lambda/8$ towards the generator, it moves onto the unit conductance circle. Hence, draw the circle of desirable admittances by transforming the unit conductance towards the load from B to A. The resulting circle is shown as (C) for this case, i.e. with a $\lambda/8$ separation between the two stubs.

Now consider a load with the normalized impedance $z = r + jx$, (1), giving the normalized admittance $y = g + jb$, (2). Transfer the admittance along the line towards the generator to the plane of A, through a distance of 0.1λ. Here the admittance of the load will be $y_1 = g_1 + jb_1$, (3). A susceptance is added by the stub at A so that the sum of the admittance of the stub and y_1, lies on the circle corresponding to the transformed unit conductance circle. Note that this last step has been *along a constant conductance circle* and gives $y_2 = g_1 + jb_2$ at (4). All points on this transformed circle (C), when moved by $\lambda/8$ towards the generator, will become the unit conductance circle. The admittance to the right of B, (5), now has the form $y_3 = 1 + jb_3$ and only requires the correct susceptance of the stub at B to bring the admittance to the characteristic admittance of the line, i.e. $y = 1 + j0$. The load has now been matched by the two stubs.

In the case that has just been considered, if the admittance of the load at A, (3), had been within the $g = 2$ circle, then the intersection point (4) could not have been realized on the (C) circle and the two stubs would not have been sufficient to carry out the matching procedure.

Example 6.4

Match a load impedance, $Z_L = (30 - j50)\,\Omega$, to the 50Ω characteristic impedance of a microstrip line using double-stub matching with 50Ω lines throughout. One stub is placed close to the load, being only 0.02λ away from it, while the distance between the stubs is 0.14λ.

Calculate the lengths of the two stubs and illustrate the region of load impedances on the Smith Chart that cannot be matched out with the stubs in this given position.

Solution:

The solution to this example is illustrated in Figure 6.13 and may be found in the following manner:

i)   Draw a circle of desirable admittances (C) that, when moved towards the generator by 0.14λ, becomes the unit conductance circle. Parallel-connected stubs are used here because they are readily fabricated as a microstrip matching circuit with a simple T-junction between the stub and the main line.

ii) Plot the normalized load impedance, $z_L = 0.6 - j1.0$ (1), and move it along a constant V.S.W.R. circle towards the generator by 0.02λ to (2). Convert to the input admittance of the load and 0.02λ line, $y = 0.543 + j0.917$ at (3).

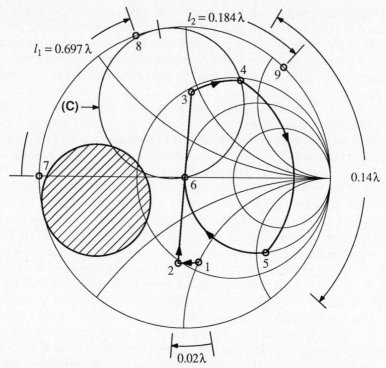

Figure 6.13 The Smith Chart solution for Example 6.4

iii) The stub that is closer to the load is used to add susceptance and move the admittance from (3) along a constant conductance circle to intersect at one of two points with the circle of desirable admittances at that plane, (C). Using an open-circuit terminated stub, the solution at (4) will lead to shorter stub lengths and will be used here. At (4), $y = 0.543 + j1.614$ and is obtained by adding a parallel stub with a normalized admittance of $+j0.697$.

iv) Moving from (4) by 0.14λ towards the generator gives $y = 1.0 - j2.278$ at (5) and, with the addition of a normalized stub admittance of $+j2.278$, the matched condition at (6) is obtained.

v) Open-circuit terminated stubs are also to be used to go from (5) to (6). The open-circuit admittance at (7) is moved *towards the generator* to $y = +j0.697$ at (8) and to $y = +j2.278$ at (9), giving lengths of 0.097λ and 0.184λ for the stubs closer to the load and generator respectively.

vi) If point (3) had had a normalized conductance greater than 1.683, then no addition of any susceptance value would have given an intersection with circle (C). Converting these unmatchable admittances to impedances and moving them by 0.02λ towards the load gives the shaded circle on Figure 6.13. This circle represents all normalized load impedances that cannot be matched with the present spacing of parallel-connected stub lines.

6.6 QUARTER-WAVE TRANSFORMER MATCHING

6.6.1 The single-section transformer

Consider a transmission line network as illustrated in Figure 6.14. In general, from (1.54), the input impedance of a line that has a characteristic impedance Z_T and length l and is terminated with a load impedance Z_L is

$$Z_{in} = Z_T \times \frac{Z_L \cos(\beta l) + j Z_T \sin(\beta l)}{Z_T \cos(\beta l) + j Z_L \sin(\beta l)} \tag{6.11}$$

Here, Z_L is the load seen by the Z_T line at its junction with the Z_0 line. If the line length l is $\lambda/4$, the line is known as a *quarter-wave transformer*. In that case $\cos(\beta l) = 0$ and, if the input to the transformer is to be matched, then

$$Z_{in} = Z_0 = \frac{Z_T^2}{Z_L} \tag{6.12}$$

i.e. $$Z_T = \sqrt{Z_{in} Z_L} = Z_0 \sqrt{z_{in} z_L} \tag{6.13}$$

As the characteristic impedances Z_0 and Z_T are in practice real quantities, the impedance at the load end of the transformer must also be real if a matched condition is to be realized. The line impedance will look purely real at the voltage maxima and minima planes. Any load may be transformed to the real impedance at such a plane through an appropriate length of transmission line. This fact, then, together with the properties of the $\lambda/4$ transformer, allows any load impedance that has a finite resistance component to be matched to the characteristic impedance of the input line.

The two unknown quantities that have to be determined for a quarter-wave transformer matching network are the distance from the load to the transformer, d, and the characteristic impedance of the transformer, Z_T.

The matching procedure is as follows:

i) Plot the normalized impedance of the load to be matched $z_L' = Z_L'/Z_0$ on a Smith Chart.

ii) Move around the chart on a constant V.S.W.R. circle, i.e. with $|\Gamma|$ constant, until the circle intersects the resistive axis, giving a plane where the impedance looking into the transmission line is real. This normalized impedance is the z_L for the quarter-wave transformer in (6.13). The rotation around the chart in a clockwise direction, towards the generator, gives the line length d.

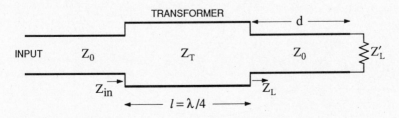

Figure 6.14 Matching with a single-section quarter-wave transformer

iii) For a matched condition requiring $z_{in} = 1 + j0$, the characteristic impedance of the transformer section becomes

$$Z_T = Z_0 \sqrt{z_L} \tag{6.14}$$

iv) As the constant $|\Gamma|$ circle of part (ii) intersects the real axis at two points, there are two possible solutions for quarter-wave transformer matching, one with $Z_T > Z_0$ and the other with $Z_T < Z_0$. Naturally, the line length d will be different for the two cases.

Example 6.5

i)   Calculate the position and characteristic impedance of a quarter-wave transformer that will match a load impedance, $(15 + j25)\,\Omega$, to a 50Ω input line.

ii) What is the magnitude of the reflection coefficient within the transformer?

Solution:

i)   This example uses the variables given in Figure 6.14 and is solved using the Smith Chart, as is illustrated in Figure 6.15. The two unknown quantities for this matching network are the length of the 50Ω line that connects the load to the output of the transformer and the characteristic impedance of the transformer. Thus

a)   Plot the load impedance normalized to the 50Ω line, (1).

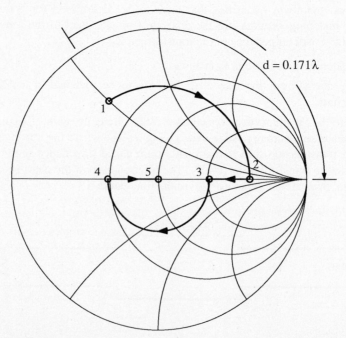

Figure 6.15   The quarter-wave transformer solution for Example 6.5

b) Move towards the generator along the $50\,\Omega$ line to a plane where the impedance is real, (2). Here $Z_L = 50 \times z_L = 211.5\,\Omega$ and the distance moved is 0.171λ.

c) The quarter-wave transformer impedance

$$Z_T = \sqrt{50 \times 211.5} = 102.8\,\Omega$$

ii) The magnitude of the reflection coefficient within the transformer is

$$|\Gamma| = \left| \frac{Z_L - Z_T}{Z_L + Z_T} \right| = 0.346$$

The magnitude and phase of the reflection coefficient within the transformer are further illustrated by following through the plots on the Smith Chart to the input of the network. Point (2) is the impedance normalized to a $50\,\Omega$ line at the load end of the transformer. Since the absolute impedance at this plane remains constant, renormalizing to the transformer characteristic impedance gives

$$\text{Point (3)} = \text{Point (2)} \times \frac{50}{102.8} = 2.057 + j\,0$$

Moving along the transformer towards the generator gives (4). This impedance, when renormalized to the $50\,\Omega$ input line, provides the perfect match at (5). Along the line from (3) to (4) there is a constant V.S.W.R. of 2.057 with $|\Gamma|$ equal to 0.346.

6.6.2 Multiple-section transformers

Multiple-section quarter-wave transformers are most often used when a low residual mismatch is required over a broad band of frequencies between transmission lines that have different characteristic impedances.

The simplified theory below assumes that the complex reflection coefficients, caused by each step in transmission line characteristic impedance in an otherwise matched system, may be added together at the input as vector quantities to give the resultant mismatch reflection coefficient. With only small magnitude reflections from each transition, second order reflections of the form $\Gamma_1\Gamma_2$ are ignored. At any change of characteristic impedance between two matched transmission lines from Z_A to Z_B, the voltage reflection coefficient as seen from the Z_A line is

$$\Gamma = \frac{Z_B - Z_A}{Z_B + Z_A} = \frac{(Z_B/Z_A) - 1}{(Z_B/Z_A) + 1} \tag{6.15}$$

Substituting $w = (Z_B/Z_A) - 1$ gives $\tag{6.16}$

$$\Gamma = \frac{w}{2 + w} \approx \frac{1}{2}\left\{ w - \frac{w^2}{2} + \frac{w^3}{4} - \cdots \right\} \tag{6.17}$$

Provided that w is small, which is the case if the impedance change is small, the first two terms of (6.17) are identical to those for the series expansion of

$$\tfrac{1}{2}\ln(1+w) \approx \frac{1}{2}\left\{w - \frac{w^2}{2} + \frac{w^3}{3} - \cdots \right\}, \quad |w| < 1 \tag{6.18}$$

i.e. $\Gamma \approx \tfrac{1}{2}\ln(Z_B/Z_A)$ \hfill (6.19)

The benefit of this approximation, where even the third term of the expansion in (6.17) differs by only $w^3/12$ from (6.18), will be appreciated when the range between the two characteristic impedances Z_0 and Z_L is subdivided to give intermediate characteristic impedances for the specific ratios of the reflection coefficients at the junctions.

Consider a two-section quarter-wave transformer, as illustrated in Figure 6.16, with $l = \lambda/4$. With no loss of generality, it will be assumed that $Z_L > Z_0$. At each of the transitions between the lines, the reflection coefficients are given by

$$\Gamma_3 \approx \tfrac{1}{2}\ln(Z_L/Z_B)$$
$$\Gamma_2 \approx \tfrac{1}{2}\ln(Z_B/Z_A)$$

and $\Gamma_1 \approx \tfrac{1}{2}\ln(Z_A/Z_0)$ \hfill (6.20)

The three reflection coefficients will be real and positive if $Z_L > Z_B > Z_A > Z_0$. If the individual reflection coefficients are small, the resultant reflection coefficient at the input does not contain any terms such as $\Gamma_1\Gamma_2$, so that

$$\Gamma = \Gamma_1 + \Gamma_2 e^{-2j\beta l} + \Gamma_3 e^{-4j\beta l} \tag{6.21}$$

For quarter-wave sections of transmission line with $l = \lambda/4$

$$\Gamma = \Gamma_1 - \Gamma_2 + \Gamma_3 \tag{6.22}$$

and it can be seen that, if $\Gamma_1 : \Gamma_2 : \Gamma_3 = 1 : 2 : 1$, there will be no resultant reflection coefficient at the input. In general, the required individual reflection coefficient ratios will be given by binomial coefficients, [6.1].

At the input and at the design or center frequency, f_0, the phase change between adjacent reflection coefficients is $\phi = 180°$. At a frequency of, say, $0.8 \times f_0$, the phase change is $\phi = 144°$. As illustrated in Figure 6.17, the resultant reflection coefficient at the design frequency is zero. At $0.8 \times f_0$ the magnitude of the resultant reflection coefficient is still small with a magnitude of $0.382\,|\Gamma_1|$, i.e. less than 10% of $4\,|\Gamma_1|$, which would be the reflection coefficient caused by an abrupt step from Z_L to Z_0. A multiple-section quarter-wave transformer matching circuit has bandpass

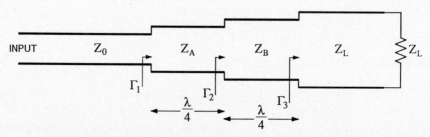

Figure 6.16 A two-section quarter-wave transformer network

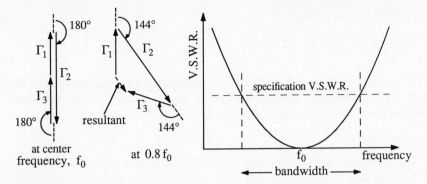

Figure 6.17 The derivation of the resultant reflection coefficient for a multiple-section quarter-wave transformer

characteristics and it is usual to specify the maximum reflection coefficient magnitude, or input V.S.W.R., that may be tolerated, together with the bandwidth over which the specification is to apply.

The curve shown in Figure 6.17 is typical of a maximally-flat response, where any further broadening of it would result in pass-band ripples. If, in the two-section transformer that has been discussed, the transformer impedances are recalculated such that

$$\Gamma_1 = \Gamma_3 = k\Gamma_2, \quad \text{where } k > 0.5 \tag{6.23}$$

then there will not be a perfect match at the center frequency. However, there will be two frequencies, one on each side of the center frequency f_0, at which a perfect match is achieved. For all frequencies in the immediate vicinity of the two matched points, the resultant V.S.W.R. will be less than that for the maximally-flat case, [6.1]. This approach is the basic design philosophy of Chebyshev transformers, where increased bandwidth is achieved at the expense of a V.S.W.R. ripple of controlled levels within the design band. A comprehensive treatment for the design of multiple-section quarter-wave transformers is given by Matthaei et al. [6.1], with detailed expressions for both maximally-flat and Chebyshev responses.

For a rigorous analysis, let Φ be the ratio of characteristic impedances of the input and output lines, i.e. $\Phi = Z_L/Z_0$. For maximally-flat response filters, expressions in [6.1] lead to the impedances for the two-section transformers as

$$Z_A = Z_0 \Phi^{\frac{1}{4}} \tag{6.24}$$

and

$$Z_B = Z_0 \Phi^{\frac{3}{4}} \tag{6.25}$$

while for the three-section case, with transformer impedances, Z_A, Z_B and Z_C, the equation

$$\psi^4 + 2\sqrt{\Phi}\psi^3 - 2\sqrt{\Phi}\psi - \Phi = 0 \tag{6.26}$$

is solved first for ψ, from which

$$Z_A = \psi Z_0 \tag{6.27}$$

$$Z_B = \sqrt{\Phi} Z_0 \tag{6.28}$$

and
$$Z_C = \frac{Z_L}{\psi} = \frac{\Phi Z_0}{\psi} \tag{6.29}$$

It is interesting to note that, for the particular case of a maximally-flat *two-section* transformer, the approximations of (6.20) actually give the same line impedances as those for the rigorous solution.

Example 6.6

Calculate the characteristic impedances for a three-section quarter-wave transformer with a maximally-flat response that will match a 200Ω load to a 50Ω input line. Compare the results from the exact and approximate theories.

Solution:

By letting the length of the 200Ω characteristic impedance line between the load and the transformer tend to zero, the 200Ω load may be connected directly to the output of the transformers. For the exact solution, $\Phi = Z_L/Z_0 = 4.0$. Hence (6.26) becomes

$$\psi^4 + 4\psi^3 - 4\psi - 4 = 0$$

giving $\psi = 1.1907$

and $Z_A = 59.54\,\Omega$, $Z_B = 100.00\,\Omega$ and $Z_C = 167.97\,\Omega$.

The other real solution, $\psi = -3.8$, yields negative impedance lines. For the approximate solution, the voltage reflection coefficients of the impedance steps considered in isolation are in the ratio

$$\Gamma_1 : \Gamma_2 : \Gamma_3 : \Gamma_4 = 1 : 3 : 3 : 1$$

being the coefficients of the binomial expansion of $(1+x)^3$. With equations of the form of (6.20), the total reflection coefficient for a step change between Z_0 and Z_L is divided into four individual reflection coefficients having the above ratios.

Thus $\Gamma_1 = \frac{1}{8} \left\{ \frac{1}{2} \ln \left(\frac{Z_L}{Z_0} \right) \right\} = 0.08664$

giving $Z_A = Z_0 e^{2\Gamma_1} = 59.46\,\Omega$

while $\Gamma_2 = 3\Gamma_1$ leads to $Z_B = 100.00\,\Omega$

and $\Gamma_3 = 3\Gamma_2$ leads to $Z_C = 168.18\,\Omega$

The exact and approximate solutions for matching this ratio of impedances are almost identical, close enough for the fabrication of microstrip circuits.

6.7 IMPEDANCE MATCHING WITH A TAPERED LINE

Broadband matching between two transmission lines with characteristic impedances Z_0 and Z_L may be achieved to some extent by using multiple-section quarter-wave transformers. However, it will always be the case that at twice the design frequency the sections become $\lambda/2$ long and the matching performance is no better than that obtained by joining the two lines directly together. For a broader bandwidth, a length of transmission line in which the cross-sectional dimensions gradually change from the characteristic impedance Z_0 to one of Z_L may be used [6.2]. The manner in which the characteristic impedance of the tapered line changes with distance gives the type of taper, e.g. exponential, hyperbolic, parabolic, etc., that is used.

Unless there are other very good reasons, it may be most expedient to use a taper type that has the simplest of geometries and is the easiest to make. Most tapered transmission line transformers that are one wavelength or more long will reduce the reflection coefficient to less than 20% of that from a step transition between the two lines. An increase in transformer length will tend to improve the broadband match between the two impedances.

The performance of a microstrip taper on $\varepsilon_r = 2.32$ substrate with a linear width variation of the line between 15Ω and 50Ω characteristic impedances has been analyzed for this book by subdividing the overall length of the taper into 100 short, but uniform, sections of line. The characteristic impedance and effective relative permittivity were evaluated for each section. The V.S.W.R. for this model of the taper, with the output line terminated by a matched load, is illustrated as curve (a) in Figure 6.18. The performance of this taper, which is of simple construction, is compared with the best of the continuous microstrip tapers (a Chebyshev taper) described by Khilla [6.3]. Curve (b) in Figure 6.18 illustrates the Chebyshev taper performance for the same impedance transformation on an $\varepsilon_r = 2.32$ substrate.

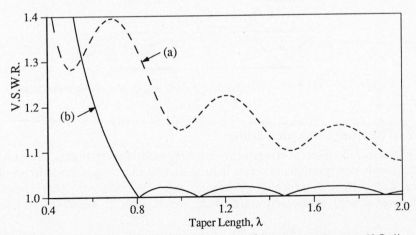

Figure 6.18 The performance of a microstrip taper between 15Ω and 50Ω lines on $\varepsilon_r = 2.32$ substrate for (a) a linear variation in the width of the taper section in comparison with (b) a Chebyshev taper from Khilla [6.3]

6.8   MATCHING WITH LUMPED LOSSLESS ELEMENTS

The lumped lossless elements may be either discrete components or approximations to lumped inductance and capacitance obtained from very short transmission line lengths.

6.8.1   Short lengths of transmission line

The reader will recall from §2.2.2 that there are two important approximations that may be used for short lengths of transmission line. A short length of high characteristic impedance transmission line may be represented by an equivalent series inductance, while that of a low characteristic impedance may be approximated by a shunt capacitance. If the short length approximations are not good enough for matching purposes, then the methods for single- and double-section line matching that are described in §6.9 and §6.10 could be used.

The short line equivalent circuit derived in §2.2.2 is reproduced in Figure 6.19 for the case of a high impedance line where the inductance, $L = Ll$. For the case of a low impedance line, the series inductance is replaced by a shunt capacitance, $C = Cl$. In terms of the respective characteristic impedances of the lines

$$L = \frac{Z_{HIGH}\sqrt{\varepsilon_{eff}}}{c} l \tag{6.1}$$

$$C = \frac{\sqrt{\varepsilon_{eff}}}{Z_{LOW} c} l \tag{6.2}$$

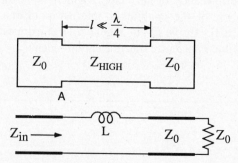

Figure 6.19   The equivalent circuit of a short length of high impedance line

6.8.2   The matching procedure

In this section, the discrete component matching techniques will only be considered for the simple combinations of series inductance and shunt capacitance illustrated in Figure 6.20. Similar matching procedures may be used for other component combinations, such as the series and shunt combination of capacitors in Exercise 6.15.

Consider case (a) in Figure 6.20. If the normalized admittance of the load with series inductance is of the form $y = 1.0 - jb$, then the capacitor may be used to add the necessary positive susceptance to give a matched condition. For the matching

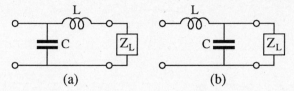

Figure 6.20   Matching networks that have series L and shunt C elements

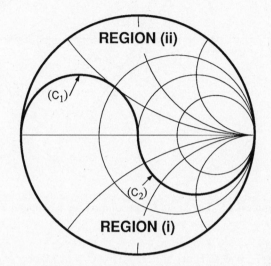

Figure 6.21   Impedance regions for the lumped element matching networks of Figure 6.20

procedure, transfer the negative susceptance half of the unit conductance circle to the impedance chart, i.e. (C_2) becomes (C_1) on Figure 6.21. Provided a load impedance lies in Region (i), i.e. the lower half of the chart bounded by (C_1) and (C_2), an inductance can always be found with a positive reactance which, when added in series with the load, will produce an impedance on (C_1) and thus an admittance on (C_2). On the other hand, in order to match with a network of type (b), the load admittance must lie within Region (i), i.e. the load impedance is in Region (ii).

Example 6.7

Using the Smith Chart, calculate the component values in Figure 6.20b that will match a $150\,\Omega$ load to a $50\,\Omega$ source impedance at $100\,\text{MHz}$.

Solution:

The Smith Chart solution is illustrated in Figure 6.22. Plot the normalized load impedance, $z_L = 3.0 + j0.0$ as Point (1), and convert to the admittance chart to obtain $y_L = 0.333 + j0.0$ at (2). As the series inductance will add positive reactance to complete the matching process, the negative reactance half of the unit resistance circle is transferred to the admittance diagram to become (C_1). Along a constant conductance path, add positive susceptance to the load admittance to give an admittance that lies on (C_1), i.e. $y = 0.333 + j0.471$ at (3).

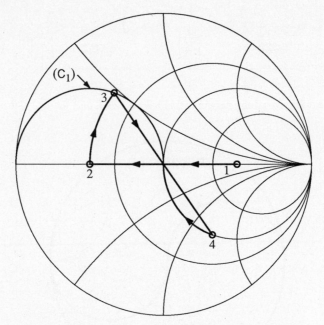

Figure 6.22   The Smith Chart solution for Example 6.7

Thus $\qquad \dfrac{\omega C}{Y_0} = 0.471$ or, at 100 MHz, $C = 15.0$ pF.

Transfer (3) back to the impedance diagram at (4), where $z = 1.0 - j1.414$. A normalized positive reactance of $+j1.414$ for the series inductance completes the matching procedure.

Hence $\qquad \dfrac{\omega L}{Z_0} = 1.414$ or, at 100 MHz, $L = 113$ nH.

The resultant matching network at 100 MHz is as shown in Figure 6.23.

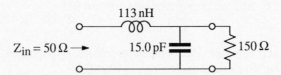

Figure 6.23   The matching network at 100 MHz for Example 6.7

6.9   SINGLE-SECTION LINE MATCHING

The representation of a short length of transmission line by a lumped L or C is, of course, only an approximation. The Π-equivalent circuit of §2.2.1, on the other hand, is an exact equivalence for all lengths of line. This Π network could also be used for

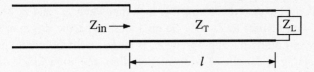

Figure 6.24   Impedance transformation with an arbitrary length of transmission line

matching purposes, but the matching procedure is actually better carried out in terms of the transmission line itself.

When the line is $\lambda/4$ long, the particular case of the quarter-wave transformer occurs, as considered in §6.6. In the present section, the matching properties of a transmission line of arbitrary length [6.4] are discussed, with its line length and characteristic impedance as variables.

Consider the network illustrated in Figure 6.24 where, in a 50Ω characteristic impedance system, a normalized input impedance z_{in} is to be derived from the load impedance, z_L. For conjugate matching to a source impedance, as discussed in §2.1.5, z_{in} may be complex, while when no reflections at the input are required, $z_{in} = 1 + j0$. Plot z_L and z_{in} on the Smith Chart, Figure 6.25, as (1) and (2) respectively. When these two impedances are each renormalized to the characteristic impedance of the transformer section, they must lie on a circle of constant $|\Gamma|$, if a transformation along the transformer length is to convert one impedance to the other.

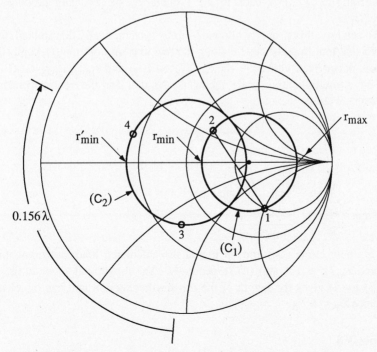

Figure 6.25   The Smith Chart for single-section line matching

The problem is thus to find a value of Z_T, such that when z_{in} and z_L are the result of normalization with respect to Z_T they lie on a circle with the origin at the center. A systematic way of finding the appropriate Z_T is based on the following property:

On traversing a length of transmission line, the impedance along the line continues to traverse a circle in the Γ plane — that is when plotted on the normalized resistance and reactance curves — even when the normalization is done with respect to an impedance that is not the characteristic impedance of the line. If the circles are normalized to real impedances, the center of each new circle will continue to lie on the real axis.

In Figure 6.25, (C_2) is the circle that results when the normalization is with respect to Z_T and (C_1) when the normalization is with respect to Z_0, both circles nevertheless referring to the impedance variation along the Z_T line. The center for (C_1) is at the intersection of the real axis and the perpendicular bisector of the line from (1) to (2). That (C_1) is a circle follows from the properties of the bilinear transformation

$$z = \frac{1 + \Gamma}{1 - \Gamma} \tag{6.32}$$

which transforms between the normalized impedance plane and the reflection coefficient plane. A bilinear transformation transforms circles into circles, with some circles becoming straight lines in limiting cases [6.5]. Since (C_2) is a circle in the Γ-plane, z is also a circle in the impedance plane. On multiplying z by the constant (Z_T/Z_0), the resulting impedance continues to be a circle in the impedance plane. Transforming $(Z_T/Z_0) \times z$ back to the Γ-plane via (6.32) again produces a circle, namely (C_1).

To see how this property allows Z_T to be systematically determined, first it is to be noted that real impedances lie along the real axis on both the (C_1) and (C_2) circles and their normalized values — r'_{min}, r'_{max} on (C_2) and r_{min}, r_{max} on (C_1) — only differ by a multiplicative scale factor, k, so that for the renormalization to the transformer section

$$k r'_{min} = r_{min} \tag{6.33}$$

$$k r'_{max} = r_{max} \tag{6.34}$$

with $k = Z_T/Z_0$. Thus

$$k^2 r'_{min} r'_{max} = r_{min} r_{max} \tag{6.35}$$

Since $r'_{min} = (r'_{max})^{-1}$, then

$$k = \sqrt{r_{min} r_{max}} \tag{6.36}$$

Draw the new circle (C_2) and on it plot the calculated load and input impedances z_L/k and z_{in}/k as (3) and (4) respectively. The movement towards the generator from (3) to (4) gives the length of the transformer section that has the characteristic impedance $Z_T = k Z_0$.

Example 6.8

A single-section matching line is required to transform a load impedance of

$(100 - j100)\,\Omega$ to an input impedance of $(65 + j30)\,\Omega$. Calculate the length and characteristic impedance of the matching line.

Solution:

It is common for many transmission line systems to have 50Ω characteristic impedance for the majority of transmission lines and it is to this value that the load and input impedances are normalized here. However, the final answer will be independent of the impedance to which these values are normalized and it is only necessary to choose a value that gives a clear presentation on the Smith Chart. The solution is the one that is specifically illustrated in Figure 6.25, with

$$z_L = 2 - j2, \qquad \text{Point (1)}$$

$$z_{in} = 1.3 + j0.6 \qquad \text{Point (2)}$$

The center of (C_1) is at $r = 2.52$, giving $r_{min} = 1.24$ and $r_{max} = 7.26$. Thus the renormalization factor, k, equals 3.00 and the transformer line impedance is 150.0Ω. The r'_{min} and r'_{max} values for (C_2) are 0.413 and 2.42 respectively and the renormalized load and input impedances of $0.667 - j0.667$ and $0.433 + j0.2$ are plotted as points (3) and (4). Hence the electrical length of the 150Ω transformer is 0.156λ, going from (3) to (4).

6.10 DOUBLE-SECTION LINE MATCHING

A pair of discrete matching elements, in particular in the form of series inductance and shunt capacitance, have been used in §6.8. While these elements may be approximated by short lengths of high and low impedance lines respectively, a more general solution [6.6] for matching with the finite line lengths, as shown in Figure 6.26, is presented here.

The high and low impedance lines, Z_{HIGH} and Z_{LOW}, are chosen to be typical of practical lines. For this discussion, the following typical values $Z_{HIGH} = 100\,\Omega$ and $Z_{LOW} = 30\,\Omega$ are taken. The matching procedure is again one where an intersection is found between two plots, representing impedances that can be obtained from the load and those that are desirable, in the sense that they can be matched to the source.

Working from the 50Ω impedance source towards the load with $Z_1 = Z_{HIGH} = 100\,\Omega$ and $Z_2 = Z_{LOW} = 30\Omega$, for an input match the following

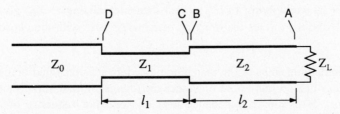

Figure 6.26 The impedance planes and selection of transformer impedances for double-section line matching

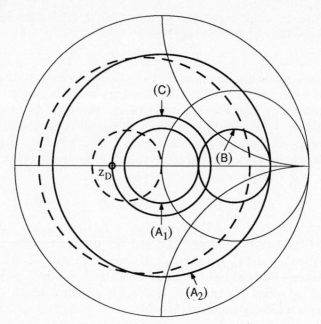

Figure 6.27 Load impedances that lie between the dashed circles for matching with a double-section transformer, $Z_1 = Z_{HIGH} = 100\Omega$ and $Z_2 = Z_{LOW} = 30\Omega$

impedances are required at the planes D to A, always looking towards the load. They are illustrated in Figure 6.27. At each plane the impedance is normalized to the characteristic impedance at that plane.

At D: $z_D = 0.5 + j0.0$, for a 50Ω impedance at the input end of a 100Ω line.

At C: the impedance lies on a constant $|\Gamma|$ circle that, through the length l_1, will transform C to D. Thus z_C lies on circle (C).

At B: z_B lies on circle (B), i.e. the circle of normalized impedances that when multiplied by Z_2/Z_1 gives the z_C on (C).

At A: z_A must lie within the region bounded by the circles A_1 and A_2, if a transformation along the low impedance line is to intersect with circle (B).

At the load: Renormalizing A_1 and A_2, i.e. converting A_1 and A_2 to the actual load impedances by multiplying by Z_{LOW} and then normalizing to Z_0, gives the region between the dashed circles that may be matched to 50Ω with the chosen impedance lines.

The remaining steps of the matching procedure are illustrated in the context of a numerical example. Figure 6.28 illustrates the regions of load impedance that may be matched to a 50Ω source impedance with one or other sequence of the 30Ω and 100Ω transformer line impedances and the shortest overall length of matching elements. In Figure 6.28, $Z_{HIGH} = mZ_0$ and $Z_{LOW} = nZ_0$, with $m = 2$ and $n = 0.6$.

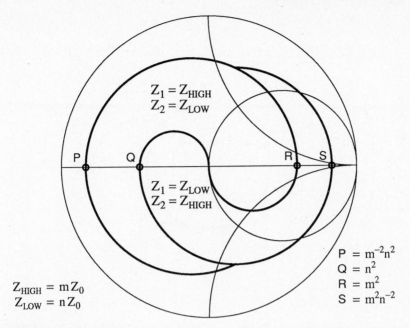

Figure 6.28 Load impedances to be matched to $50\,\Omega$ with $m=2$, $n=0.6$, using a double-section matching transformer

Example 6.9

Using $100\,\Omega$ and $30\,\Omega$ transformer sections, find the line lengths that match a $(60 - j60)\,\Omega$ load to a $50\,\Omega$ input transmission line.

Solution:

From Figure 6.28, it is seen that, to match a normalized impedance of $1.2 - j1.2$, the high impedance line should be closer to the load, leading to a matching network, as shown in Figure 6.26, with $Z_1 = 30\,\Omega$ and $Z_2 = 100\,\Omega$. The matching procedure is illustrated in Figure 6.29.

i) Plot the load impedance z_L, normalized to $50\,\Omega$ (1) and renormalized to $100\,\Omega$ to give z_A at (2).

ii) z_B lies on the constant $|\Gamma|$ circle (C_1) that passes through (2).

iii) Renormalizing (C_1) to the $30\,\Omega$ line gives obtainable values of z_C as the circle (C_2).

iv) To match to the $50\,\Omega$ source impedance, $z_D = (50 + j0)/30 = 1.67$, (3), is required.

v) The required values of z_C lie on the constant $|\Gamma|$ circle (C_3) that passes through (3).

vi) Solutions will be given by the intersections of (C_2) and (C_3). In particular for minimal line lengths, a solution is given at (4).

vii) (4) is renormalized back to (C_1), giving the solution for z_B at (5).

viii) l_2 is the movement towards the generator from (2) to (5) along the $100\,\Omega$ line.

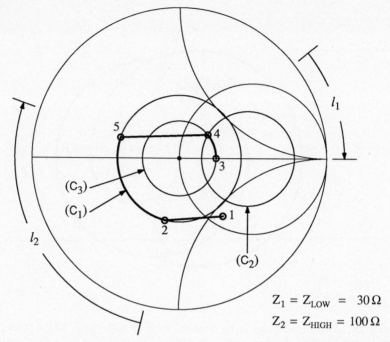

Figure 6.29 Double-section transformer matching of $(60 - j60)\,\Omega$ to a $50\,\Omega$ source impedance

ix) l_1 is the movement towards the generator from (4) to (3) along the $30\,\Omega$ line. For the solution, the following specific values are obtained:

Point (1)	z_L	$1.2 - j1.2$
Point (2)	z_A	$0.6 - j0.6$
Point (3)	z_D	$1.667 + j0$
Point (4)	z_C	$1.391 + j0.467$
Point (5)	z_B	$0.417 + j0.140$
	l_1	0.054λ
	l_2	0.133λ

6.11 IMPEDANCE SYNTHESIS

Most of this chapter has dealt with the situation where a given load impedance is transformed through a matching network to become an impedance equal to the characteristic impedance of the input line. There are situations, e.g. in Chapter 11, where the converse is required and a specified impedance has to be derived from a matched load. While in principle this may be achieved by reversing the steps of any of the previous techniques, it is better to take a direct approach that makes use of both

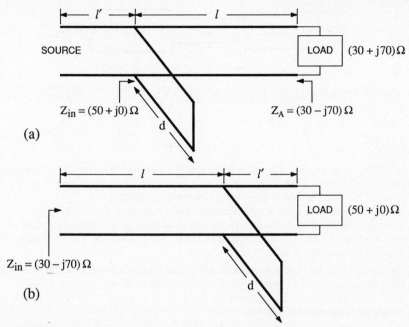

Figure 6.30 Illustrating impedance synthesis with $50\,\Omega$ lines, showing (a) matching a load $(30 + j70)\Omega$ to a $50\,\Omega$ source, and (b) the reinterpretation of the network in (a) as converting a $50\,\Omega$ matched load into an input impedance of $(30 - j70)\Omega$

the conjugate matching properties of a lossless network, discussed in §2.1.5, and the matching procedures just presented.

Consider the single-stub matching network that was derived for Example 6.2. This network is redrawn in Figure 6.30a. As there is conjugate matching ($50\,\Omega$ to $50\,\Omega$) at the source end, there will also be conjugate matching at the load end. Consequently, the source impedance as seen from the load will be $Z_A = Z_L^* = (30 - j70)\Omega$. In Figure 6.30b, the network is redrawn in the reverse order, i.e. the original matched source is now the load, and an input impedance $(30 - j70)\Omega$ is obtained. Thus, to derive an impedance Z_{in} starting with a matched termination, (i) choose a load equal to Z_{in}^*, (ii) carry out the matching process as in Figure 6.30a, and (iii) reverse the order of the matching elements as in Figure 6.30b. Note that the stub is now placed at a matched load plane and that the length l is required between it and the desired impedance plane. The length l must not be confused with l', the latter length being arbitrary in both Figures 6.30a and b.

EXERCISES

6.1 The characteristic impedances of the $\lambda/8$ lines in the figure below may be selected such that $25\,\Omega < Z_{LOW} < 50\,\Omega$ and $50\,\Omega < Z_{HIGH} < 100\,\Omega$. Using a Smith Chart, construct the region of load impedances, normalized to $50\,\Omega$, that may be matched using this network.

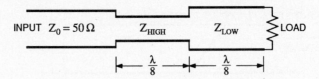

6.2 What normalized load impedances give the following voltage reflection coefficients?

i) $0.7 \underline{/-30°}$

ii) $0.5 \underline{/90°}$

iii) $0.1 \underline{/180°}$

iv) $1.0 \underline{/0°}$

6.3 The V.S.W.R. for a standing wave pattern set up by an unknown load is 2.0 with a voltage minimum at 102.5 mm. When the load is replaced by a short circuit, there are voltage minima at 25.1 and 123.3 mm, the smaller value being closer to the generator.

i)   What is the normalized load impedance referred to the attached short-circuit plane?

ii) If the previous readings had been obtained when the plane of the replacement short circuit was 5.0 mm in front of the actual load plane, what would be the normalized load impedance at the load plane?

6.4 i)   Illustrate on a Smith Chart the region of normalized load impedances that may be matched using the lumped element network illustrated in the figure below.

ii) What capacitance values are required for the network to match a $(30 + j50)\,\Omega$ load to a $50\,\Omega$ source impedance at 1.0 GHz?

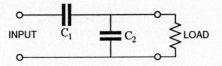

6.5 In Example 6.8, the load and source impedances were normalized to $50\,\Omega$ in order to carry out a Smith Chart solution. Verify that the same answer is obtained if the impedances are normalized to $100\,\Omega$ instead.

6.6 Calculate the length and characteristic impedance of a single-section impedance transformer, Figure 6.24, that will match a $(70 - j50)\,\Omega$ load to a $50\,\Omega$ source impedance.

6.7 In Example 6.4 for double-stub matching, there are two sets of stub admittances that may be used. Solve the example for the second solution and, having found the required normalized stub admittances, determine the stub lengths assuming short-circuit terminations.

6.8 The circuit elements shown in the figure below are expressed as normalized admittances. What is the distance in wavelengths, l, between the two susceptances if there is to be maximum power transfer from the matched input line to the load? Is there a voltage maximum or minimum at plane A, midway between the two susceptances? Deduce the results from a Smith Chart.

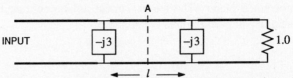

6.9 Determine the length and position of a single-stub matching network, using $50\,\Omega$ lines, that will match a load, $(30 + j20)\,\Omega$, to a $50\,\Omega$ input line. With the stub matching network, *estimate* the approximate percentage range or bandwidth for the V.S.W.R. to remain less than 1.2, assuming that the load remains $(30 + j20)\,\Omega$ over the frequency range of interest. **Hint:** Observe on the Smith Chart how the input impedance varies for a small frequency change.

6.10 A 100Ω characteristic impedance line with a 100Ω load is to be matched to a 50Ω line using sections of 30Ω and 120Ω lines as illustrated in the figure below. Using the Smith Chart, determine the lengths, l_1 and l_2 as fractions of a wavelength, that are required for the input match to be achieved.

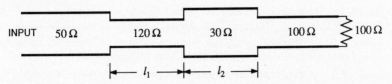

6.11 The figure below illustrates the conductor pattern for a microstrip network that is to be used to match a load, $(12.5 - j37.5)\,\Omega$, to a 50Ω input line. Determine (i) the length l of the open-circuit terminated stub that has a characteristic impedance of 100Ω and (ii) the characteristic impedance of the quarter-wave transformer necessary to achieve the matched condition. Ignore discontinuity effects.

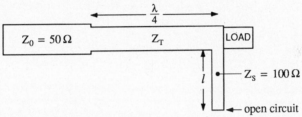

6.12 With line lengths given in wavelengths where appropriate, match a $(15 - j25)\,\Omega$ load to a 50Ω source with the following methods and constraints:
i)   single-stub matching with 50Ω lines and an open-circuit stub termination,
ii)   double-stub matching with 50Ω lines and open-circuit terminated stubs at 0.2λ and 0.33λ from the load plane,
iii) quarter-wave transformer matching with the transformer impedance less than 50Ω,
iv)   a series inductance, shunt capacitance network with the component values calculated at 1.0GHz,
v)   a double-section transformer with 25Ω and 100Ω impedance lines.

6.13 50Ω and 100Ω characteristic impedance lines may be matched together at a spot frequency by alternating lengths of each line type as shown in the figure below. Calculate the line lengths for a matched condition.

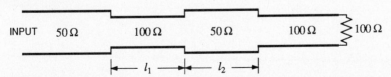

6.14 i)   Using the approximate theory for multiple-section quarter-wave transformers, calculate the transformer impedances between a 20Ω resistive load and a 50Ω input line for the maximally-flat 1-, 2- and 4-section cases.
ii)   For the single-section case, show that the result derived in (i) corresponds to the normal quarter-wave transformer solution.
iii) For the 2-section case, estimate from the Smith Chart the V.S.W.R. at 0.8 times the design center frequency.

6.15 Indicate on a Smith Chart those impedances that may be matched using a series and shunt combination of two capacitors.

REFERENCES

[6.1]   Matthaei, G. L., Young, L. and Jones, E. M. T., *Microwave Filters, Impedance-matching Networks, and Coupling Structures*, McGraw-Hill, New York, 1964.

[6.2]   Hall, A. H., "Impedance matching by tapered or stepped transmission lines", *Microwave Journal*, Vol. 9, No. 3, March 1966, pp. 109-14.

[6.3]   Khilla, A. M., "Optimum continuous microstrip tapers are amenable to computer-aided design", *Microwave Journal*, Vol. 26, No. 5, May 1983, pp. 221-4.

[6.4]   Somlo, P. I., "A logarithmic transmission line chart", *IEEE Trans. Microwave Theory and Techniques*, Vol. MTT-8, No. 4, July 1960, p. 463.

[6.5]   Churchill, R. V. and Brown, J. W., *Complex Variables and Applications*, McGraw-Hill, New York, 4th edn, 1984.

[6.6]   French, G. N. and Fooks, E. H., "Double section matching transformers", *IEEE Trans. Microwave Theory and Techniques*, Vol. MTT-17, No. 9, September 1969, p. 719.

7 Hybrid-line couplers

7.1 INTRODUCTION

Hybrid-line couplers represent the class of couplers that simulate the properties of the hybrid-coil transformer, Figure 7.1, in common use in telephony. With each port correctly terminated, a signal input at port 1 will give in-phase signals equally split between ports 2 and 3 and no output from port 4. Conversely, a signal input at port 4 will give out-of-phase but equal amplitude signals at ports 2 and 3 with no output at port 1. The correct choice of loads ensures a perfectly matched four-port device. Hybrid-line couplers are often simply referred to as hybrids.

An ideal directional coupler may be defined as a lossless reciprocal four-port network that appears perfectly matched looking into any one port when the others are terminated with matched loads, i.e. it is matched in every port. It can then be proved that there must be complete isolation between two distinct pairs of ports, say ports 1, 4 and ports 2, 3 (see Exercise 2.7). When an ideal directional coupler is inserted into a transmission line with the input connection at port 1 and the output at port 2, as illustrated in Figure 7.2, then, for a forward propagating wave from port 1 to port 2, some portion of it will be coupled out at port 3 with nothing being coupled to port 4. When waves propagate along other port directions, the coupled waves are as shown in Figure 7.2. Thus, a reverse traveling wave that is incident at port 2 will result in the coupled signal appearing at port 4, while port 3 is now isolated. The directional properties of the coupler are now evident, since the signals that are measured on ports 3 and 4 represent a measure of the forward and reverse traveling waves respectively on the through transmission line, i.e. from port 1 to 2 in this case. For each type of coupler that is discussed, the ports will be consistently numbered in this manner for comparison purposes.

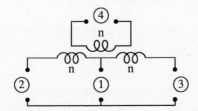

Figure 7.1 The hybrid-coil transformer

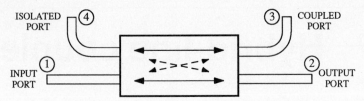

Figure 7.2   Power flow in a directional coupler

Hybrid-line couplers may be regarded as directional couplers which can produce an equal power split between the output ports. This definition is good in the main, but it is possible to have hybrids with an unequal power split. Inherently, a hybrid and a directional coupler are the same type of device. The only difference is really one of construction, with the hybrid-line coupler historically arising out of structures which aimed to produce equal power splitting with good isolation, directional coupling aspects not being of interest.

The hybrid-line directional couplers that are described in this chapter are thus constructed from an interconnected symmetrical network of transmission lines. The characteristic impedances of the interconnecting lines are chosen to give the input matched conditions for each port as well as giving the desired amount of coupling to the output ports. Directional couplers that derive their coupling from the action of closely spaced parallel transmission lines, without any direct connection between the lines, are the subject of the next chapter.

Consider a practical coupler with matched terminations at all ports. The insertion loss, I dB, between the input and output ports is dependent on both the dissipation loss and the power that is transmitted to the other ports. The level of the coupled signal, C, at port 3 is used in the description of the coupler, e.g. a 20dB coupler where 1% of the incident power comes out of that port. Thus, with matched loads at ports 2 to 4

$$C \; = \; -10 \, log_{10} \left\{ \frac{P_{out} \text{ at port 3}}{P_{in} \text{ at port 1}} \right\} \quad dB \tag{7.1}$$

Any unwanted signal strength at port 4 is compared with the coupled signal strength to give the directivity, D, as

$$D \; = \; -10 \, log_{10} \left\{ \frac{P_{out} \text{ at port 4}}{P_{out} \text{ at port 3}} \right\} \quad dB \tag{7.2}$$

If the unwanted signal is compared with the incident power at port 1, then the term "isolation" is used.

Example 7.1

At a certain frequency, a directional coupler has equal power outputs at -3.3dB below the incident signal level at the input. If the input V.S.W.R. is 1.2 and the directivity is -19.7dB, calculate the dissipation loss in the coupler.

Solution:

With an incident power to the coupler at port 1 of 1.0W, the following powers flowing out into matched loads are obtained:

At port 1: 0.0083 W, since V.S.W.R. = 1.2
At port 2: 0.4667 W
At port 3: 0.4667 W

The power at the isolated port is 19.7 dB below the power output at port 2. Thus the power is 23.0 dB below the input power level.

At port 4: 0.0050 W

The total output power is 0.9487 W, leaving 0.0513 W of power being dissipated within the coupler. Hence the dissipation loss = 0.23 dB.

7.2 EVEN- AND ODD-MODE ANALYSIS

A symmetrical four-port network is illustrated in Figure 7.3. This network may be analyzed in terms of even and odd modes with respect to the plane of symmetry.

For the even mode, two signals of equal amplitude and in-phase with each other are the inputs at the two ports reflected in the plane of symmetry, in this case at ports 1 and 4. On any line that crosses the plane of symmetry, there will be zero current flow at the plane together with a maximum in the magnitude of the voltage. This situation is equivalent to there being an open-circuit termination on any transmission line as it crosses the plane of symmetry. Thus, with no power flow across the symmetry plane, the circuit may be separated into two parts and analyzed accordingly. For the even mode, the four-port network now appears as two identical two-port networks. The detailed structure of the two-port network with an input signal at port 1 will determine the reflected signal amplitude and phase at port 1 and the transmitted signal amplitude and phase at port 2.

The odd mode is characterized by the input of two equal-amplitude signals at ports 1 and 4 that have a phase difference of 180°. There will be zero potential along the plane of symmetry, equivalent to a short-circuit termination across each transmission line at this plane. As for the even mode, the four-port network may be analyzed as two identical two-port networks.

An incident signal at port 1 with zero incident signal at port 4 may now be considered as the superposition of equal amplitude even- and odd-mode components with a phase relationship between them that gives a zero input signal at port 4. In general, there will now be a reflected signal at port 1 with transmitted signals to all the other ports. The ideal properties of a directional coupler are achieved if there is

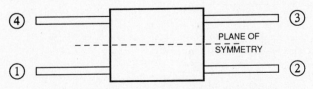

Figure 7.3 A symmetrical four-port network

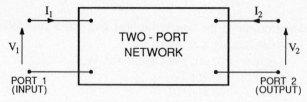

Figure 7.4 The voltages and currents for a two-port network

no reflected signal at port 1 and no transmitted signal to one of the other ports. This may be provided, for example, if both the equivalent two-port networks for the even and odd modes are themselves matched at their respective inputs. The direct and coupled signals to ports 2 and 3 may now be evaluated from a phasor addition of the transmitted signals for the two-port equivalent circuits.

The ABCD or transmission parameters for a two-port network have been discussed in Chapter 2 and are defined in conjunction with the sign convention for the terminal voltages and currents, illustrated in Figure 7.4, as

$$
\begin{bmatrix} V_1 \\ I_1 \end{bmatrix} = \begin{bmatrix} A & B \\ C & D \end{bmatrix} \begin{bmatrix} V_2 \\ -I_2 \end{bmatrix}
\tag{7.3}
$$

The reader will recall from Example 2.3 that the s_i and s_f scattering parameters for the two-port network, normalized to Z_0, are

$$
s_i = \frac{(A - D) + (BY_0 - CZ_0)}{(A + D) + (BY_0 + CZ_0)}
\tag{7.4}
$$

and

$$
s_f = \frac{2}{(A + D) + (BY_0 + CZ_0)}
\tag{7.5}
$$

Consider again the symmetrical four-port network with the ports as illustrated in Figure 7.3. The four-port network is analyzed in terms of two equivalent two-port networks, one for the even and the other for the odd mode. Associated with each mode will be the even- and odd-mode s_i, s_f parameters, denoted by $s_i^{(e)}$, $s_i^{(o)}$ and $s_f^{(e)}$, $s_f^{(o)}$ respectively. With an incident wave, a_1, at port 1, there will be reflected waves at all four ports, i.e. b_1 to b_4 inclusive. Now, in terms of the two-port equivalent networks, the incident wave at port 1 may be reconstructed as the sum of equal magnitude even- and odd-mode components, with their difference giving a zero incident wave at port 4. Thus, the four reflected waves are given by

$$
b_1 = \left[\frac{s_i^{(e)} + s_i^{(o)}}{2} \right] a_1
\tag{7.6a}
$$

$$
b_2 = \left[\frac{s_f^{(e)} + s_f^{(o)}}{2} \right] a_1
\tag{7.6b}
$$

$$
b_3 = \left[\frac{s_f^{(e)} - s_f^{(o)}}{2} \right] a_1
\tag{7.6c}
$$

$$
b_4 = \left[\frac{s_i^{(e)} - s_i^{(o)}}{2} \right] a_1
\tag{7.6d}
$$

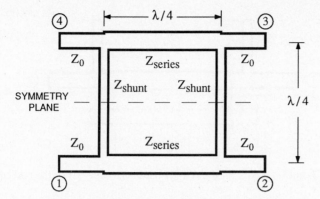

Figure 7.5 A branch-line coupler

7.3 THE BRANCH-LINE COUPLER

The fundamental structure for the branch-line coupler consists of four quarter-wavelength transmission lines that are connected together in a square format as illustrated in Figure 7.5.

The limitations of permissible microstrip line characteristic impedances make the branch-line coupler suitable only for tightly coupled requirements with a typical range of 2dB to 9dB. As a directional coupler, it has a narrow bandwidth which, it will be seen, can be increased either by having multiple sections or by using external matching elements. The continuous connection of the microstrip line to all ports gives d.c. coupling and, if implemented with low-loss transmission lines, is suitable for higher power applications with little danger of breakover that may otherwise occur at narrow gaps between adjacent transmission lines.

The even- and odd-mode analysis of the previous section will be used here to evaluate the performance of the basic coupler. The two-port network for the odd mode is illustrated in Figure 7.6, with the short-circuit terminations at the original plane of symmetry.

The ABCD-parameter or transmission parameter matrix, [T], is used to find the overall transmission and reflection characteristics of the network, [7.1]. Let $Z_{series} = Z_A = 1/Y_A$ and $Z_{shunt} = Z_B = 1/Y_B$. The circuit is split up into its three parts, representing each stub and the series $\lambda/4$ connecting line, as illustrated in Figure 7.7.

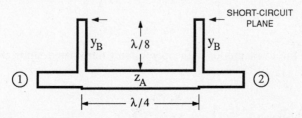

Figure 7.6 Odd-mode symmetry for the branch-line coupler

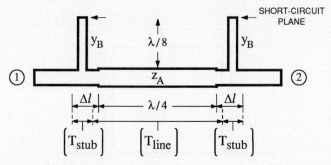

Figure 7.7 Circuit segments for the ABCD-parameter analysis of the odd-mode case

For an infinitesimally short length of through line, Δl, to which a $\lambda/8$ short-circuit terminated stub with input admittance $-jY_B$ is connected, and ignoring the T-junction effects described in §5.5

$$\left[T_{stub}\right] = \begin{bmatrix} 1 & 0 \\ -jY_B & 1 \end{bmatrix}$$

(7.7)

For the length of transmission line, l, between the two stubs, (2.31) gives

$$\left[T_{line}\right] = \begin{bmatrix} \cos\beta l & jZ_A\sin\beta l \\ jY_A\sin\beta l & \cos\beta l \end{bmatrix}$$

(7.8)

When this line length is a quarter-wavelength long line, with $\beta l = \dfrac{\pi}{2}$, then

$$\left[T_{line}\right] = \begin{bmatrix} 0 & jZ_A \\ jY_A & 0 \end{bmatrix}$$

(7.9)

The complete transmission parameter matrix for the odd-mode network in Figure 7.6 is found from the multiplication in the correct order of the individual matrices for the network. Thus

$$\left[T_{odd}\right] = \left[T_{stub}\right]\left[T_{line}\right]\left[T_{stub}\right]$$

(7.10)

Substituting from (7.7) and (7.9), and evaluating, gives

$$\left[T_{odd}\right] = \begin{bmatrix} 1 & 0 \\ -jY_B & 1 \end{bmatrix}\begin{bmatrix} 0 & jZ_A \\ jY_A & 0 \end{bmatrix}\begin{bmatrix} 1 & 0 \\ -jY_B & 1 \end{bmatrix}$$

(7.11)

$$= \begin{bmatrix} Y_B Z_A & jZ_A \\ j(Y_A - Y_B^2 Z_A) & Y_B Z_A \end{bmatrix}$$

(7.12)

Likewise for the even mode, with open-circuit terminated stubs that have an input admittance of $+jY_B$

$$\left[T_{even}\right] = \begin{bmatrix} -Y_B Z_A & jZ_A \\ j(Y_A - Y_B^2 Z_A) & -Y_B Z_A \end{bmatrix}$$

(7.13)

Note that $(A - D)_{odd} = (A - D)_{even} = 0$, $B_{odd} = B_{even}$ and $C_{odd} = C_{even}$. In view of (7.4), it follows that if $BY_0 = CZ_0$, then both $s_i^{(e)}$ and $s_i^{(o)}$ are zero and for the complete four-port network $b_1 = b_4 = 0$. In other words, the condition $BY_0 = CZ_0$ ensures that with matched loads at ports 2 to 4 the four-port network is perfectly matched looking into port 1 and that port 4 is completely isolated. Thus, for matching and isolation, and in terms of normalized quantities

$$j z_A = j \left[y_A - y_B^2 z_A \right] \tag{7.14}$$

i.e.

$$y_A^2 = 1 + y_B^2 \tag{7.15}$$

From this equation, it is seen that the characteristic impedance of the series connecting $\lambda/4$ line must always be less than the characteristic impedance of the external connecting transmission lines.

Again, with $A = D$ and $BY_0 = CZ_0$, and substituting for the odd mode into (7.5), the transmission coefficient becomes

$$s_f^{(o)} = \frac{1}{z_A(y_B + j)} = \frac{y_B - j}{y_A} \tag{7.16}$$

The transmission coefficient magnitude is unity in view of (7.15), as expected for a lossless two-port network that is perfectly matched at the input. Retaining the quadrant information in the signs of the real and imaginary parts, the phase angle for the odd-mode transmission coefficient is given by

$$\Theta_o = \tan^{-1} \left[\frac{-1}{+y_B} \right] \tag{7.17}$$

Similarly for the even-mode, the transmission coefficient magnitude is also unity and

$$\Theta_e = \tan^{-1} \left[\frac{-1}{-y_B} \right] \tag{7.18}$$

The relative phase angles for the incident and transmitted waves for the two-port networks are illustrated in Figure 7.8. The phase angles of the even- and odd-mode transmission coefficients are symmetrical with respect to the $-90°$ phase shift associated with a single quarter-wavelength line of unit characteristic impedance. This phase angle will now be taken as the reference angle.

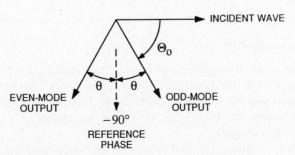

Figure 7.8 Two-port network phase relationships

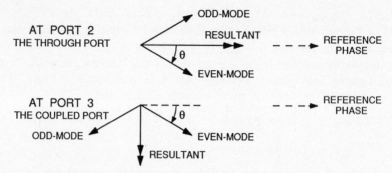

Figure 7.9 Output signal components for the branch-line coupler

Returning now to the complete four-port network, an incident wave at port 1 is treated as the sum of two equal amplitude components of the even and odd modes added together in-phase. The direct wave out of port 2 is the phasor sum of the two mode components as illustrated in Figure 7.9, with an amplitude given by

$$|b_2| = \cos \theta \tag{7.19}$$

where $\theta = |\Theta_e - 90°| = |90° - \Theta_o|$. The direct wave is thus no longer unity. Now there is power flow across the plane of symmetry.

The output at the coupled port, port 3, is given in a similar manner but being produced by the odd-mode incident wave out-of-phase with the even-mode component. The coupled signal amplitude

$$|b_3| = \sin \theta \tag{7.20}$$

From the definition of coupling given in (7.1)

$$C = -10 \, log_{10}(\sin^2\theta) \quad \text{dB} \tag{7.21}$$

Therefore

$$C = -20 \, log_{10}\left\{\frac{y_B}{\sqrt{1 + y_B^2}}\right\} \tag{7.22}$$

Solving for y_B gives

$$y_B = \frac{c}{\sqrt{1 - c^2}} \tag{7.23}$$

where

$$c = 10^{-C/20} \tag{7.24}$$

The results for a range of single-section branch-line couplers in a 50Ω characteristic impedance transmission line system are presented in Table 7.1. Referring to Figure 7.8, it is also seen that the outputs at ports 2 and 3 are 90° out-of-phase with each other. For this reason, the branch-line coupler is a member of a class of couplers known as *quadrature couplers*.

The power-split ratio, P, is used to express the coupling of a directional coupler in terms of the ratio of powers to the coupled and direct ports, i.e.

$$P = -20 \, log_{10} \left[\frac{|b_3|}{|b_2|} \right] = C - I \tag{7.25}$$

For a lossless coupler, the coupling coefficient may be expressed as

$$c = \frac{p}{\sqrt{1 + p^2}} \quad \text{where } p = 10^{-P/20} \tag{7.26}$$

Table 7.1 Line impedances for a single-section branch-line coupler, designed for 50Ω external connecting impedances

Coupling C, dB	y_B	y_A	Z_{shunt} Ω	Z_{series} Ω
2	1.308	1.646	38.24	30.37
3	1.002	1.416	49.88	35.31
3.01	1.000	1.414	50.00	35.36
4	0.813	1.289	61.48	38.79
5	0.680	1.209	73.52	41.35
6	0.579	1.156	86.33	43.27
7	0.499	1.118	100.15	44.73
8	0.434	1.090	115.21	45.87
9	0.380	1.070	131.75	46.75
10	0.333	1.054	150.00	47.43

There are advantages in using the power-split ratio when directional couplers are being used in a network to provide appropriate power ratios at several output ports. The relationship between the power-split ratio and the coupling ratio is illustrated in Figure 7.10. The asymptote, representing a plot of the coupling ratio on both axes, clearly shows how the power-split ratio approaches the coupling ratio for high values.

The behavior of the branch-line coupler when the input is applied to another port can be deduced in a straightforward manner from knowledge of what happens with an input to port 1. It is readily observed that every port has an identical environment, with two lines of different characteristic impedances branching from it. The higher impedance line leads to the isolated port.

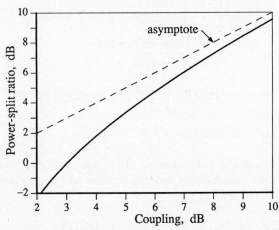

Figure 7.10 Power-split ratio and coupling for an ideal directional coupler

At this point, it is worth remarking that for any type of coupler, there is no fundamental difference between the through and coupled ports. Both are output ports, but it is usual to take the port with the smaller amount of power as the coupled port.

Example 7.2

Calculate the power-split ratio in dB for a 6.02dB directional coupler.

Solution:

From (7.26)

$$p = \frac{c}{\sqrt{1 - c^2}}$$

For a coupler with $C = 6.02$dB, then from (7.24) $c = 0.50$, giving $p = 0.577$ and a power-split ratio, $P = 4.77$dB.

Example 7.3

A single-section 3.01dB branch-line coupler has shunt and series line impedances of 50.0 and 35.36Ω respectively. To observe one possible effect of manufacturing tolerances, consider the case where the series line impedance is 37.0Ω while all other aspects of the coupler may be considered to be unaltered. Calculate the input V.S.W.R., coupling and directivity for this modified coupler.

Solution:

The circuit segments for ABCD-parameter analysis are shown in Figure 7.7. The line immittances are $Y_B = 0.02$S and $Z_A = 37.0\Omega$. From (7.12) and (7.13)

$$\left[T_{even}\right] = \begin{bmatrix} -0.74 & j37.0 \\ j0.01223 & -0.74 \end{bmatrix}$$

and

$$\left[T_{odd}\right] = \begin{bmatrix} 0.74 & j37.0 \\ j0.01223 & 0.74 \end{bmatrix}$$

From (7.4), $s_i^{(o)} = \dfrac{(A - D) + (BY_0 - CZ_0)}{(A + D) + (BY_0 + CZ_0)}$

$$= \frac{j(0.74 - 0.6114)}{1.48 + j1.3514} = 0.0642 \underline{/47.6°}$$

while $s_i^{(e)} = 0.0642 \underline{/-47.6°}$

Each mode scattering parameter is for a unit incident wave. Now, for the complete coupler with half the incident wave at port 1 associated with each of the modes, from (7.6a) with $a_1 = 1$

$$b_1 \equiv \Gamma_{in} = \frac{s_i^{(e)} + s_i^{(o)}}{2} = 0.0433$$

i.e.   input V.S.W.R. $= \dfrac{1 + |\Gamma_{in}|}{1 - |\Gamma_{in}|} = 1.09$

At the coupled port, from (7.6c)

$$b_3 = \frac{s_f^{(e)} - s_f^{(o)}}{2}$$

i.e. $b_3 = \left[\dfrac{1}{(A + D) + (BY_0 + CZ_0)} \right]_{even} - \left[\dfrac{1}{(A + D) + (BY_0 + CZ_0)} \right]_{odd}$

giving $b_3 = \dfrac{1}{1.48 + j1.3514} - \dfrac{1}{-1.48 + j1.3514} = 0.737$

Hence, the coupled signal level $= -2.65\,dB$.

From (7.6d), the reflected wave at the isolated port is given by

$$b_4 = \frac{s_i^{(e)} - s_i^{(o)}}{2} = j0.0474$$

representing a power level of -26.48 dB with respect to the input power. Now, with a coupled signal of $-2.65\,dB$, the directivity of this branch-line coupler is $23.8\,dB$.

7.4   THE BRANCH-LINE COUPLER
— with improved coupling performance

The single-section branch-line coupler was designed so that, when it was inserted into a matched system of transmission lines, a perfect match was provided at each port at the design frequency. This was achieved by maintaining a match for both the even- and odd-mode two-port equivalent networks. However, it is important to consider the performance of the coupler as the frequency of operation is varied. Ideally, the input should remain as near to a perfect match for both modes over as wide a frequency band as possible. How well this is achieved may be seen with the aid of the transformations through the two-port networks at the design frequency, f_0, described in conjunction with Figure 7.11 using the line parameters for a 3 dB branch-line coupler.

The intermediate admittances are labeled at the respective planes commencing at port 2, which is terminated in a matched load, through to the input at port 1. Consider the case of the even mode with the $\lambda/8$ open-circuit terminated parallel-stub transmission lines. Each admittance is highlighted by a black dot on the expanded central section of a Smith Chart, Figure 7.11b. Commencing from the matched load termination, y_1, the stub susceptance of $+j1.0$, normalized to a 0.02 S characteristic admittance (50 Ω characteristic impedance), is added to give y_2. Renormalizing to 35.36 Ω transforms the normalized admittance to y_3. Note that this admittance is situated on the line that passes through the center of the chart and perpendicular to the

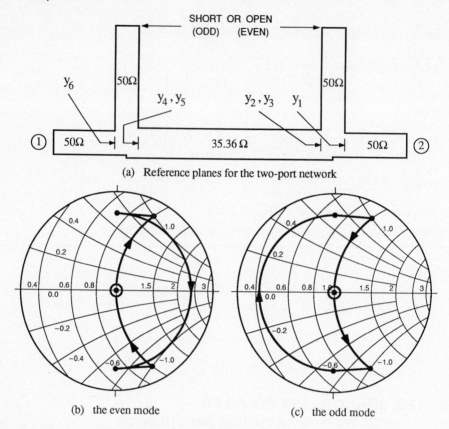

(a)   Reference planes for the two-port network

(b)   the even mode   (c)   the odd mode

Figure 7.11   The transformation of admittance through even- and odd-mode two-port networks

zero susceptance axis. As will be seen when the odd mode is considered, this is necessary if both the even- and odd-mode equivalent two-port networks are to be matched. The $\lambda/4$ line transforms the admittance to y_4, while renormalization back to the 50Ω characteristic impedance line gives the admittance, y_5. At this plane, the stub susceptance of $+j1.0$ completes the perfect match, y_6.

For the odd mode, with the short-circuit terminated stub, the normalized stub admittance is $-j1.0$. Thus, using the same order of admittance points, the matching process is illustrated for the odd-mode case in Figure 7.11c.

The perfect match at the design frequency will not be maintained as the frequency changes. Let the frequency be reduced by 10% to $0.9f_0$. Each transmission line, with its fixed physical length, now has an electrical length that is reduced by 10%. Thus the normalized input admittance of the short-circuit terminated transmission line will become $-j1.171$, while that with the open-circuit termination will become $+j0.854$. The series line that was a quarter-wavelength at the design frequency is now reduced, in effect, to 0.225λ. With these changes, the even- and odd-mode admittance plots at $0.9f_0$ are illustrated in Figure 7.12.

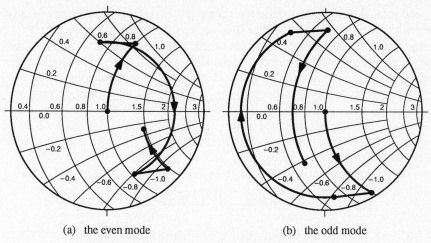

      (a) the even mode   (b) the odd mode

Figure 7.12   Admittance transformations at $0.9\,f_0$

The input V.S.W.R. is evaluated from the input reflection coefficient

$$\Gamma_{in} = \frac{s_i^{(e)} + s_i^{(o)}}{2} \tag{7.27}$$

At $0.9\,f_0$, the individual mode reflection coefficients from Figure 7.12 are

$$s_i^{(e)} = 0.223\,\underline{/-25.4°}, \qquad s_i^{(o)} = 0.298\,\underline{/-111.6°}$$

giving

$$|\Gamma_{in}| = 0.192 \quad \text{and} \quad \text{V.S.W.R.} = 1.475.$$

For a minimum V.S.W.R. at the input, it is not necessary for both even and odd components to be minimum, provided that the two components are out-of-phase with each other.

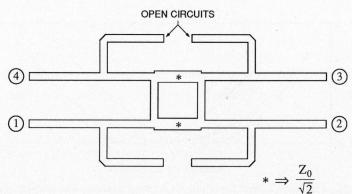

Figure 7.13 The 3.01 dB branch-line coupler with four external matching circuits and all line impedances of Z_0 unless otherwise specified, adapted from Riblet [7.2] (© 1978, IEEE)

Additional networks, external to the ports of the two-port network, may be added without influencing the center frequency input impedance at each port for both modes, if at this frequency the additional networks appear as open circuits across the lines. An open-circuit terminated parallel-stub line with a 50Ω characteristic impedance and $\lambda/2$ long exhibits some of the necessary circuit requirements. This line has an infinite input impedance for both modes and may, in principle, be connected at any plane along the transmission line. However, an improved performance is obtained when the stub is connected at $\lambda/4$ from each of the shunt arms of the branch-line coupler, as illustrated in Figure 7.13. A compact structure is achieved by folding the $\lambda/2$ stubs.

For this network at $0.9\,f_0$, (7.27) gives

$$\Gamma_{in} = \frac{0.244\,\underline{/15.2°} \ + \ 0.015\,\underline{/105.9°}}{2} \tag{7.28}$$

The resulting V.S.W.R. of 1.278 is an improvement over the value obtained earlier.

It will be seen later in this section that, by correctly selecting the characteristic impedances, precise lengths of the stubs and their positions relative to each port, [7.2], improved broadband matching may be obtained. A theoretical comparison for a 3 dB coupler is presented in Figure 7.14. Curve (a) shows the performance of the basic branch-line coupler and (b) that of the same coupler with $\lambda/2$ open-circuit terminated stubs placed at $\lambda/4$ from each port.

Instead of having a perfect match at the design frequency, f_0, it is possible to match at two frequencies on either side of f_0. The matching networks used are illustrated in Figure 7.15a. They are connected in tandem with each port. For a microstrip line it is more practical to use the open-circuit terminated stub transmission line, Figure 7.15b. With $\theta \approx \pi/2$, a line of length 2θ and normalized characteristic admittance $y_2/2$ with an open-circuit termination will have approximately the same input admittance as the short-circuit terminated line of length, θ, and admittance, y_2. The $\lambda/4$ sections of the branch-line coupler also have the new electrical length, θ.

From the results presented by Riblet [7.2], if $\Delta f/f_0$ is the required frequency

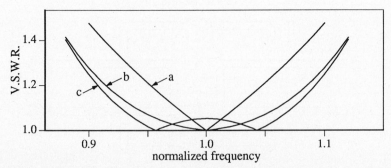

Figure 7.14 The input VSWR for a 3.01 dB improved branch-line coupler, showing (a) the basic branch-line coupler, (b) the coupler shown in Figure 7.13, and (c) using broadband matching elements of Figure 7.15

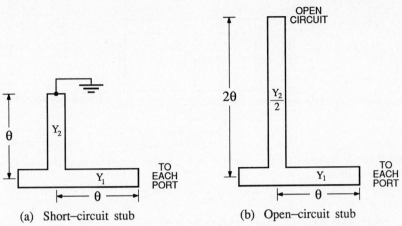

(a) Short–circuit stub   (b) Open–circuit stub

Figure 7.15 Broadband matching elements, attached to all ports of the hybrid coupler, from Riblet [7.2] (© 1978, IEEE)

separation for the two matched frequencies, then θ is given approximately by

$$\theta = \cos^{-1}\left\{\frac{1}{\sqrt{2}}\cos\left[\frac{\pi}{2}\left(1+\frac{\Delta f}{2f_0}\right)\right]\right\} \tag{7.29}$$

If $\zeta = y_1/y_2$ and ξ is a free parameter that is chosen such that $\xi\zeta \approx 1$ in order to give a realizable range of microstrip line admittances, then the two equations that must be satisfied are

$$\frac{\sin^2\theta}{y_1^2} + (1+1/\zeta)^2\cos^2\theta = \frac{\sin\theta}{\xi\sqrt{\zeta^2-1}} \tag{7.30}$$

$$\frac{\sin\theta}{y_1} - (1+1/\zeta)\left[y_1\sin\theta - y_2\frac{\cos^2\theta}{\sin\theta}\right] = -\left\{\frac{\zeta+1}{\zeta-1}\right\}^{\frac{1}{2}} \tag{7.31}$$

These two equations are symmetrical in θ and $(180° - \theta)$, giving two solutions where the design frequency, f_0, is the lower or upper matched frequency respectively. The particular case described in detail earlier and plotted as curve (b) in Figure 7.14 is given if $\Delta f/f_0 = 0$, i.e. the matched condition is maintained at f_0. However, whereas 50Ω characteristic impedance lines were used, an improved performance as predicted by (7.29), (7.30) and (7.31) is given if the line parameters are altered to $y_1 = 1.026$ ($48.73\,\Omega$) and $y_2 = 1.195$ ($41.84\,\Omega$) respectively. In order to have the perfectly matched frequencies on either side of a central design frequency, the transmission line lengths become a quarter- or half-wavelength long at the center frequency. With this change, the V.S.W.R. at 0.9 and $1.1 \times f_0$ becomes 1.25; a small improvement, as seen by comparing curves (b) and (c) in Figure 7.14.

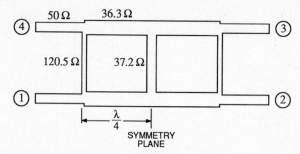

Figure 7.16   A symmetrical three-branch 3dB coupler, based on data from Levy and Lind [7.3] (© 1968, IEEE)

7.5 THE BRANCH-LINE COUPLER
— with multiple sections

Multiple-section branch-line couplers appear to offer the desirable coupler characteristics of constant coupling, low input V.S.W.R. and high isolation over a wider bandwidth than the single-section (two-branch) coupler. Levy and Lind [7.3] derive the theory for the synthesis of multiple-section couplers by considering a frequency dependent function that represents both the input reflection coefficient and isolation. Minimizing the function with maximally-flat or Chebyshev characteristics leads to the required line impedances for each of the coupler's quarter-wave lines.

Further design details are not given here because, from the tables of line impedances given in [7.3] for couplers having from 3 to 9 branches, it is seen that almost all couplers would be unachievable using microstrip transmission line circuits as a result of the high impedance shunt lines that are required. Limiting the characteristic impedance of all lines to the range of 25 to 150Ω, maximally-flat coupler designs are restricted to the range of 2 to 4dB coupling at the center frequency. A 3-branch coupler with 3dB coupling at the center frequency is illustrated in Figure 7.16, where all the line lengths are nominally λ/4 at the center frequency.

Broadband designs for multiple-branch 3dB couplers that do no require excessively high line impedances have been described by Muraguchi et al. [7.4]. The

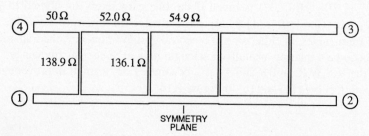

Figure 7.17   An optimized 3dB coupler, based on Example 4.9 of Muraguchi et al. [7.4] (© 1983, IEEE)

impedance problem is solved by applying a computer-aided design that limits the maximum allowable line impedances. A penalty function, in terms of all the line impedances, is summed at intervals across the desired frequency band. The function for a 3dB coupler has terms of the form $|s_{11}|^2$, $|s_{21}|^2 - 0.5$, $|s_{31}|^2 - 0.5$ and $|s_{41}|^2$ and their sum across the frequency band is minimized by an appropriate search through possible line impedances. From their results, a four-branch 3dB coupler with the impedance values rounded to one decimal place is illustrated in Figure 7.17.

7.6 THE HYBRID-RING COUPLER

7.6.1 Introduction

The branch-line coupler described in the previous section may be constructed with the series and shunt transmission lines in a circular shape. However, this should not be confused with the hybrid-ring coupler that has $\lambda/4$ and $3\lambda/4$ interconnecting line lengths as illustrated in Figure 7.18. The hybrid-ring coupler is also known as the "rat-race" coupler. Unlike the 3dB branch-line coupler where the phase of the coupled signals ideally differs by 90°, for the 3dB hybrid-ring coupler the output signal pairs are either in-phase or out-of-phase depending upon which port of the coupler is used for the input. For this reason the "rat-race" coupler is closer to the action of an ideal hybrid, Figure 7.1, and the waveguide junction "magic tee" than the branch-line hybrid. The facility of having two equal amplitude in-phase signals that are isolated from each other, and the improved bandwidth performance of this coupler compared with the single-section branch-line coupler, makes it an ideal choice for a matched power splitter in an antenna feed system (Reed and Wheeler [7.1]).

It will be noticed that there is a plane of symmetry in Figure 7.18, which will allow the four-port network to be split into equivalent two-port networks with even- and odd-mode symmetry, as was done for the branch-line coupler. With the isolated

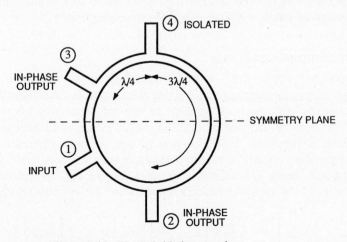

Figure 7.18 The hybrid-ring coupler

port still being designated as port 4, the port arrangement with respect to the plane of symmetry differs from that in Figures 7.3 and 7.5 by having ports 3 and 4 interchanged. In Figure 7.3, the isolated port 4 was at the input end of the structure, whereas in Figure 7.18 it is port 3 that has that position. It will be seen in a later analysis that input matching is achieved not with $s_i^{(e)} = s_i^{(o)} = 0$, as for the branch-line coupler, but with $s_i^{(e)} + s_i^{(o)} = 0$. This also means that the resultant reflection coefficient at port 3 is not zero but is equal to $s_i^{(e)} - s_i^{(o)}$, representing a transmitted signal flowing out of that port.

7.6.2 A qualitative description

The following simplified view of a 3 dB hybrid-ring coupler leads to an understanding of which port is isolated as well as giving the correct characteristic impedance for the circular connecting line. Consider an input signal at port 1 that splits into two equal waves traveling in opposite directions around the ring. At port 2, the two waves have traveled distances $\lambda/4$ and $5\lambda/4$ respectively and arrive at port 2 in-phase with each other. The same occurs at port 3, making the signal at port 3 in-phase with that at port 2. However, at port 4 the two waves that have traveled distances of $\lambda/2$ and λ, arrive out-of-phase and will cancel. Thus port 4 is isolated from port 1. On the ring between ports 2 and 3 via port 4, there will be a standing wave present with zero electric field at port 4. The transmission line may be short circuited at this plane. This short circuit, through the $\lambda/4$ and $3\lambda/4$ lengths of line, presents an open-circuit impedance at each of the junctions to the ring of ports 2 and 3 respectively. With each port connected into a matched $50\,\Omega$ line, the connections from ports 2 and 3 should each appear as a $100\,\Omega$ load at port 1 so that their parallel combination presents a matched load to the input. The $100\,\Omega$ impedance level is provided by transforming each $50\,\Omega$ load of the coupled output ports through a $70.7\,\Omega$ characteristic impedance quarter-wave transformer. A similar argument when port 3 is taken as the input to the hybrid-ring coupler leads to the complete ring being found to have a uniform $70.7\,\Omega$ characteristic impedance. Further analysis will show that for an input wave at any port there will be equal power split to adjacent ports, either in-phase or out-of-phase, with isolation at the remaining port.

7.6.3 A complete analysis

The two-port network for even- and odd-mode analysis is illustrated in Figure 7.19. The characteristic impedance of the feed line at each port is assumed to be $50\,\Omega$ while the characteristic impedances of appropriate sections of the ring become Z_{stub} and Z_{series}. For the even and odd modes, the stub lines are terminated with open and short circuits respectively. It will be noticed that the only difference between the two-port equivalent circuits of branch-line and hybrid-ring couplers is the length of the second stub, being $\lambda/8$ for the branch-line coupler but $3\lambda/8$ in this case. The ABCD-matrix of the two-port equivalent circuit for the hybrid-ring coupler is now given by

$$\left[T_{e,o} \right] = \left[T_{e,o} \right]_{\frac{\lambda}{8}\text{-stub}} \times \left[T_{e,o} \right]_{\frac{\lambda}{4}\text{-line}} \times \left[T_{e,o} \right]_{\frac{3\lambda}{8}\text{-stub}} \quad (7.32)$$

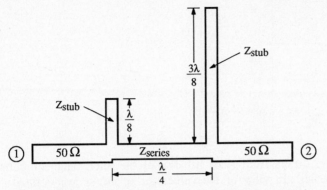

Figure 7.19 The hybrid-ring two-port equivalent network

Thus

$$
\left[T_{e,o} \right] = \begin{bmatrix} 1 & 0 \\ \pm \dfrac{j}{Z_{stub}} & 1 \end{bmatrix} \begin{bmatrix} 0 & jZ_{series} \\ \dfrac{j}{Z_{series}} & 0 \end{bmatrix} \begin{bmatrix} 1 & 0 \\ \mp \dfrac{j}{Z_{stub}} & 1 \end{bmatrix}
\tag{7.33}
$$

$$
= \begin{bmatrix} \pm \dfrac{Z_{series}}{Z_{stub}} & jZ_{series} \\ j\left(\dfrac{1}{Z_{series}} + \dfrac{Z_{series}}{Z_{stub}^2} \right) & \mp \dfrac{Z_{series}}{Z_{stub}} \end{bmatrix}
\tag{7.34}
$$

In order to evaluate s_i and s_f via (7.4) and (7.5), certain symmetries inherent in (7.34) are to be noted, namely

$$
(A+D)_{even} = (A+D)_{odd} = 0
\tag{7.35a}
$$

$$
(A-D)_{even} = -(A-D)_{odd} = 2A_{even} = -2A_{odd}
\tag{7.35b}
$$

$$
B_{even} = B_{odd} = B
\tag{7.35c}
$$

$$
C_{even} = C_{odd} = C
\tag{7.35d}
$$

so that

$$
s_i^{(e,o)} = \frac{\pm 2A_{even} + (BY_0 - CZ_0)}{BY_0 + CZ_0}
\tag{7.36}
$$

and

$$
s_f^{(e)} = s_f^{(o)} = \frac{2}{BY_0 + CZ_0}
\tag{7.37}
$$

Returning now to the four-port network, the reflected waves out of the four ports for an incident wave at port 1 are

$$
b_1 = \left[\frac{s_i^{(e)} + s_i^{(o)}}{2} \right] a_1
\tag{7.38a}
$$

$$b_2 = \left[\frac{s_f^{(e)} + s_f^{(o)}}{2}\right] a_1 \tag{7.38b}$$

$$b_3 = \left[\frac{s_i^{(e)} - s_i^{(o)}}{2}\right] a_1 \tag{7.38c}$$

$$b_4 = \left[\frac{s_f^{(e)} - s_f^{(o)}}{2}\right] a_1 \tag{7.38d}$$

This equation differs from (7.6) because of the interchange of ports 3 and 4, as explained earlier. In view of the values for s_i and s_f just derived as well as the symmetry relations (7.35), the reflected power wave expressions in (7.38) simplify to

$$b_1 = \frac{(BY_0 - CZ_0)}{BY_0 + CZ_0} a_1 \tag{7.39a}$$

$$b_2 = \frac{2}{BY_0 + CZ_0} a_1 \tag{7.39b}$$

$$b_3 = \frac{2 A_{even}}{BY_0 + CZ_0} a_1 \tag{7.39c}$$

$$b_4 = 0 \tag{7.39d}$$

Isolation at port 4 is thus always achieved, i.e. for all values of Z_{series} and Z_{stub}, but it is necessary that $BY_0 = CZ_0$ for input matching, giving

$$b_1 = b_4 = 0 \tag{7.40}$$

$$b_2 = \frac{1}{BY_0} a_1 \tag{7.41}$$

and $$b_3 = \frac{A_{even}}{BY_0} a_1 \tag{7.42}$$

In terms of normalized quantities and substituting from (7.34)

$$b_2 = \frac{-j}{z_{series}} a_1 \tag{7.43}$$

$$b_3 = \frac{-j}{z_{stub}} a_1 \tag{7.44}$$

and the condition $BY_0 = CZ_0$ becomes

$$\frac{1}{z_{series}^2} + \frac{1}{z_{stub}^2} = 1 \tag{7.45}$$

In view of (7.43) and (7.44), it will be recognized that (7.45) simply states that the power out of ports 2 and 3 equals the power into port 1. For an equal power split between the two output ports, $z_{series} = z_{stub} = \sqrt{2}$, giving $70.7\,\Omega$ for the characteristic impedance of the ring in a $50\,\Omega$ system.

Unequal power division may be obtained by alternating the characteristic

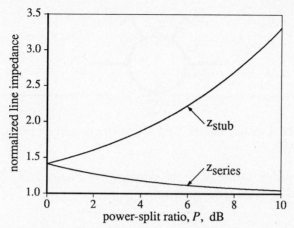

Figure 7.20 Normalized line impedances for a hybrid-ring coupler with an output power-split ratio, P

impedances of the ring lines between the two different values, as shown by Pon [7.5]. The output signals at ports 2 and 3 retain the same phase relationship even for an unequal power split. When inputs are applied to other ports, an identical analysis reveals that each port is itself matched. Thus the hybrid-ring coupler, even with unequal power split, satisfies the conditions for an ideal directional coupler as outlined in §7.1. From (7.25)

$$P = -10 \, log_{10} \left[\frac{|b_3|^2}{|b_2|^2} \right] \tag{7.46}$$

Substituting for b_2 and b_3, and using (7.45), gives

$$z_{stub} = \left\{ 1 + 10^{P/10} \right\}^{\frac{1}{2}} \text{ and } z_{series} = \left\{ 1 + 10^{-P/10} \right\}^{\frac{1}{2}} \tag{7.47}$$

The two normalized line impedances are plotted as a function of P in Figure 7.20.

Example 7.4

The hybrid-ring coupler, illustrated schematically in Figure 7.21, has power fed into port 1 and twice the power out from port 3 compared with that from port 2. If perfect matching and isolation are assumed in a $50\,\Omega$ system, calculate the line impedance around the ring.

Solution:

The coupler has a power-split ratio of 3.01 dB. However, since the coupled power out from port 3 is greater than the transmitted power to port 2, the power-split ratio is actually -3.01 dB and in effect gives a directional coupler with $C = 1.76$ dB. From (7.47)

$$z_{stub} = 1.225 \times 50 = 61.2\,\Omega$$

and

$$z_{series} = 1.732 \times 50 = 86.6\,\Omega$$

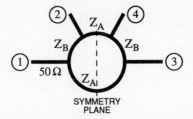

Figure 7.21 Line impedances of the hybrid-ring coupler for Example 7.4

Noting the position of the plane of symmetry in Figure 7.21, Z_A is the stub impedance and Z_B is the series impedance.

7.7 THE HYBRID-RING COUPLER
— with modified ring impedances

With an upper limit on line characteristic impedance of about $150\,\Omega$ for many microstrip transmission lines, the hybrid-ring coupler is limited to typical power-split ratios from 0 to 9 dB. In the modified ring structure, analyzed by Agrawal and Mikucki [7.6], the $3\lambda/4$ section is divided into three separate $\lambda/4$ sections of transmission line, as illustrated in Figure 7.22. Increased power-split ratios may now be obtained while the line impedances remain below $150\,\Omega$. The output signals from ports 2 and 3 remain in-phase for the input at port 1 and out-of-phase for the input at port 4. The improved performance is, however, at the expense of the input match of each port and the isolation between ports 1 and 4, and ports 2 and 3.

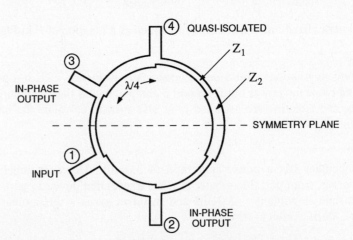

Figure 7.22 The modified hybrid-ring coupler with alternating line impedances around the ring

Modified ring impedances of $Z_1 = 118.6\,\Omega$ and $Z_2 = 52.1\,\Omega$ in a $50\,\Omega$ system give a power-split ratio of 10.0 dB at the center frequency, with most of the power going to port 2. This modified hybrid-ring coupler has a theoretical input V.S.W.R. = 1.072 and directivity (port 4 with respect to port 3) = 8.5 dB. These line impedances satisfy the requirements [7.6] that

$$P = 20\,log_{10}\left\{\frac{2Z_1}{Z_2(1 + Z_2/Z_1)}\right\} \tag{7.48}$$

and for a good input V.S.W.R. condition, in terms of normalized quantities

$$\frac{1}{z_1^3} + \frac{1}{z_2^3} = \frac{1}{z_2} \tag{7.49}$$

In the conventional hybrid-ring coupler, an impedance $Z_1 = 118.6\,\Omega$ (with $Z_2 = 55.1\,\Omega$) would only give a power-split ratio of 6.6 dB.

EXERCISES

7.1 Calculate the level of the transmitted and coupled signals from a directional coupler that has a total dissipation loss of 0.5 dB, while retaining a 0 dB power-split ratio. The input return loss and directivity for the coupler are both 20 dB.

7.2 In the first stage of a three-way equal power-split coupler, an ideal directional coupler is required with a power-split ratio of 3.01 dB, i.e. the transmitted signal at port 2 has twice the power of the coupled port. Calculate (i) the coupling, CdB required, and (ii) the line impedances for a two-branch branch-line coupler.

7.3 In the design of a lossless 2.0 dB branch-line coupler, the quarter-wave line impedances are 38.24 and $30.37\,\Omega$, see Table 7.1. Retaining the $38.24\,\Omega$ line but changing the $30.37\,\Omega$ line to $59.30\,\Omega$, gives the necessary line impedance values for a 3.81 dB coupler. Sketch the two coupler designs, label the ports appropriately and comment on the similarities and/or differences of the performance of the two designs.

7.4 Calculate the input V.S.W.R. in a $50\,\Omega$ system for a nominal 3 dB hybrid-ring coupler, if the ring impedance is $75\,\Omega$ instead of the designed value of $70.7\,\Omega$. Is a voltage maximum or minimum expected at the input plane to the ring?

7.5 Calculate the impedance parameters for a hybrid-ring coupler, to be used in a $50\,\Omega$ system, that has in-phase output signals with a 0.5 dB power-split ratio.

7.6 Consider the three-branch directional coupler that is illustrated in Figure 7.16. Using ABCD-parameter analysis for the even and odd modes, calculate (i) the input V.S.W.R., (ii) the coupled signal level, and (iii) the directivity at $0.9 \times f_0$ and f_0, where f_0 is the center frequency for the coupler.

7.7 A 3 dB branch-line coupler with $Z_{series} = Z_0/\sqrt{2}$ and $Z_{shunt} = Z_0$ in Figure 7.5 is driven by two inputs at ports 1 and 4 respectively. Identical load impedances, Z_L, are connected to the two output ports, 2 and 3, and produce the identical impedances Z_{in} looking into the inputs at 1 and 4. In all cases, Z_{in} and Z_L are evaluated at the junctions of the connecting lines with the hybrid proper. Z_{in} is a function not only of Z_L but also depends on whether the input excitation is of the even- or odd-mode type. Show that

$$Z_{in} = jkZ_0 \quad \text{when } Z_L = \infty$$

and $$Z_{in} = -jkZ_0 \quad \text{when } Z_L = 0$$

where $k = +1$ for even-mode excitation and $k = -1$ for odd-mode excitation.

7.8 Express the insertion loss I, coupling C, and directivity D of the directional coupler in Figure 7.2 in terms of the scattering parameters of the four-port network.

REFERENCES

[7.1]   Reed, J. and Wheeler, G. J., "A method of analysis of symmetrical four-port networks", *IRE Trans. Microwave Theory and Techniques*, Vol. MTT-4, No. 4, October 1956, pp. 246-52.

[7.2]   Riblet, G. P., "A directional coupler with very flat coupling", *IEEE Trans. Microwave Theory and Techniques*, Vol. MTT-26, No. 2, February 1978, pp. 70-4. Correction: *IEEE Trans. Microwave Theory and Techniques*, Vol. MTT-26, No. 9, September 1978, p. 691.

[7.3]   Levy, R and Lind, L. F., "Synthesis of symmetrical branch-guide directional couplers", *IEEE Trans. Microwave Theory and Techniques*, Vol. MTT-16, No. 2, February 1968, pp. 80-9.

[7.4]   Muraguchi, M., Yukitake, T. and Naito, Y., "Optimum design of 3-dB branch-line couplers using microstrip lines", *IEEE Trans. Microwave Theory and Techniques*, Vol. MTT-31, No. 8, August 1983, pp. 674-8.

[7.5]   Pon, C. Y., "Hybrid-ring directional coupler for arbitrary power division", *IRE Trans. Microwave Theory and Techniques*, Vol. MTT-9, No. 6, November 1961, pp. 529-35.

[7.6]   Agrawal, A. K. and Mikucki, G. F., "A printed-circuit hybrid-ring directional coupler for arbitrary power divisions", *IEEE Trans. Microwave Theory and Techniques*, Vol. MTT-34, No. 12, December 1986, pp. 1401-7.

Parallel-coupled lines and directional couplers

8.1 INTRODUCTION

The performance of hybrid-line couplers was analyzed in Chapter 7 in terms of the even- and odd-mode properties of the transmission line structures. For the hybrid-line couplers, there were transmission line connections between all the ports of the network. In this chapter, analysis in terms of the even and odd modes will again be considered, but this time for a network where there is continuous coupling — but no d.c. connection — between two transmission lines that are parallel and in close proximity to each other. Figure 8.1 illustrates a typical pair of parallel-coupled microstrip lines producing a directional coupler, with the coupling taking place over an electrical length θ.

The microstrip lines are in the one plane and parallel to each other; for this reason this circuit is also described as an edge-coupled circuit. The separation between the line is s and the width of the lines in the coupling region is w. In general, this width is different from the width of the connecting lines at the ports, which will be denoted by $w^{(50)}$, for a $50\,\Omega$ characteristic impedance system.

Following the convention that port 1 is the input port with ports 2, 3 and 4 being the direct, coupled and isolated ports respectively, it is seen from the labeling of the

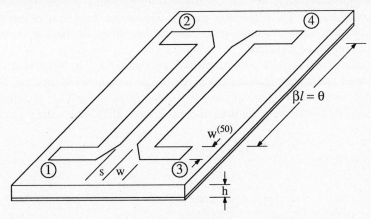

Figure 8.1 The parallel-coupled lines of an edge-coupled directional coupler

175

ports in Figure 8.1 that a solution with the coupled signal in the opposite direction to that in a branch-line coupler is anticipated.

For the development of the basic coupled line theory [8.1, 8.2], it will be necessary to make three assumptions:

i)   that the quasi-static TEM-mode of operation holds and voltage and current are thus meaningful quantities on the lines,

ii) that the transmission lines have the same ε_{eff} for both the even and odd modes. This is equivalent to the lines being immersed in a uniformly filled dielectric space with relative permittivity ε_{eff}. With this assumption, the even and odd modes will have the same phase velocity. In practice, the velocities for the two modes will be slightly different. This will produce some degradation in the performance of the coupler.

iii) that the transmission lines possess an appropriate plane of symmetry.

8.2 EVEN- AND ODD-MODE ANALYSIS

The two coupled lines make a four-port device. The total voltages that result when port 1 is excited by an input signal are shown in Figure 8.2a. If the two ports at one end of the structure are driven with the same phase and magnitude voltages, the even-mode configuration as illustrated in Figure 8.2b results, with four voltages and four currents at the ports. These variables are distinguished by the superscript "(e)". For the odd mode, Figure 8.2c, port 3 is driven antiphase to port 1 and the odd-mode voltages and currents are denoted by the superscript "(o)".

The total voltages at each port are given by superposition of the even- and odd-mode voltages. Thus

$$V_1 = V_i^{(e)} + V_i^{(o)} \qquad V_2 = V_o^{(e)} + V_o^{(o)}$$
$$V_3 = V_i^{(e)} - V_i^{(o)} \qquad V_4 = V_o^{(e)} - V_o^{(o)} \tag{8.1}$$

The transverse electric field patterns for the even and odd modes for a microstrip line are illustrated in Figure 8.3. As was the case for the single microstrip line, rather than evaluating the inductance per unit length for each mode, each of the even- and odd-mode impedances, Z_{0e} and Z_{0o}, is found from both air-filled and dielectric-filled line capacitances for the respective mode. The mode impedances represent the propagating wave voltage/current ratio on each line when the pair of coupled lines have been appropriately excited.

Consider the line with a characteristic impedance Z_0, as illustrated in Figure 8.4. The load and source impedances give voltage reflection coefficients, Γ_L and Γ_S respectively, at the ends of the transmission line. In Exercise 2.11, the forward and reverse traveling waves were considered for this case, giving the voltages at each end of the line as

$$V_i = \frac{V_S}{2}\left[1 - \frac{\Gamma_S - \Gamma_L e^{-j2\theta}}{1 - \Gamma_S \Gamma_L e^{-j2\theta}}\right] \tag{8.2}$$

and

$$V_o = \frac{V_S}{2}\left[\frac{(1 - \Gamma_S)(1 + \Gamma_L)e^{-j\theta}}{1 - \Gamma_S \Gamma_L e^{-j2\theta}}\right] \tag{8.3}$$

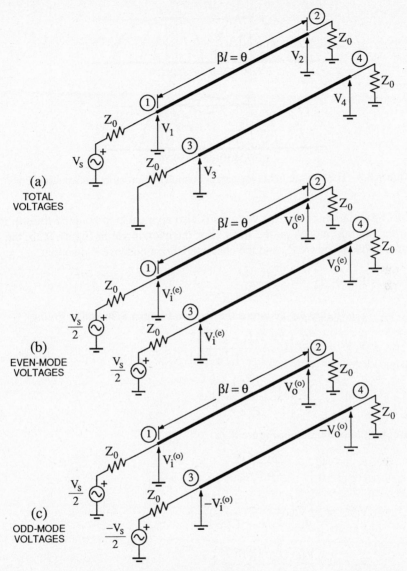

Figure 8.2 Parameters for coupled transmission lines. (a) The total voltages that result when port 1 is excited by an input signal. The voltages in (a) are decomposed into two sets, (b) the even-mode voltages, and (c) the odd-mode voltages.

When $\Gamma_S = \Gamma_L = \Gamma$, say, (8.2) and (8.3) become

$$V_i = \frac{V_S}{2} \left[1 - \frac{\Gamma(1 - e^{-j2\theta})}{1 - \Gamma^2 e^{-j2\theta}} \right] \tag{8.4}$$

and

$$V_o = \frac{V_S}{2} \left[\frac{(1 - \Gamma^2) e^{-j\theta}}{1 - \Gamma^2 e^{-j2\theta}} \right] \tag{8.5}$$

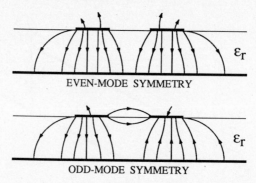

EVEN-MODE SYMMETRY

ODD-MODE SYMMETRY

Figure 8.3 The electric fields for parallel-coupled microstrip transmission lines

This formulation of the line voltages is also applicable to coupled lines in terms of the even and odd modes on the lines. For the even mode in Figure 8.2b, the load and source impedances are Z_0 and the characteristic impedance Z_{0e} so that

$$\Gamma \rightarrow \Gamma_e = \frac{Z_0 - Z_{0e}}{Z_0 + Z_{0e}}$$

(8.6)

Further, the V_S in Figure 8.4 becomes the $V_S/2$ in Figures 8.2b and c, giving

$$V_i^{(e)} = \frac{V_S}{4}\left[1 - \frac{\Gamma_e(1 - e^{-j2\theta})}{1 - \Gamma_e^2 e^{-j2\theta}}\right]$$

(8.7)

and

$$V_o^{(e)} = \frac{V_S}{4}\left[\frac{(1 - \Gamma_e^2)e^{-j\theta}}{1 - \Gamma_e^2 e^{-j2\theta}}\right]$$

(8.8)

Likewise, for the odd mode in Figure 8.2c

$$\Gamma \rightarrow \Gamma_o = \frac{Z_0 - Z_{0o}}{Z_0 + Z_{0o}}$$

(8.9)

giving

$$V_i^{(o)} = \frac{V_S}{4}\left[1 - \frac{\Gamma_o(1 - e^{-j2\theta})}{1 - \Gamma_o^2 e^{-j2\theta}}\right]$$

(8.10)

and

$$V_o^{(o)} = \frac{V_S}{4}\left[\frac{(1 - \Gamma_o^2)e^{-j\theta}}{1 - \Gamma_o^2 e^{-j2\theta}}\right]$$

(8.11)

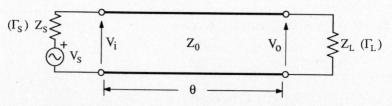

Figure 8.4 Parameters for a terminated transmission line

Note that θ in (8.10) and (8.11) is identical to that in (8.7) and (8.8), by virtue of the assumption that the even- and odd-mode velocities are equal.

Input match condition

Apply an input signal to port 1 only. The input voltage $V_1 = V_i^{(e)} + V_i^{(o)}$, so that

$$V_1 = \frac{V_s}{4}\left\{1 + 1 - (1 - e^{-j2\theta})\left[\frac{\Gamma_e}{1 - \Gamma_e^2 e^{-j2\theta}} + \frac{\Gamma_o}{1 - \Gamma_o^2 e^{-j2\theta}}\right]\right\} \quad (8.12)$$

The condition for a matched input at port 1 in Figure 8.2a is $V_1 = V_s/2$, given when

$$\frac{\Gamma_e}{1 - \Gamma_e^2 e^{-j2\theta}} + \frac{\Gamma_o}{1 - \Gamma_o^2 e^{-j2\theta}} = 0 \quad (8.13)$$

To satisfy this condition in a frequency independent manner, i.e. for all θ, $\Gamma_e = -\Gamma_o$ is obviously *sufficient*, but it is left to show in Exercise 8.9 that it is also necessary. Now, from (8.6) and (8.9)

$$\Gamma_e = -\Gamma_o \quad \Rightarrow \quad \frac{Z_0}{Z_{0e}} = \frac{Z_{0o}}{Z_0} \quad (8.14)$$

i.e. $\qquad Z_{0e}Z_{0o} = Z_0^2 \quad (8.15)$

Isolation

The voltage at port 4, where good isolation is required, is given by $V_o^{(e)} - V_o^{(o)}$, yielding

$$V_4 = \frac{V_s}{4}\left[\frac{(1 - \Gamma_e^2)e^{-j\theta}}{1 - \Gamma_e^2 e^{-j2\theta}} - \frac{(1 - \Gamma_o^2)e^{-j\theta}}{1 - \Gamma_o^2 e^{-j2\theta}}\right] \quad (8.16)$$

Again, $\Gamma_e = -\Gamma_o$ is clearly *sufficient*, and may likewise be shown to be necessary, to make $V_4 = 0$.

The coupled port

At port 3, the coupled port

$$V_3 = \frac{V_s}{4}\left[1 - \frac{\Gamma_e(1 - e^{-j2\theta})}{1 - \Gamma_e^2 e^{-j2\theta}} - 1 + \frac{\Gamma_o(1 - e^{-j2\theta})}{1 - \Gamma_o^2 e^{-j2\theta}}\right] \quad (8.17)$$

With $\Gamma_e = -\Gamma_o$

$$V_3 = -\frac{V_s}{2}\left[\frac{\Gamma_e(e^{j\theta} - e^{-j\theta})}{e^{j\theta} - \Gamma_e^2 e^{-j\theta}}\right] \quad (8.18)$$

giving $\qquad V_3 = -\frac{V_s}{2}\left[\frac{2j\Gamma_e\sin\theta}{(1 - \Gamma_e^2)\cos\theta + j\sin\theta(1 + \Gamma_e^2)}\right] \quad (8.19)$

$$= -\frac{V_s}{2}\left[\frac{2\Gamma_e}{1 + \Gamma_e^2}\right]\frac{j\sin\theta}{\left[\frac{1 - \Gamma_e^2}{1 + \Gamma_e^2}\right]\cos\theta + j\sin\theta} \quad (8.20)$$

Now it is readily shown from (8.6) and (8.15) that

$$-\frac{2\Gamma_e}{1+\Gamma_e^2} = \frac{Z_{0e}-Z_{0o}}{Z_{0e}+Z_{0o}} \equiv \mathbf{c}, \text{ by definition}$$

$$(8.21)$$

where $\mathbf{c}$ will appear later as a coupling coefficient, consistent with its use in §7.3. It is simple to show that

$$1-\mathbf{c}^2 = \left\{\frac{1-\Gamma_e^2}{1+\Gamma_e^2}\right\}^2$$

$$(8.22)$$

Thus

$$V_3 = \frac{V_s}{2}\left[\frac{j\mathbf{c}\sin\theta}{\sqrt{1-\mathbf{c}^2}\cos\theta + j\sin\theta}\right]$$

$$(8.23)$$

Transmission

With the same condition $\Gamma_e = -\Gamma_o$, the transmitted signal at port 2 may be similarly found as

$$V_2 = \frac{V_s}{2}\left[\frac{\sqrt{1-\mathbf{c}^2}}{\sqrt{1-\mathbf{c}^2}\cos\theta + j\sin\theta}\right]$$

$$(8.24)$$

Summary and discussion

When V_s is set as 2V for each of the two modes, a unit input voltage at port 1 results and the following relationships are found, namely

$$V_1 = 1$$
$$V_2 = \frac{\sqrt{1-\mathbf{c}^2}}{\sqrt{1-\mathbf{c}^2}\cos\theta + j\sin\theta}$$
$$V_3 = \frac{j\mathbf{c}\sin\theta}{\sqrt{1-\mathbf{c}^2}\cos\theta + j\sin\theta}$$
$$V_4 = 0$$

$$(8.25)$$

The maximum coupling from port 1 to port 3 occurs when the coupling length is one quarter-wavelength, i.e. $\theta = \pi/2$. Under these conditions

$$V_1 = 1$$
$$V_2 = -j\sqrt{1-\mathbf{c}^2}$$
$$V_3 = \mathbf{c}$$
$$V_4 = 0$$

$$(8.26)$$

To summarize, with the assumptions that have been previously stated and including condition (8.15), the following points should be noted with respect to (8.25) and (8.26):

i) Port 4 always has zero output, irrespective of the electrical length of the coupling region. In practical circuits, a major cause of the poor isolation may be unequal even- and odd-mode phase velocities.

ii) The input at each port is matched to the feed line characteristic impedance, Z_0, again irrespective of the electrical length of the coupling region.

iii) The total output power equals the input power (see Example 8.1).
iv) The maximum coupling to port 2 occurs at the frequency that gives a quarter-wave coupling length. This will be the mid-band frequency. Because of this property, these couplers are also known as quarter-wave couplers.
v) At this maximum coupling frequency, the through-line voltage V_2 is 90° out of phase with the coupled line voltage V_3, i.e. this coupler may be described as a *quadrature coupler*. The coupled voltage V_3 is in phase with V_1 and thus V_2 lags V_1 by 90°, the latter phase difference being identical to the electrical length of the coupling region.
vi) At frequencies other than the maximum coupling frequency, the ideal frequency response is found by evaluating the terms $|V_2(\theta)|$ and $|V_3(\theta)|$, remembering that θ is a function of frequency.

The coupling for a directional coupler is generally expressed in dB, i.e.

$$\text{coupling, } C \text{ (dB)} = -20\,log_{10}(\mathbf{c}) \tag{8.27}$$

Thus, it follows from (8.21) that

$$Z_{0e} = Z_0 \left[\frac{1 + \mathbf{c}}{1 - \mathbf{c}}\right]^{\frac{1}{2}} = Z_0 \left[\frac{1 + 10^{-C/20}}{1 - 10^{-C/20}}\right]^{\frac{1}{2}} \tag{8.28}$$

and

$$Z_{0o} = Z_0^2 \div Z_{0e} = Z_0 \left[\frac{1 - 10^{-C/20}}{1 + 10^{-C/20}}\right]^{\frac{1}{2}} \tag{8.29}$$

The edge-coupled coupler may also have been analyzed in terms of the scattering parameters of the equivalent two-port networks as were the branch-line and hybrid-ring couplers in the previous chapter. For this alternative analysis, the required even- and odd-mode s-parameters can be deduced from the solution to Exercise 2.1(v).

Example 8.1

For a lossless directional coupler, there is a balance between the input and output powers. Verify this at the mid-band frequency.

Solution:

Assuming equal even- and odd-mode phase velocities, then, in the even- and odd-mode analysis, (8.26) has been obtained at the mid-band frequency for a quarter-wave parallel-line directional coupler. It is assumed that the coupler is matched to input lines with equal characteristic impedances, Z_0. With the only input signal being applied to port 1, and there being no reflected wave at that port, then

$$\text{the input power} = |V_1|^2 Z_0^{-1}$$

and

$$\text{the output power} = (|V_2|^2 + |V_3|^2 + |V_4|^2) Z_0^{-1}$$

Assuming, with no loss of generality, that $V_1 = 1$, then

$$\text{output power} = (|-j\sqrt{1 - \mathbf{c}^2}|^2 + |\mathbf{c}|^2 + 0) Z_0^{-1} = Z_0^{-1}$$

$$= \text{input power.}$$

8.3 COUPLED-LINE PARAMETERS

The quasi-static analysis of coupled microstrip transmission lines gives the capacitance in terms of the physical parameters w/h, s/h and the substrate permittivity, which in turn leads to the even- and odd-mode impedances, Z_{0e} and Z_{0o}, and the effective relative permittivity for each mode. This latter quantity relates the mode phase velocities to the phase velocity in free space by

$$v_{phase}^{(e,o)} = \frac{c}{\sqrt{\varepsilon_{eff}^{(e,o)}}} \tag{8.30}$$

For the quasi-static TEM-mode propagation, Bryant and Weiss [8.4] produced a rigorous theoretical analysis using a Green's function that expressed the discontinuity of the electric fields at the dielectric-air interface (see Appendix 2). Their published results for ε_r = 1, 9 and 16 have served as a benchmark against which later derivations have been compared. For each substrate permittivity, a graphical presentation of results gives the mode impedances as a function of w/h, with s/h as a parameter. Akhtarzad et al. [8.5] recognized the need for an easier determination of w/h and s/h when the mode impedances are known, while Ros [8.6] presented a chart for ε_r = 9.7 that had the mode impedances along the two axes with both the w/h and s/h contours drawn for interpolation.

Jansen [8.7] extended these results for coupled microstrip transmission lines with a rigorous hybrid-mode solution, that gave the frequency dependence of the derived quantities. From this work, the detailed empirical formulae of Kirschning and Jansen [8.8] were later derived, giving Z_{0e}, Z_{0o} and $\varepsilon_{eff}^{(e,o)}$ as functions, not only of the line parameters w/h, s/h and ε_r, but also of frequency. The zero-frequency equations from [8.8] are presented in Appendix 4, together with sample data values. Using the equations of [8.8], Z_{0e} and Z_{0o} are plotted in Figure 8.5 at low frequencies for ε_r = 2.5 and 10.0.

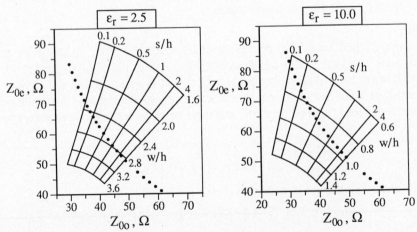

Figure 8.5 The even- and odd-mode impedances for coupled microstrip transmission lines, including the condition $Z_{0e} Z_{0o} = (50)^2$, (dotted curve)

For directional couplers that are to be matched to the input feed lines, it was seen in (8.15) that only certain combinations of Z_{0e} and Z_{0o} would be required. As the majority of systems are designed for $50\,\Omega$ characteristic impedance interconnections, the required condition (8.15) is also plotted in Figure 8.5.

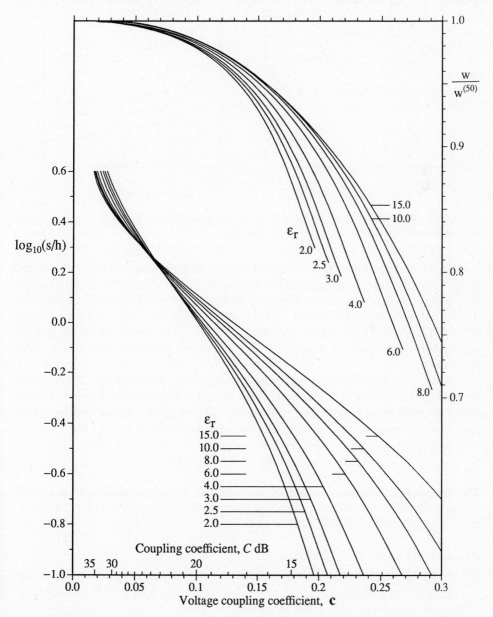

Figure 8.6 The line separation and width for a single section of an edge-coupled directional coupler, matched to $50\,\Omega$ input and output lines

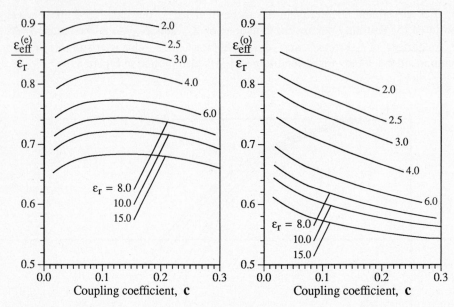

Figure 8.7 Normalized even- and odd-mode effective permittivities for directional coupler design, with the condition $Z_{0e} Z_{0o} = (50)^2$ applying

In the design of directional couplers, before the line parameters are considered, the necessary voltage coupling coefficients must be found. Figure 8.6 is derived from the equations in Appendix 4 and is plotted so that, given the coupling coefficient, the line parameters may be found. The curves cover the line-separation range $0.1 \leq s/h \leq 4.0$ and also give the width of the lines in the coupling region, normalized to the width of a single 50Ω characteristic impedance line on the same substrate.

Curves for the effective permittivity for each mode, normalized to the substrate permittivity and for use in a 50Ω system, are plotted in Figure 8.7 as a function of the coupling coefficient. An effective permittivity ε_{eff} that gives a phase velocity that is the arithmetic mean of the even- and odd-mode phase velocities is given by

$$\sqrt{\varepsilon_{eff}} = \frac{\sqrt{\varepsilon_{eff}^{(e)}} + \sqrt{\varepsilon_{eff}^{(o)}}}{2} \tag{8.31}$$

The percentage difference between the effective mode-permittivities and their average, which for the parameter range plotted in Figure 8.7 may be as great as 10%, tends to be greater for small s/h and large ε_r .

Example 8.2

i) Design a 20dB single-section parallel-line directional coupler with a center frequency of 1.0GHz on a 1.58 mm substrate for use with a 50Ω characteristic impedance system. The substrate permittivity is 2.5 .

ii) If the coupling is not to decrease by more than 1.0dB, estimate the bandwidth for the design in (i), assuming equal even- and odd-mode phase velocities.

Solution:

i) Using Figure 8.6 and the $\varepsilon_r = 2.5$ curves, C of 20 dB gives

$$w/w^{(50)} = 0.98 \quad \text{and} \quad \log_{10}(s/h) = 0.02, \quad \text{i.e. } s/h = 1.047$$

Now from Appendix 3, $w^{(50)}/h = 2.837$ giving, for the coupling region

$$w/h = 2.78$$

From Figure 8.7

$$\varepsilon_{\text{eff}}^{(e)} = 0.875 \times \varepsilon_r = 2.188, \quad \text{and} \quad \varepsilon_{\text{eff}}^{(o)} = 0.782 \times \varepsilon_r = 1.955$$

and, from (8.31)

$$\varepsilon_{\text{eff}} = 2.07$$

The wavelength along the coupling region at 1.0 GHz is

$$\lambda = \lambda_0/\sqrt{\varepsilon_{\text{eff}}} = 208.5 \text{ mm}$$

Thus, for a 1.0 GHz center frequency directional coupler on a 1.58 mm thick substrate

$$s = 1.65 \text{ mm}, \quad w = 4.39 \text{ mm}, \quad \text{and coupling length} = 52.1 \text{ mm}$$

ii) From (8.25), the coupled signal for a 1 V input is

$$V_3 = \frac{jc \sin\theta}{\sqrt{1 - c^2}\cos\theta + j\sin\theta}$$

The magnitude of this signal

$$|V_3| = \frac{c\sin\theta}{\left\{(1 - c^2)\cos^2\theta + \sin^2\theta\right\}^{\frac{1}{2}}}$$

For a 20 dB coupler, V_3 at the center frequency equals the value for c, i.e. 0.1. When the coupled signal falls by a further 1.0 dB

$$V_3 = 0.0891$$

Therefore

$$(1 - c^2)\cos^2\theta + \sin^2\theta = \left[\frac{c}{0.0891}\right]^2 \sin^2\theta$$

or

$$\tan\theta = \left[\frac{1 - c^2}{(c/0.0891)^2 - 1}\right]^{\frac{1}{2}} = \left[\frac{0.99}{0.259}\right]^{\frac{1}{2}} = 1.955$$

giving

$$\theta = 62.9°, \quad 117.1°$$

The electrical length is inversely proportional to frequency. As the center frequency coincides with an electrical path length given by $\theta = 90°$ and the coupling is not to fall by more than 1.0 dB, then

$$\text{fractional bandwidth} = \left|\frac{90°}{62.9°} - \frac{90°}{117.1°}\right| \times 100 = 66\%$$

8.4 MULTIPLE-SECTION DIRECTIONAL COUPLERS

The analytical equivalence, originally developed by Young [8.9], between the single-section of a parallel-coupled directional coupler and the forward and reverse waves on a single stepped-impedance transmission line, is developed in this section. The equivalent line concept is then extended to analyze multiple-section directional couplers.

Consider an electrical length, θ, of transmission line as illustrated in Figure 8.8, that has a characteristic impedance, Z_1, and is connected to unit impedance lines at each end. The output is terminated with a matched load so that there is no reflected signal.

In Figure 8.8a, the source and load reflection coefficients are identical. The voltages, V_A and V_B, are thus given by (8.4) and (8.5) with Γ being the reflection coefficient of a *load* Z_0 with respect to a Z_1 line, i.e.

$$\Gamma = \frac{Z_0 - Z_1}{Z_0 + Z_1} \tag{8.32}$$

With no loss of generality, $V_S/2$ can be set to unity and (8.4) and (8.5) can be rearranged to give

$$V_A = 1 + \frac{j\gamma\sin\theta}{\sqrt{1 - \gamma^2}\cos\theta + j\sin\theta} \tag{8.33}$$

and

$$V_B = \frac{\sqrt{1 - \gamma^2}}{\sqrt{1 - \gamma^2}\cos\theta + j\sin\theta} \tag{8.34}$$

with

$$\gamma = -\frac{2\Gamma}{1 + \Gamma^2} \quad \text{and} \quad \sqrt{1 - \gamma^2} = \frac{1 - \Gamma^2}{1 + \Gamma^2} \tag{8.35}$$

Now, V_A can be regarded as the sum of the voltages carried by the incident and reflected waves in the Z_0 line at A, of amplitudes 1 and d respectively, so that

$$d = \frac{j\gamma\sin\theta}{\sqrt{1 - \gamma^2}\cos\theta + j\sin\theta} \tag{8.36}$$

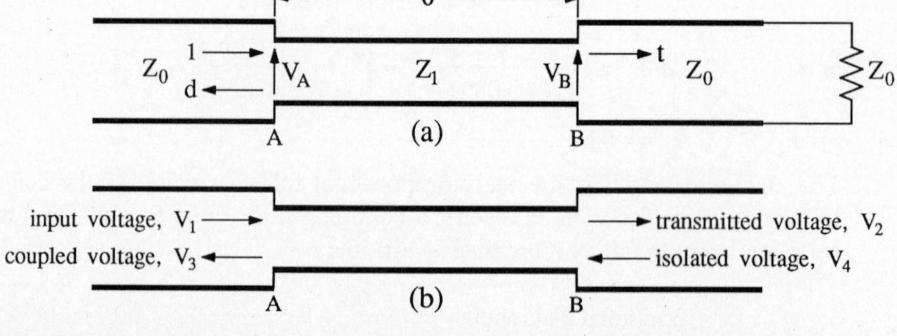

Figure 8.8 The equivalent transmission line for the coupling region of parallel-coupled lines (a) with no input to the isolated port, i.e. the isolated port terminated with just a matched load. (b) The more general analogy with inputs into both the input and isolated ports.

As there is no incident wave in Z_0 line impinging on B, in view of the matched load to the right of B, the amplitude of the wave moving right from B equals V_B and is thus given by

and
$$t = \frac{\sqrt{1 - \gamma^2}}{\sqrt{1 - \gamma^2}\cos\theta + j\sin\theta} \tag{8.37}$$

On comparing (8.36) and (8.37) with (8.25), it is apparent that d and t are identical to the values of V_3 and V_2 in (8.25) if **c** is equated with γ

i.e.
$$\mathbf{c} = -\frac{2\Gamma}{1 + \Gamma^2} = \frac{Z_1^2 - Z_0^2}{Z_1^2 + Z_0^2} \tag{8.38}$$

Thus, an analogy between the *total* voltages at the ports of the directional coupler in Figure 8.1 and the *incident* and *reflected* waves in Figure 8.8a has been established. The coupled port output is simply given by the reflection coefficient looking into A from the Z_0 line. This reflection coefficient in the analogous circuit is in some circumstances easier to calculate than the coupling coefficient in the original coupled circuit.

In Figure 8.8a there is no incident wave from the right of B, because of the matched load. This is analogous to there being no output at the isolated port 4. If there were to be a separate excitation at port 4, this would require an incident wave at B and simple linear superposition then yields the more complete equivalent circuit of Figure 8.8b. Note that the *total* voltages in a 4-port device are equated to the *traveling* waves (incident and reflected) in a 2-port device. With the help of this equivalence, it is now easy to see that cascading directional couplers, D_1 and D_2, say, with

$$\text{Port 2 of } D_1 \leftrightarrow \text{Port 1 of } D_2$$
$$\text{Port 4 of } D_1 \leftrightarrow \text{Port 3 of } D_2$$

is equivalent to cascading appropriate lengths of single transmission lines. If need be, the Z_0 line may be considered to be of zero length, e.g. as at the junctions of the Z_1 and Z_2 lines in Figure 8.11.

The equivalence just described does not particularly aid in the design of a single-section directional coupler, but is useful when multiple sections are considered. Provided that the condition, $Z_{0e}Z_{0o} = Z_0^2$, is maintained throughout the coupling region, the principle may also be used for continuously variable coupling between the lines as in the exponential coupler [8.10, 8.11].

Example 8.3

What configuration of two $\lambda/4$-long sections of coupled lines, where ports 2 and 4 of one section are fed into ports 1 and 3 of the second section, gives an equivalent 3.0dB directional coupler?

Solution:

Two identical sections of $\lambda/4$-long coupled lines joined directly together are equivalent to a $\lambda/2$ coupled length and will not give any coupled signal to port 3, as seen from (8.25).

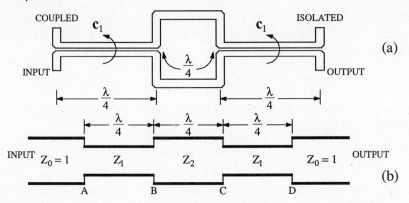

Figure 8.9 The connection of two $\lambda/4$-long directional couplers to produce increased overall coupling. (a) The circuit layout, and (b) the single line equivalent circuit.

In physical terms, the coupled signal from the second $\lambda/4$ coupler will be out-of-phase with the signal from the first and the two will cancel. To avoid cancellation, an additional quarter-wavelength line (without coupling) that gives a further 180° phase shift to the second coupled signal may be used. This is illustrated in Figure 8.9a with the single line equivalent circuit in Figure 8.9b. The center $\lambda/4$ section (from B to C) is purely to provide a connection and the appropriate phase shift between the two coupled sections. In the equivalent circuit, it becomes a $\lambda/4$-long Z_0 line. However, in order to derive results which can also be used in a more general situation later (viz. Figure 8.11), Z_2 in Figure 8.9b is left arbitrary for the time being. With no loss of generality, Z_0 is made unity.

The coupled signal level is found from the voltage reflection coefficient at the input of the equivalent single line. At the mid-band frequency where the line lengths are all $\lambda/4$, each section of line acts as a quarter-wave transformer and this fact allows easy evaluation of the total effective load impedance at plane A.

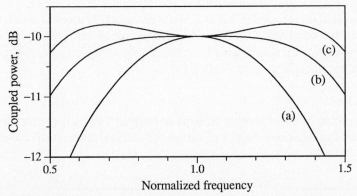

Figure 8.10 The frequency characteristics of 10dB directional couplers with (a) a single section coupler, (b) a maximally-flat 3-section coupler, and (c) a 0.2dB ripple, 3-section coupler.

Thus the impedances on the line are as follows

at D: $Z_D = 1$

at C: $Z_C = Z_1^2$

at B: $Z_B = Z_2^2/Z_1^2$

at A: $Z_A = Z_1^2/(Z_2^2/Z_1^2) = Z_1^4/Z_2^2$

and the input reflection coefficient

$$\Gamma_{in} = \frac{Z_A - 1}{Z_A + 1} = \frac{Z_1^4 - Z_2^2}{Z_1^4 + Z_2^2}$$

$$= \frac{Z_1^4 - 1}{Z_1^4 + 1} \quad \text{when } Z_2 = 1$$

For a combination that gives a 3 dB coupler

$$c \equiv \Gamma_{in} = 0.707$$

giving $$Z_1 = \left[\frac{1 + \Gamma_{in}}{1 - \Gamma_{in}}\right]^{\frac{1}{4}} = 1.554$$

and the coupling of an individual section

$$c_1 = \frac{Z_1^2 - 1}{Z_1^2 + 1} = 0.4142$$

i.e. $C_1 = 7.66 \text{ dB}$

Note

i)   A greater amount of coupling has been achieved than that given by an individual coupler, but the overall coupler will have a smaller bandwidth than one of the component couplers.

ii) A 3 dB coupler may also be achieved by another interconnection of two 8.34dB couplers [8.12], with a greater bandwidth than that for this example. However, ports 2 and 3 of one coupler have to be fed into ports 1 and 4 of the second coupler and the complete design with the 8.34dB couplers cannot be fabricated in one plane on the same substrate, without having some of the lines crossing each other.

A symmetrical coupler made from three sections may be synthesized with selected passband properties. Three typical coupler frequency characteristics are illustrated in Figure 8.10 and show (a) a single coupler for comparison, (b) a maximally-flat coupler, and (c) a coupler with 0.2dB ripple in the passband, all the couplers giving 10dB coupling at the center frequency.

A maximally-flat coupler exhibits the greatest bandwidth without passband ripple. The acceptance of a limited amount of passband ripple will give an increased bandwidth. Young [8.9] gives a detailed analysis of the coupler parameter evaluation for 3- and 5-section symmetrical couplers.

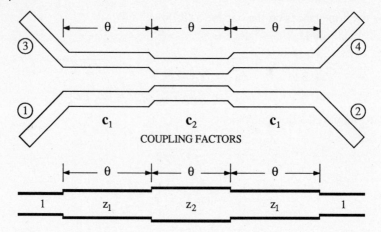

Figure 8.11   Variables for the design of a 3-section directional coupler

Figure 8.11 shows a 3-section directional coupler and its equivalent single line model, with solutions for the equivalent line impedance, Z_1, given in Figure 8.12. It has already been seen in the previous example that, from the equivalent single line model, the overall coupling coefficient

$$c = \frac{Z_1^4 - Z_2^2}{Z_1^4 + Z_2^2}$$

(8.39)

so that

$$Z_2 = Z_1^2 \left\{ \frac{1-c}{1+c} \right\}^{\frac{1}{2}}$$

(8.40)

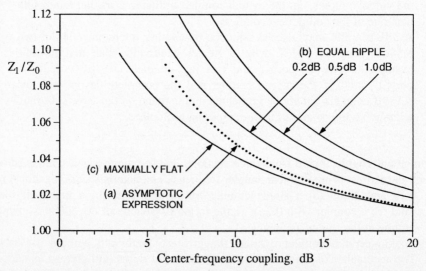

Figure 8.12   The equivalent single line normalized impedance for the end section of 3-section directional couplers as a function of the overall coupling at the center frequency

Knowing c, Z_1 and Z_2 can now be determined. The solutions are strictly true only for pure TEM modes with equal phase velocities for the even and odd modes, as is the case with symmetrical parallel-coupled striplines.

For small levels of coupling and a maximally-flat response, curve (a) gives the asymptotic expression

$$Z_1 = \frac{7 + \sqrt{\chi}}{8} \tag{8.41}$$

where $\chi = (1 + |c|)/(1 - |c|)$ and c is the mid-band overall voltage coupling coefficient.

Broad bandwidth for a 3-section directional coupler may be achieved if a ripple in the coupled signal level is permissible in the passband. The advantage of having a 0.2dB ripple has been illustrated in Figure 8.10. Design data for three different ripple levels (0.2, 0.5 and 1.0dB) are presented in Figure 8.12. In each case, the center frequency coupling level gives the equivalent single line impedance for the two end-sections, while the center-section impedance is determined from (8.40). The data for a maximally-flat design is also shown as (c) in Figure 8.12.

From (8.26), the coupled signal at the center frequency for a single-section directional coupler is in phase with the input signal. Further, it was seen in Example 8.3 that there was no coupled signal for a pair of identical couplers in tandem, in effect forming a $\lambda/2$ coupled section, as the 180° phase difference between the two parts gave complete signal cancellation. Now, for a symmetrical 3-section directional coupler, it turns out that the center section will have the highest coupling coefficient with the coupling from the two end-sections actually reducing the overall coupled signal level at the center frequency. Thus, for the overall coupler, the coupled signal at port 3 will be out-of-phase with respect to the input signal at port 1, and the mid-band coupling coefficient in (8.39) and (8.40) will be negative.

Example 8.4

Design a symmetrical 3-section maximally-flat 20dB directional coupler, giving the results in terms of the coupling coefficients of the individual quarter-wave sections.

Solution:

For a directional coupler where $C = 20$dB, the magnitude of the voltage coupling coefficient is 0.1, and thus $c = -0.1$ for the overall symmetrical 3-section coupler. Using the asymptotic expression (8.41) gives the equivalent single line normalized impedance for the end section as

$$Z_1 = 1.0132$$

while for the center section, from (8.40)

$$Z_2 = (1.0132)^2 \left[\frac{1 - (-0.1)}{1 + (-0.1)} \right]^{\frac{1}{2}}$$

i.e. $Z_2 = 1.1349$

The coupling coefficients for the individual quarter-wave directional couplers are found by working out the input reflection coefficients of sections such as in

Figure 8.8a. (Remember: the Z_0 lines may be of zero length.)

For the end section, $Z_1 = 1.0132 \rightarrow c_1 = 0.0131$

and the center section, $Z_2 = 1.1349 \rightarrow c_2 = 0.1259$

Thus, the end- and center-section couplers are 37.7dB and 18.0dB couplers respectively.

These results serve to reinforce the fact that, in a multi-section directional coupler of this form — as a trade-off for the larger bandwidth achieved — the center section will have tighter coupling when compared with a single-section coupler possessing the same mid-band coupling specification. This may be a limiting factor in the design of broadband directional couplers requiring a tight coupling specification.

8.5   THE LANGE COUPLER

A 3.0dB directional coupler has a voltage coupling coefficient of 0.707 and, in a 50Ω system, even- and odd-mode impedances of 120.7Ω and 20.7Ω respectively. For parallel-coupled microstrip lines, extrapolation of the contour plots of Figure 8.5 indicate that, while narrow line widths are required, it will be the minute separation between the coupled lines that is the limiting factor for manufacturing purposes. The small line separation is needed in order to enhance the odd-mode capacitance between adjacent lines, for the small odd-mode impedance to be achieved.

An isolated third conductor [8.13] above, but close to, the parallel-coupled pair of lines will increase the odd-mode capacitance, Figure 8.13. However, this configuration has the disadvantage of using additional dielectric material, as well as not being fabricated with all the conductor patterns in the one plane.

Increasing the number of parallel lines in the one plane for the coupling structure in place of the original two lines will give more adjacent pairs of edges and the possibility of increasing the odd-mode capacitance. A series of four parallel lines is illustrated in Figure 8.14.

For the even mode, all the lines are driven in phase, Figure 8.14(a). For the odd mode with pairs of lines driven in opposite phase, Figures 8.14(b), (c) and (d), it is the configuration (b) that gives the maximum capacitance between positive and

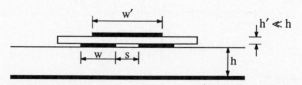

Figure 8.13 The use of a third parallel line to enhance the odd-mode capacitance, from Malherbe and Losch [8.13] (Reprinted with permission of *Microwave Journal*, from the November 1987 issue, © 1987 Horizon House, Inc.)

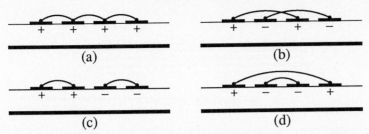

Figure 8.14 Non-degenerate modes for four parallel-coupled lines, showing (a) the even mode, (b) the odd mode with maximum odd-mode capacitance, (c) and (d) the other odd-mode configurations, from Paolino [8.16] (© 1976, *Microwaves and RF*)

negative driven lines, for fixed line widths and separations. The required odd-mode potential distribution is forced by connecting the pairs of similar polarity lines together as illustrated. It will always be necessary for some wire bonds or bridges to be used in the construction of a practical directional coupler of this type.

The Lange coupler [8.14] is a 3dB coupler design that was the first to utilize this approach. The coupler is illustrated in Figure 8.15, where it will be observed that the layout of the ports differs from that for a conventional two-line coupler, with the direct and isolated ports interchanged. The odd-mode potential configuration is that of Figure 8.14b. Waugh and LaCombe [8.15] unfolded the Lange coupler design, Figure 8.16a, to give a design that requires fewer wire cross-connections, as well as interchanging the physical position of the direct and isolated ports. A further line configuration, Figure 8.16b, that possesses the same port geometry as the original Lange coupler, has been suggested by Paolino [8.16]. This design, as with the original Lange design, may be used if a 3dB equivalent directional coupler is to be made from a pair of 8.34dB directional couplers, since the output ports of the first coupler feed directly into ports 1 and 4 of the second coupler.

Initially, the four-line Lange couplers and their derivatives were developed using an intuitive approach and it was some time before Ou [8.17] provided the basis for a synthesis approach. Ou considered the capacitances of a system of an even number of lines, n, with identical widths and separations, Figure 8.17. Only the capacitances between adjacent lines were considered with the capacitances between non-adjacent lines being neglected.

For any strip, m

$$C_{m,m+1} = C_{12} \tag{8.42}$$

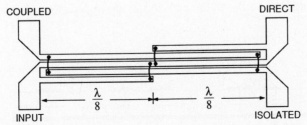

Figure 8.15 The four-strip Lange coupler, from Lange [8.14] (© 1969, IEEE)

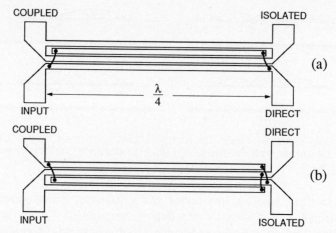

COUPLED   ISOLATED

(a)

INPUT   DIRECT

COUPLED   DIRECT

(b)

INPUT   ISOLATED

Figure 8.16 Alternative layouts for the Lange coupler, showing (a) the "unfolded" coupler of Waugh and LaCombe [8.15] (© 1972, IEEE), and (b) the Lange coupler variation of Paolino [8.16] (© 1976, *Microwaves and RF*)

$$C_{m0} = C_{10} \quad \text{if } m = 1,n \tag{8.43}$$
$$= C_{20} \quad \text{if } m \neq 1,n \tag{8.44}$$

and
$$C_{20} \approx C_{10} - \frac{C_{10}C_{12}}{C_{10} + C_{12}} \tag{8.45}$$

In [8.17], equations are developed to achieve (i) the matched conditions necessary for a $\lambda/4$-long directional coupler, and (ii) the required coupled power to port 3. When a system of only an adjacent pair from the n-conductors (i.e. a two-line system) is considered, the line capacitances for the pair may be derived from the known even- and odd-mode impedances for the composite structure as in Figure 8.14. It is thus possible to present the final equations for the n-conductor system in terms of the established parameters for a two-line coupler. This approach has been further developed and verified experimentally by Presser [8.18]. Results for multiconductor couplers are given by Tajima and Kamihashi [8.19], in particular for the six-line couplers that are required for 1.5dB coupling levels in the center section of a three-section 3dB broadband coupler. For the important case of four coupled lines, further manipulation of the equations given by Presser leads to the following design formulae.

Let Z_{0e} and Z_{0o} be the even- and odd-mode impedances of a pair of coupled

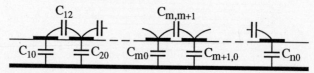

Figure 8.17 Interline capacitances for a system of n parallel-coupled lines

lines that have w/h and s/h equal to that of the final four-line coupler. The ratio, Φ, is defined by

$$\Phi = \frac{Z_{0o}}{Z_{0e}} \tag{8.46}$$

and is always less than unity. From the equations of [8.18], Φ may be shown to be a function of c, viz.

$$\Phi = \frac{-c + \sqrt{9 - 8c^2}}{3c + 3} \quad \text{when } n = 4 \tag{8.47}$$

c here is the voltage coupling coefficient for the *complete* four-line coupler. For the complete coupler to be matched to the input connecting lines of impedance Z_0, then Z_{0o} for the two-line coupler must be

$$Z_{0o} = Z_0 \frac{\sqrt{3\Phi^3 + 10\Phi^2 + 3\Phi}}{1 + \Phi} \tag{8.48}$$

Hence, given Z_0 and c for the four-line coupler, Z_{0e} and Z_{0o} for a two-line coupler are derived, giving the necessary values for w/h and s/h. Constructing a four-line coupler with these dimensions and a $\lambda/4$ coupling length will provide the necessary coupling and matched conditions.

Example 8.5

Design a 6.0dB four-line interdigitated microstrip directional coupler on a 1.58 mm thick, $\varepsilon_r = 2.5$ substrate.

Solution:

The voltage coupling coefficient, c, for a 6.0dB directional coupler is 0.5. For a coupler with only two lines, it is seen from (8.28) and (8.29) that the even- and odd-mode impedances will be $86.8\,\Omega$ and $28.9\,\Omega$ respectively. A rough extrapolation of the curves in Figure 8.5 would indicate that a value for s/h $\ll 0.1$ is required if there are only two coupled lines in the directional coupler. From (8.47), the odd- to even-mode impedance ratio, Φ, is 0.4768, with (8.48) giving the odd-mode impedance of a pair of coupled lines with the same w/h and s/h as $Z_{0o} = 67.96\,\Omega$. To give the desired impedance ratio, the even-mode impedance for the same pair of lines is $Z_{0e} = 142.5\,\Omega$. Using the equations from [8.8], reproduced here in Appendix 4, for these impedance values, gives w/h = 0.664 and s/h = 0.283 while, from (8.31), $\varepsilon_{\text{eff}} = 1.89$.

The treatment given here so far has assumed zero thickness conductors. Finite conductor thickness for given w/h and s/h will have the main effect of increasing the odd-mode capacitance, i.e. Z_{0o} is reduced, leading via (8.21) to overcoupled characteristics for the coupler.

From detailed experimental studies with $\varepsilon_r = 6.6$ and 10.0, Presser [8.18] concluded that a line-separation correction term, modified from one in Wheeler [8.20]

$$\frac{\Delta s}{h} = \frac{t/h}{\pi\sqrt{\varepsilon_{\text{eff}}^{(0)}}} \left\{ 1 + \ln\left[\frac{4\pi w/h}{t/h}\right] \right\} \tag{8.49}$$

was applicable. For a practical circuit, the line separation has to be increased by Δs over the value calculated for zero thickness lines. The increased gap is achieved in practice by reducing the widths of the coupled lines, keeping the overall lateral dimension constant.

EXERCISES

8.1   Two $50\,\Omega$ characteristic impedance microstrip lines, forming different parts of a microwave circuit, are in parallel with each other for a distance of 50 mm with a separation of 5 mm between the lines. The lines are on a 1.58 mm thick, $\varepsilon_r = 2.5$ substrate. Estimate the maximum level of interference that may be expected between the lines and the frequency (frequencies) at which this maximum level may occur.

8.2   Design a single-stage parallel-line coupler on a 3.2 mm thick, $\varepsilon_r = 2.5$ substrate, with a power split ratio of 15 dB at a center frequency of 1.5 GHz.

8.3   Verify that two 8.34 dB parallel-coupled directional couplers, connected so that ports 2 and 3 of one coupler are connected either directly or with equal path lengths to ports 1 and 4 of the second coupler, give an overall 3.0 dB coupler performance.

8.4   What value directional couplers are required if a pair of identical couplers are to be connected together as in the previous exercise to give an overall 6.0 dB coupler performance?

8.5   Design a maximally-flat 3-section symmetrical 13 dB coupler that is to be fabricated on a 1.0 mm thick, $\varepsilon_r = 10.0$ substrate.

8.6   Verify that, for any coupling length θ of two lossless parallel-coupled transmission lines, there is a balance between the input and output powers. Assume that the even- and odd-mode impedances have been chosen so that $Z_{0e} Z_{0o} = Z_0^2$ and that the mode phase velocities are equal.

8.7   The direct port, i.e. port 2, of an ideal parallel-line 10 dB directional coupler is open-circuited at the plane of the junction between the coupler and the external line to that port. At the coupler's mid-band frequency, what are the input V.S.W.R. to the coupler and the signal levels at the remaining two matched ports?

8.8   Design an 8.34 dB interdigital-line directional coupler for a center frequency of 2.0 GHz on a 1.58 mm thick, $\varepsilon_r = 2.5$ substrate.

8.9   Show that $Z_{0e} Z_{0o} = Z_0^2$ is a necessary condition for an input match to a pair of coupled lines, i.e. to satisfy (8.13), and to produce a zero voltage to the isolated port, in (8.16).

8.10   Calculate the bandwidth of an ideal 3 dB quarter-wave coupler, if the direct and coupled powers are not to differ by more than 0.5 dB.

REFERENCES

[8.1]   Oliver, B. M., "Directional electromagnetic couplers", *Proc. IRE*, Vol. 42, No. 11, November 1954, pp. 1686-92.

[8.2]   Jones, E. M. T. and Bolljahn, J. T., "Coupled-strip-transmission-line filters and directional couplers", *IRE Trans. Microwave Theory and Techniques*, Vol. MTT-4, No. 2, April 1956, pp. 75-81.

[8.3]   Edwards, T. C., *Foundations for Microstrip Circuit Design*, Wiley, Chichester, 1981, Appendix A.

[8.4]   Bryant, T. G. and Weiss, J. A., "Parameters of microstrip transmission lines and of coupled pairs of microstrip lines", *IEEE Trans. Microwave Theory and Techniques*, Vol. MTT-16, No. 12, December 1968, pp. 1021-7.

[8.5] Akhtarzad, S., Rowbotham, T. R. and Johns, P. B., "The design of coupled microstrip lines", *IEEE Trans. Microwave Theory and Techniques*, Vol. MTT-23, No. 6, June 1975, pp. 486-92.

[8.6] Ros, A. E., "Design charts for inhomogeneous coupled microstrip lines", *IEEE Trans. Microwave Theory and Techniques*, Vol. MTT-26, No. 6, June 1978, pp. 394-400.

[8.7] Jansen, R. H., "High-speed computation of single and coupled microstrip parameters including dispersion, high-order modes, loss and finite strip thickness", *IEEE Trans. Microwave Theory and Techniques*, Vol. MTT-26, No. 2, February 1978, pp. 75-82.

[8.8] Kirschning, M. and Jansen, R. H., "Accurate wide-range design equations for the frequency-dependent characteristic of parallel coupled microstrip lines", *IEEE Trans. Microwave Theory and Techniques*, Vol. MTT-32, No. 1, January 1984, pp. 83-90. Corrections: *IEEE Trans. Microwave Theory and Techniques*, Vol. MTT-33, No. 3, March 1985, p. 288.

[8.9] Young, L., "The analytical equivalence of TEM-mode directional couplers and transmission-line stepped-impedance filters", *Proc. IEE*, Vol. 110, No. 2, February 1963, pp. 275-81.

[8.10] Sharpe, C. B., "An equivalence principle for nonuniform transmission-line directional couplers", *IEEE Trans. Microwave Theory and Techniques*, Vol. MTT-15, No. 7, July 1967, pp. 398-405.

[8.11] Sobhy, M. I. and Hosny, E. A., "The design of directional couplers using exponential lines in inhomogeneous media", *IEEE Trans. Microwave Theory and Techniques*, Vol. MTT-30, No. 1, January 1982, pp. 71-6.

[8.12] Shelton, J. P., Wolfe, J. and Van Wagoner, R. C., "Tandem couplers and phase shifters for multi-octave bandwidth", *Microwaves*, Vol. 4, No. 4, April 1965, pp. 14-9.

[8.13] Malherbe, J. A. G. and Losch, I. E., "Directional couplers using semi-re-entrant coupled lines", *Microwave Journal*, Vol. 30, No. 11, November 1987, pp. 121-2, 124, 126, 128.

[8.14] Lange, J., "Interdigitated stripline quadrature hybrid", *IEEE Trans. Microwave Theory and Techniques*, Vol. MTT-17, No. 12, December 1969, pp. 1150-1.

[8.15] Waugh, R. and LaCombe, D., "Unfolding the Lange coupler", *IEEE Trans. Microwave Theory and Techniques*, Vol. MTT-20, No. 11, November 1972, pp. 777-9.

[8.16] Paolino, D. D., "Design more accurate interdigitated couplers", *Microwaves*, Vol. 15, No. 5, May 1976, pp. 34-8.

[8.17] Ou, W., "Design equations for an interdigitated directional coupler", *IEEE Trans. Microwave Theory and Techniques*, Vol. MTT-23, No. 2, February 1975, pp. 253-5.

[8.18] Presser, A., "Interdigitated microstrip coupler design", *IEEE Trans. Microwave Theory and Techniques*, Vol. MTT-26, No. 10, October 1978, pp. 801-5.

[8.19] Tajima, Y. and Kamihashi, S., "Multiconductor couplers", *IEEE Trans. Microwave Theory and Techniques*, Vol. MTT-26, No. 10, October 1978, pp. 795-801.

[8.20] Wheeler, H. A., "Transmission-line properties of parallel strips separated by a dielectric sheet", *IEEE Trans. Microwave Theory and Techniques*, Vol. MTT-13, No. 2, March 1965, pp. 172-85.

⑨ Filters

9.1 INTRODUCTION

This chapter introduces the reader to the problems of microstrip filter design and presents some useful design techniques. The chapter thus is not meant to be a comprehensive treatise on microstrip filters; such a treatment would exceed the scope of this book. For a more comprehensive treatment of filter designs, the reader is referred to specialized texts, e.g. [9.1, 9.2].

In the next section, the injection and blocking of d.c. bias voltages is described using a filter characteristic that is ideally suited to the d.c. source requirement, i.e. appearing either as a d.c. open or short circuit, while at the same time appearing transparent to the microwave signals. Bias injection networks are simple structures in which little attention is paid to the detail of the transmission characteristics across the frequency band.

The extraction of a lower frequency signal, such as a 70MHz intermediate frequency signal from a microwave mixer circuit, requires a low-pass filter that needs to be more carefully designed than a bias network. Low-pass filters can be designed from the classical lumped-element low-pass prototype circuits and may either be fabricated in lumped element form or transformed into equivalent transmission line networks.

Band-pass filters require precise transmission characteristics that allow a desired band of signals to pass through the two-port network. Thus, between a transmitter and the transmitting antenna, a band-pass filter may be used to attenuate unwanted signals and harmonic components that may cause interference to other users of the electromagnetic spectrum. Conversely, between an antenna and a receiver, a band-pass filter will reject out-of-band signals that may cause interference within the receiver, especially if they are at a high signal level in comparison with the desired signals.

Band-stop filters reflect signals over a limited range of frequencies while allowing all others to pass through the network. They are used to minimize the transmission of possible high-level signals, e.g. the local oscillator of an upconverter where only the upper-sideband is desired, and as tuned reflective elements in oscillator circuits.

It will be assumed here that the reader has a basic knowledge of classical lumped filter designs. Prototype designs are available in classical textbooks and handbooks [9.3–9.5]. Common filters include maximally-flat (Butterworth), equi-ripple (Chebyshev), Bessel and elliptic filters, where each name is descriptive of the filter characteristics. It will be assumed that the reader is able to use published filter tables intelligently for the design of filters.

There are two basic mechanisms for achieving filter action, that is, for obtaining variation of signal transmission through a circuit. If the filter is lossless, the only way in which a transmission of less than unity may be achieved is by reflection at the input. Most filters are of this type and, in their case, attenuation must imply a high reflection coefficient at the input. Thus it is impossible to maintain a good match across the attenuation band for lossless filters. Conversely, matching the input for no reflections with lossless elements, as in Chapter 6, will automatically give complete transmission from input to output.

Reduction in transmission can also be achieved if lossy absorbing elements are inserted inside the filter. In this case, at least in principle, a good match across the attenuation band could be maintained, as the reduction in transmission is no longer predicated on the presence of reflections at the input. An example of a filter depending on a lossy element is the wavemeter, where a resonator, coupled to a waveguide or a transmission line, couples energy into itself at resonance and dissipates it within the resonator losses. This is really a form of band-stop filter. Another example occurs when a diplexer is formed by connecting a low-pass filter in parallel with its complementary high-pass network. If the outputs of the two parallel networks are combined, say with a hybrid, an all-pass network is obtained. However, with completely separate loads a diplexer is obtained, with low frequencies going to one load and high frequencies to the other, while maintaining a matched input at all times for the combined network. If, say, only the low-pass function is of interest, then the low-pass filter section that is matched in both the pass and attenuation bands may be used, with the matched load terminating the high-pass network simply being an internal lossy element as far as the low-pass network is concerned.

While microstrip filters may take many forms, a useful range of filter designs may be obtained by taking classical lumped filter designs and converting them to microstrip form, using the equivalence of short lengths of transmission line to inductance or capacitance. An extension of this procedure, applicable to the case of band-pass and band-stop filters, is not to take individual inductances and capacitances, but to replace whole resonating L-C sections of the lumped-element prototype with resonant lengths of microstrip line. To achieve the required range of L or C values it may be necessary to use quite long lines, due to the limited range of characteristic impedances that are available in practice. The short length line approximations then may no longer hold and a correction to the classical filter design process may be required. Also, corrections due to losses in resonator elements may be needed, especially in very narrowband designs. However, this is not a problem unique to microstrip filters, as it is similar to that found in classical lumped filters.

Microstrip equivalents exist for both series-resonant and parallel-resonant circuits. However, it is often desirable to use resonators of one type only, in which case immittance inverters may be employed to convert between the two types of resonant circuit. Immittance inversion in lumped circuit filters is achieved with active circuits to produce filters with only one type of reactive element, as in active filter arrangements with only capacitors. In microstrip circuits, immittance inversion is often conveniently achieved with quarter-wave directional coupler arrangements.

Given a particular low-pass lumped filter design, a complementary high-pass filter is obtained with the transformation

$$\frac{\omega'}{\omega_1'} \; \rightarrow \; -\frac{\omega_1}{\omega} \tag{9.1}$$

while

$$\frac{\omega'}{\omega_1'} \; \rightarrow \; \frac{1}{\Omega}\left[\frac{\omega}{\omega_0} - \frac{\omega_0}{\omega}\right] \tag{9.2}$$

transforms from the low-pass to band-pass form. In both equations, ω' and ω_1' are the frequency variable and band-edge frequency respectively for the low-pass filter. In standard filter tables, ω_1' is normally $1.0 \, \text{radian.s}^{-1}$. Likewise ω_1 is the band-edge frequency for the high-pass filter with a frequency variable, ω. For the band-pass filter transformation in (9.2), ω_0 is now the center frequency and Ω is the fractional bandwidth given by $\Omega = (\omega_2 - \omega_1)/\omega_0$, ω_2 and ω_1 being the band-edge frequencies. Structurally, (9.1) is equivalent to replacing every inductor in the low-pass filter with a capacitor and every capacitor with an inductor. In the low-pass to band-pass transformation (9.2)

| | L | $\rightarrow$ | series L-C resonator |
| and | C | $\rightarrow$ | parallel L-C resonator. |

The transformation (9.2) applied to a high-pass filter will yield a band-stop filter. This is equivalent to taking a low-pass design and letting

| | L | $\rightarrow$ | *parallel* L-C resonator |
| and | C | $\rightarrow$ | *series* L-C resonator. |

Filters in microstrip form may also be constructed with dielectric resonators. They offer the advantages of improved pass- and stop-band filter characteristics at the expense of increased circuit complexity in design and fabrication. A discussion of dielectric resonators will be found in Chapter 10.

9.2   BIAS NETWORKS

Microwave diodes and transistors require appropriate d.c. voltages to be applied to their terminals for correct circuit operation. The necessary connections of d.c. power supplies and earth return paths to the main microstrip transmission line are made using biasing networks. As a filter, a bias injection network is one that is designed to combine both microwave and low frequency (d.c.) signals without any transmission of energy between the microwave and low frequency ports. Thus, not only should there be no low frequency path to the microwave input branch, but the d.c. input port must also appear as a microwave open circuit at the junction to the through transmission line. A circuit diagram of a bias network that has these functions is illustrated in Figure 9.1. The network may be considered in two parts:

i) For the microwave input branch, an infinite capacitance is ideally required to provide a d.c. open circuit and a microwave short circuit.

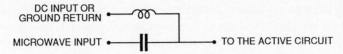

Figure 9.1 A bias injection network

ii) For the d.c. input branch, an infinite inductance will give the d.c. short circuit and will appear as a microwave open circuit at the plane of the through transmission line.

9.2.1 The d.c. open circuit

Miniature chip ceramic capacitors with dimensions of the order of 1 mm and with a capacitance of 1000 pF make ideal d.c. blocking elements, since 1000 pF at 1.0 GHz gives an impedance of 0.16 Ω. Such an impedance as a series element in a 50 Ω characteristic impedance transmission line gives a negligible reflection coefficient of magnitude 0.0016. This value naturally reduces in magnitude with increasing frequency. However, a detailed study by Ingalls and Kent [9.6] of this type of capacitor mounted as a series element in a 50Ω microstrip transmission line has shown the existence of resonances, giving high impedance values at resonance, associated with the equivalent folded transmission line of its structure. The separation between resonances depends on the capacitor size, and insertion loss maxima of up to 2dB have been observed at gigahertz intervals.

A narrowband design in the form of a λ/4 open-circuited stub, series-connected in the line, will give the desired result of a d.c. open circuit. The physical realization, with the "stub" folded along the main line as illustrated in Figure 9.2, is a more complex structure and needs rigorous analysis in terms of coupled transmission lines.

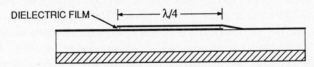

Figure 9.2 A tightly coupled quarter-wave line forming a d.c. open circuit

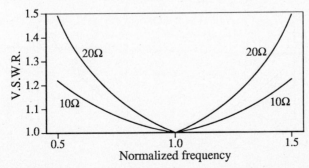

Figure 9.3 The performance of a λ/4 open-circuited stub series-connected in a 50Ω line, with the stub characteristic impedance as a parameter. The stub is exactly λ/4 at a normalized frequency of unity.

A low characteristic impedance, compared with the $50\,\Omega$ microstrip transmission line, is required for broadband operation, as is seen from Figure 9.3.

An appraisal of the complete microwave network in the vicinity of a solid-state element may reveal that a specific d.c. block is not required for the bias injection network. This will be the case if, for example, a parallel-coupled directional coupler or band-pass filter automatically provides the required d.c. blocking function.

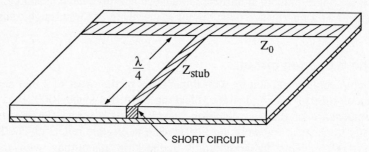

Figure 9.4   A d.c. return network

9.2.2   The d.c. return and r.f. block

A quarter-wave shunt stub line with a short-circuit termination, as illustrated in Figure 9.4, will transform to an open circuit at the main through line for the design frequency. The microwave short-circuit termination on the stub line may be either a true short circuit for a d.c. ground return or in the form of a capacitive feed-through element. The short-circuit connection is shown in Figure 9.4 as a connection between the strip and the ground plane at the edge of the substrate. Since the circuit geometry may preclude this arrangement, metal posts may be inserted through the substrate and be soldered to both the strip and the ground plane. End-effect corrections for the short-circuit posts are not normally required since such an element would only be used in frequency-insensitive situations. Of importance is the broadband performance of the circuit, as illustrated in Figure 9.5 for two possible stub characteristic impedances on a $50\,\Omega$ line.

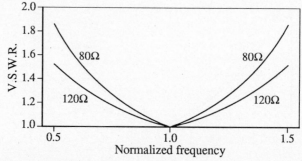

Figure 9.5   The performance of a $\lambda/4$ short-circuited stub parallel-connected in a $50\,\Omega$ line, with the stub characteristic impedance as a parameter. The stub is exactly $\lambda/4$ long at a normalized frequency of unity.

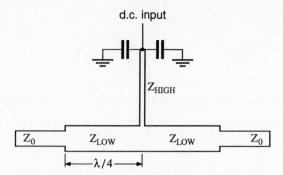

Figure 9.6 An improved d.c. injection network with λ/4 sections

The arrangement of Figure 9.4 gives better performance for a high ratio of Z_{stub}/Z_0. Therefore, if the main line is transformed through a low impedance quarter-wave transformer to give an even lower impedance at the plane of the stub, Z_T say, as is illustrated in Figure 9.6, then the higher ratio Z_{stub}/Z_T leads to even better circuit performance. With the high impedance stub line at 120Ω, Figure 9.7 illustrates typical performance curves with the characteristic impedance of the main line quarter-wave transformers as a parameter.

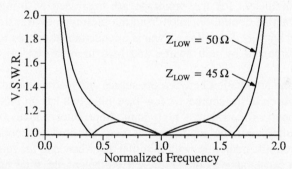

Figure 9.7 The input V.S.W.R. of the network of Figure 9.6, as a function of frequency and normalized to the frequency where line lengths are λ/4 long. The stub-line impedance is maintained at 120Ω with $Z_0 = 50Ω$.

9.3 LOW-PASS FILTERS

The bias network for a microwave circuit that has just been considered is one form of a low-pass filter. This filter, taking the form of a quarter-wave high impedance line with feed-through capacitors, however, may be quite unsuitable for the injection of i.f. signals into a circuit. A filter with more precise low-pass characteristics is required.

Consider an n-section lumped element prototype low-pass filter, illustrated in Figure 9.8. For κ from 1 to n, the prototype elements represent an alternating

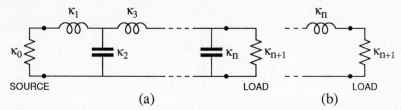

Figure 9.8 A prototype n-section low-pass filter with an inductor κ_1 as the first element, showing (a) the case of an even n, with κ_{n+1} a resistance and (b) the case of an odd n, with κ_{n+1} a conductance

network of series inductances and shunt capacitances, with either type of component as an end-element in the filter network. In Figure 9.8, κ_1 is shown as a series inductance, but κ_1 could equally be a shunt capacitance, with κ_2 a series inductance, etc. The termination elements, κ_0 and κ_{n+1}, when considered as a continuation of the sequence of series and shunt elements, will be resistances if adjacent to a capacitive element and conductances if adjacent to an inductive element. The other κ-values are in henries or farads as the case may be. κ_0 may equal unity without loss of generality, but some filters, in particular those with pass-band ripple and an even number of sections, will not necessarily have $\kappa_{n+1} = 1$. In these cases, the correct interpretation of κ_{n+1} as either a resistance or a conductance is important.

Frequency parameters for the prototype are expressed as primed quantities, while those for an actual circuit, obtained from the prototype by transformation, will be expressed by an unprimed notation. Typical attenuation responses for a prototype low-pass filter, terminated by unit source and load immittances, are illustrated in Figure 9.9.

The out-of-band attenuation between the source and load is due to the reflection of the input signal back to the source. A low-pass filter that has an inductance as the first reactive element will present a high input impedance above the pass band. Connected in parallel with the main line, it may be used to couple out a low frequency signal from the main line, with minimal disturbance to the microwave path. Conversely, with a capacitance as the first reactive element, the filter has a low input impedance above the pass band. This latter configuration may be useful for a low frequency path to a diode that otherwise requires a low impedance to ground for a microwave return path at that terminal.

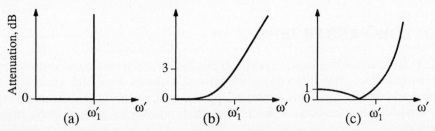

Figure 9.9 Low-pass filter attenuation/frequency responses showing (a) the ideal low-pass characteristic, (b) a maximally-flat or Butterworth response with two sections and (c) the 1.0dB equi-ripple or Chebyshev response with two sections

For the ideal low-pass characteristic, Figure 9.9a, there is no loss at any frequency below the edge of the pass band at ω_1', but infinite loss above that frequency. A maximally-flat or Butterworth response, Figure 9.9b, cannot be made any broader in bandwidth without attenuation ripples appearing in the pass band. A small level of pass-band ripple, Figure 9.9c, may indeed be desirable since there is the added benefit of a much steeper transition region between the pass and stop bands. For the equi-ripple filter, the pass-band edge at ω_1' is the frequency above which the attenuation will be greater than the maximum pass-band ripple attenuation. However, as such a definition will not be meaningful for a maximally-flat filter, ω_1' is taken at the 3dB attenuation level for this case. An allowance must be made if the low-pass filter specification calls for an attenuation other than 3dB at some nominated band-edge frequency.

9.3.1 Terminology

The transmission properties of filters will be described by P_{out}/P_{avs}, where P_{out} is the power delivered by the filter output to the load and P_{avs} is the available power from the source. P_{out}/P_{avs} is formally known as the transducer power gain, G_t, of the two-port network. More will be said about G_t in Chapter 11, but for the present it is to be noted that G_t is a function of both the source impedance Z_S and the load impedance Z_L. Thus G_t is not meaningfully defined unless both Z_S and Z_L are specified. It is usual practice to make $Z_S = Z_L = Z_0$ and, for prototype filters, Z_0 is often made 1Ω.

In the context of filters, $G_t \leq 1$ and it is more convenient to speak of an attenuation or a transmission loss that is greater than unity, by taking $1/G_t$. Thus attenuation in dB will be defined by $10\,log_{10}(P_{avs}/P_{out})$. Insertion loss is also a term that is often used. However, care must be exercised in using this term unless $Z_S = Z_L = Z_0$, in which case the insertion loss or gain is identical to $1/G_t$ or G_t respectively, as will be seen in Exercise 11.11.

Attenuation, insertion loss, and transmission loss will continue to be used interchangeably in the remainder of the book, with the assumption that they are evaluated with $Z_S = Z_L = Z_0$, unless specifically indicated otherwise.

9.3.2 The maximally-flat response

The output power for an n-section low-pass filter that possesses a maximally-flat response is given [9.1] by

$$P_{out} = \frac{1}{1 + (\omega'/\omega_1')^{2n}} \times P_{avs} \tag{9.3}$$

The output power is one half the input power at $\omega' = \omega_1'$, where ω_1' is the 3dB frequency. The filter attenuation, given by

$$L = 10\,log_{10}\left[1 + (\omega'/\omega_1')^{2n}\right] \quad \text{dB} \tag{9.4}$$

is plotted in Figure 9.10 and may be used as a guide to estimate the required number

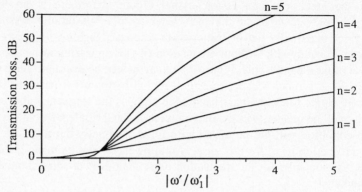

Figure 9.10 The normalized response for n-section maximally-flat low-pass filters

of sections for a practical low-pass filter. The curves may also be used in §9.4.5 for an estimation of the number of sections that are required for a narrowband band-pass filter. Now, with $\kappa_0 = \kappa_{n+1} = 1$, and with $\omega_1' = 1$, the prototype element values [9.1] are computed from

$$\kappa_k = 2 \sin\left\{\frac{(2k-1)\pi}{2n}\right\}, \quad k = 1, \cdots, n \tag{9.5}$$

The numerical values for κ_k with filter orders from 1 to 7 are presented in Table 9.1.

Table 9.1 Prototype element values for a maximally-flat low-pass filter

n	1	2	3	4	5	6	7
κ_1	2.0000	1.4142	1.0000	0.7654	0.6180	0.5176	0.4450
κ_2		1.4142	2.0000	1.8478	1.6180	1.4142	1.2470
κ_3			1.0000	1.8478	2.0000	1.9319	1.8019
κ_4				0.7654	1.6180	1.9319	2.0000
κ_5					0.6180	1.4142	1.8019
κ_6						0.5176	1.2470
κ_7							0.4450

9.3.3 The equi-ripple response

A complete exposition of the theory for equi-ripple or Chebyshev response filters [9.1] is beyond the scope of this book and only the relevant equations that are used to generate the filter prototype coefficients are presented here.

Let the magnitude of the pass-band ripple be L_r and the frequency at the band edge where the filter loss is equal to the pass-band ripple be ω_1'. Between a unit immittance for the source and an appropriate load immittance — that is later seen to be other than unity for an even number of sections — the output power is given by

$$P_{out} = \frac{1}{1 + F_0 C_n^2(\omega'/\omega_1')} \times P_{avs} \tag{9.6}$$

where F_0 is a constant, related to the pass-band ripple L_r (dB) by

$$F_0 = 10^{L_r/10} - 1 \qquad (9.7)$$

and $C_n(x)$ is the Chebyshev polynomial of order n and argument x. The generating equation for the Chebyshev polynomial is

$$C_n(x) = 2x\,C_{n-1}(x) - C_{n-2}(x) \qquad (9.8)$$

with $\qquad C_0(x) = 1, \; C_1(x) = x$ and $C_n(1) = 1$

$\omega' = \omega'_1$ makes $x = 1$ and the polynomial becomes unity for all n. When $x > 1$, the Chebyshev polynomial increases monotonically with increasing x while for $x < 1$, it oscillates as a function of the polynomial order within the range zero to unity. It follows from (9.6), that the attenuation at $\omega' = \omega'_1$ is also the maximum pass-band attenuation, which in fact occurs at every frequency within the passband for which $C_n(x) = 1$.

Let the source immittance be unity, i.e. $\kappa_0 = 1$. The prototype element values are then given [9.1] by

$$\kappa_1 = \frac{a_1}{F_2}$$

$$\kappa_k = \frac{a_{k-1} a_k}{b_{k-1} \kappa_{k-1}}, \qquad k = 2, \cdots, n$$

$$\begin{aligned} \kappa_{n+1} &= 1, & \text{for n odd} \\ &= \coth^2(F_1), & \text{for n even} \end{aligned} \qquad (9.9)$$

where

$$F_1 = \frac{1}{4} \, ln \left\{ \coth \left[\frac{L_r}{17.372} \right] \right\} \qquad (9.10)$$

$$F_2 = \sinh \left\{ \frac{2F_1}{n} \right\} \qquad (9.11)$$

$$a_k = 2 \sin \left\{ \frac{(k-1)\pi}{2n} \right\}, \qquad k = 1, \cdots, n \qquad (9.12)$$

$$b_k = F_2^2 + \sin^2 \left\{ \frac{k\pi}{n} \right\}, \qquad k = 1, \cdots, n \qquad (9.13)$$

Table 9.2 gives the prototype element values for equi-ripple filters with $L_r = 0.1\,\text{dB}$ and $0.25\,\text{dB}$ and with the filter order, n, from 1 to 7. The termination element, κ_{n+1}, is unity and of the same dimension as κ_0 for the symmetrical odd-order filter but, for the even-order filters, there is a dimensional inversion between the two ends of the filter, i.e. if κ_0 is a resistance, then κ_{n+1} is a conductance and vice versa. Further, there is an immittance magnitude transformation in the even-order case. For this reason, the odd-order equi-ripple filter is more common in practice.

Table 9.2 Prototype element values for an equi-ripple low-pass filter

Pass-band ripple = 0.1 dB, $F_0 = 0.0233$, $F_1 = 1.2894$							
n	1	2	3	4	5	6	7
κ_0	1.0000	1.0000	1.0000	1.0000	1.0000	1.0000	1.0000
κ_1	0.3052	0.8430	1.0316	1.1088	1.1468	1.1681	1.1812
κ_2		0.6220	1.1474	1.3062	1.3712	1.4040	1.4228
κ_3			1.0316	1.7704	1.9750	2.0562	2.0967
κ_4				0.8181	1.3712	1.5171	1.5734
κ_5					1.1468	1.9029	2.0967
κ_6						0.8618	1.4228
κ_7							1.1812
κ_{n+1}	1.0000	1.3554	1.0000	1.3554	1.0000	1.3554	1.0000
Pass-band ripple = 0.25 dB, $F_0 = 0.0593$, $F_1 = 1.0603$							
n	1	2	3	4	5	6	7
κ_0	1.0000	1.0000	1.0000	1.0000	1.0000	1.0000	1.0000
κ_1	0.4868	1.1132	1.3034	1.3782	1.4144	1.4345	1.4468
κ_2		0.6873	1.1463	1.2693	1.3180	1.3422	1.3560
κ_3			1.3034	2.0558	2.2414	2.3126	2.3476
κ_4				0.8510	1.3180	1.4279	1.4689
κ_5					1.4144	2.1737	2.3476
κ_6						0.8858	1.3560
κ_7							1.4468
κ_{n+1}	1.0000	1.6196	1.0000	1.6196	1.0000	1.6196	1.0000

Example 9.1

i) Evaluate the prototype component values for a 3-section Chebyshev response low-pass filter that has a pass-band ripple of 0.05 dB.

ii) Compare the filter frequency response with a maximally-flat filter that has the same bandwidth to the 0.05 dB attenuation level.

Solution:

i) An equi-ripple filter that has an odd number of sections has symmetrical component values about the center element. Thus, in this case only κ_1 and κ_2 need to be evaluated.

From (9.10 – 9.13), the required terms are found in the following order:

$$F_1 = 1.4626 \qquad F_2 = 1.1371$$

$$a_1 = 1.0000 \qquad a_2 = 2.0000$$

$$b_1 = 2.0430$$

and the prototype element values are calculated from (9.9) as

$$\kappa_1 = 0.8794 = \kappa_3$$

$$\kappa_2 = 1.1132$$

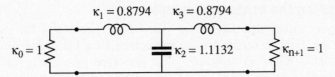

Figure 9.11 The 0.05 dB Chebyshev prototype network for Example 9.1

The final network is illustrated in Figure 9.11.

ii) For the Chebyshev response, from (9.6) and (9.7), the attenuation

$$L = 10\,log_{10}\left\{1 + \left(10^{L_r/10} - 1\right) \times C_n^2(\omega'/\omega_1')\right\}$$

with L_r = 0.05 dB and, from (9.8), the Chebyshev polynomial

$$C_3(x) = 4x^3 - 3x \quad \text{with } x \equiv \omega'/\omega_1'$$

This response is plotted as curve (a) in Figure 9.12. A comparison is made with a three-section maximally-flat filter that has the same low-pass bandwidth to the 0.05 dB attenuation level. While ω_1' is unity for the band edge in the Chebyshev response, the same parameter represents the 3 dB frequency in the maximally-flat filter. Solving the maximally-flat response equation, (9.4)

$$L = 0.05 = 10\,log_{10}\left\{1 + \left[\frac{\omega'}{\omega_1'}\right]^6\right\}$$

to find the normalized frequency for 0.05 dB attenuation, gives $\omega' = 0.4756\,\omega_1'$. For a direct comparison with the Chebyshev response, the expression

$$L = 10\,log_{10}\left(1 + (0.4756\,\omega')^6\right)$$

is now plotted as curve (b) in Figure 9.12.

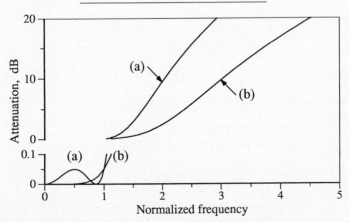

Figure 9.12 A comparison of equal bandwidth low-pass filter responses showing (a) the 0.05 dB equi-ripple (Chebyshev) response, and (b) the maximally-flat response

9.3.4 Scaling the prototype values

The prototype values that have been derived for maximally-flat and equi-ripple responses have element values that are appropriate for a 1Ω source impedance and a cut-off frequency ω_1' of $1.0 \, \text{radian.s}^{-1}$. The prototype values may now be scaled in two steps to give the practical values for a low-pass filter. For the first stage of scaling, the prototype values are scaled such that the impedance of each element increases by the same ratio. This gives derived values that are now adjusted to a new impedance level, say 50Ω, while the cut-off frequency still remains at ω_1'. The second stage of scaling is carried out to adjust the frequency properties of the network while maintaining the impedance values of all the elements at the new frequency.

Let the impedance go to x ohm and the cut-off frequency to $y \, \text{radian.s}^{-1}$, then

$$R \rightarrow xR, \quad L \rightarrow \left[\frac{x}{y}\right]L \text{ and } C \rightarrow \left[\frac{1}{xy}\right]C \tag{9.14}$$

Example 9.2

i) Calculate the inductance and capacitance values for a maximally-flat low-pass filter that has a 3dB bandwidth of 400MHz. The filter is to be connected between 50Ω source and load impedances. It must present a high input impedance at 1.0GHz and, at that frequency, have an attenuation greater than 20dB.

ii) Up to what frequency will the attenuation be less than 0.1dB?

Solution:

i) From (9.4), with a 3dB frequency f_1 of 400 MHz, the attenuation at 1.0GHz is given by

$$L = 10 \, log_{10}\left\{1 + \left[\frac{1000}{400}\right]^{2n}\right\}$$

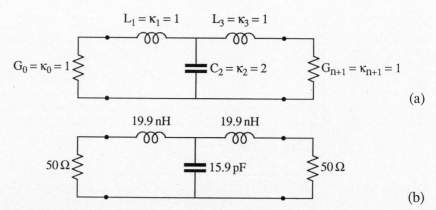

Figure 9.13 The 3-section maximally-flat filter for Example 9.2, showing (a) prototype values and (b) the actual circuit values

Thus, with $L = 20\,dB$, the minimum number of sections is

$$n = \frac{1}{2\,log_{10}2.5} \times log_{10}(10^{L/10} - 1) = 2.51$$

i.e. the filter must have a practical minimum of three sections. This will give an attenuation of 23.9 dB at 1.0 GHz.

The maximally-flat low-pass prototype with three sections and an inductance as the reactive element at each end in order to satisfy the high-frequency high impedance requirement is illustrated in Figure 9.13. Each of the prototype values is scaled to allow for a transformation first from $1\,\Omega$ to a $50\,\Omega$ characteristic impedance and then from a cut-off frequency of $1.0\,radian.s^{-1}$ to 400 MHz. The results are presented in Table 9.3 and in Figure 9.13b.

Table 9.3   The component values for a maximally-flat low-pass filter

	Prototype		Impedance scaling		Frequency scaling
L_1, H	1.0	$\Rightarrow$	50.0	$\Rightarrow$	19.9×10^{-9}
C_2, F	2.0	$\Rightarrow$	0.04	$\Rightarrow$	15.9×10^{-12}
L_3, H	1.0	$\Rightarrow$	50.0	$\Rightarrow$	19.9×10^{-9}
Z_0, Ω	1.0	$\Rightarrow$	50.0	$\Rightarrow$	50.0

ii) From (9.4), the insertion loss is less than 0.1 dB up to a frequency f_0 MHz, given by

$$0.1 = 10\,log\left\{1 + \left[\frac{f_0}{400}\right]^6\right\}$$

i.e. $f_0 = 213\,MHz$

9.3.5 Practical considerations

A low-pass filter may be constructed from a tandem connection of alternating high and low characteristic impedance lines. Characteristic impedance values that are as extreme as possible will lead to the best approximations to lumped inductive and capacitive elements. However, because of the transmission line nature of the elements, the lumped component approximations for a particular length of line are not as accurate when the frequency increases and there will be a degradation of the stop-band performance. This degradation for a stripline low-pass filter has been calculated by Howe [9.7, Figure 6.12] as a function of line impedance selections. Howe recommends (i) that additional sections should be included as a safety measure and (ii) that the filter 3 dB cut-off frequency should not be too low; a value about one-third of the frequency that has to be blocked should be considered. This second recommendation implies that the L and C values for the low-pass filter will be kept small and thus may be modeled using the short line approximation.

Consider a practical low-pass filter with alternating high and low impedance sections. There will be fringing fields from the ends of the low impedance lines. The finite lengths of line further require that the short line approximations are modified by the inclusion of the full Π- and T-equivalent circuits, as discussed in §2.2.1, for the high and low impedance sections respectively. These considerations necessitate an iterative approach, with the correction elements modifying the initial ideal design. For a high impedance line, the line length l in terms of the required inductance is given (2.38) by

$$l = \frac{\lambda_H}{2\pi} \sin^{-1}(\omega L / Z_H) \tag{9.15}$$

with the two shunt-capacitance elements

$$C_L = \frac{1}{\omega Z_H} \tan\left[\frac{\pi l}{\lambda_H}\right] \tag{9.16}$$

In these two equations, the values of λ_H and Z_H are the wavelength and characteristic impedance associated with the high impedance microstrip line.

In a similar manner for the low impedance line, the line length for the desired capacitance is given by

$$l = \frac{\lambda_L}{2\pi} \sin^{-1}(\omega C Z_L) \tag{9.17}$$

and the two series-inductance elements of the T-equivalent model are

$$L_C = \frac{Z_L}{\omega} \tan\left[\frac{\pi l}{\lambda_L}\right] \tag{9.18}$$

The transmission line lengths of a practical low-pass filter are found using an iterative procedure. The initial element lengths are found and the circuit values are modified to allow for the fringing capacitance, the capacitive components of the high impedance lines and, to a lesser extent, the inductive components of the low impedance lines. New element lengths are now determined and the process repeated until a sufficiently stable solution is found. This process is illustrated in the following example.

Example 9.3

Design a fifth-order maximally-flat low-pass filter with a 3 dB cut-off frequency of 2.0 GHz for use in a 50 Ω characteristic impedance system. The filter is to be constructed on a 1.58 mm thick, $\varepsilon_r = 2.5$ substrate.

Solution:

The prototype filter elements are found in Table 9.1 and are scaled for impedance level and frequency such that

$$L_k = \frac{\kappa_k Z_0}{\omega} \text{ H}, \quad C_k = \frac{\kappa_k}{\omega Z_0} \text{ F} \tag{9.19}$$

where $Z_0 = 50\,\Omega$ and $\omega = 4\pi \times 10^9$ radian.s^{-1}. The circuit element values that result are presented in Table 9.4. Data for the different values of characteristic impedance lines that have been selected to be used on the 1.58 mm thick, $\varepsilon_r = 2.5$ substrate are presented in Table 9.5.

Table 9.4 The component values for a fifth-order maximally-flat filter

k	κ_k	L_k, nH	C_k, pF
1	0.618	2.46	-
2	1.618	-	2.58
3	2.000	7.96	-
4	1.618	-	2.58
5	0.618	2.46	-

Table 9.5 Transmission line data for the low-pass filter construction

Z_0, Ω	$\frac{w}{h}$	w, mm	ε_{eff}	λ, mm at 2.0GHz
25	7.28	11.50	2.226	99.8
50	2.837	4.48	2.090	103.8
130	0.404	0.64	1.901	108.8

The 130 Ω line is used for each inductive component, with the line length and correction capacitance derived from (9.15) and (9.16) respectively. Hence

$$L_1 = 2.46\,\text{nH}, \ Z_H = 130\,\Omega \ \Rightarrow \ l_1 = 0.038\lambda_H = 4.2\,\text{mm}, \ C_L = 0.073\,\text{pF}$$

$$L_3 = 7.96\,\text{nH}, \ Z_H = 130\,\Omega \ \Rightarrow \ l_3 = 0.140\lambda_H = 15.2\,\text{mm}, \ C_L = 0.287\,\text{pF}$$

A T-equivalent circuit with a shunt capacitance as the major component is used for a short length of low impedance line. From (9.17) and (9.18)

$$C_2 = 2.58\,\text{pF}, \ Z_L = 25\,\Omega \ \Rightarrow \ l_2 = 0.150\lambda_L = 15.0\,\text{mm}, \ L_C = 1.02\,\text{nH}$$

The desired values of the circuit inductances and capacitances are reduced by the values of the adjacent line correction terms. Thus

$$\begin{aligned}
L_1 &= 2.46 - 1.02 &&= 1.44\,\text{nH} \\
C_2 &= 2.58 - 0.073 - 0.287 &&= 2.22\,\text{pF} \\
L_3 &= 7.96 - 2 \times 1.02 &&= 5.92\,\text{nH}
\end{aligned}$$

An iteration is now undertaken to recalculate l_1, l_2, and l_3 from L_1, C_2, and L_3 just calculated, giving

$$\begin{aligned}
l_1 &= 2.42\,\text{mm} \ \ \text{with} \ C_L = 0.043\,\text{pF} \\
l_2 &= 12.26\,\text{mm} \ \ \text{with} \ L_C = 0.81\,\text{nH}
\end{aligned}$$

and $\qquad l_3 = 10.55\,\text{mm} \ \ \text{with} \ C_L = 0.192\,\text{pF}$

From the filter symmetry, $l_4 = l_2$ and $l_5 = l_1$. Further iterations may not be necessary, but if they are carried out then these correction terms just calculated are subtracted from the initial L and C values given in Table 9.4.

Allowances must now be made for
i) the fringing capacitance at each end of the low impedance sections,

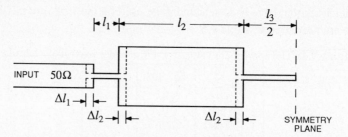

Figure 9.14 One half of the microstrip circuit for the 2.0GHz low-pass filter

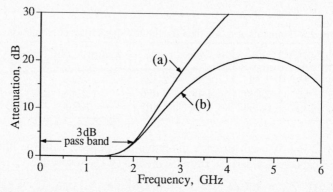

Figure 9.15 A comparison of the responses of the low-pass filter in Example 9.3, showing (a) the lumped-element circuit response with five sections, and (b) the transmission line equivalent circuit response

ii) the additional 0.047 pF capacitance associated with the Π-equivalent network for l_1 at the step junction between the high impedance line and the 50 Ω input lines. A similar correction is made between l_5 and the output line.

In both cases, the effects will be compensated by lengthening the high impedance lines, with correction lengths Δl_1 and Δl_2 as shown in Figure 9.14.

A plot of the lumped element circuit response compared with the transmission line equivalent network with finite line lengths obtained after one iteration above, but still assuming ideal lines that do not have any dispersion or step-junction fringing capacitance, is given in Figure 9.15 and clearly illustrates the reduced stop-band attenuation for practical circuits.

9.4 BAND-PASS FILTERS

A resonating structure that forms the basis of many microstrip filters is a half-wavelength section of line that ideally is terminated at each end by an open circuit. The physical line will be reduced in length from a half wavelength when allowances are made for the open-circuit fringing capacitances discussed in §5.2.

There are two basic forms of coupled-resonator band-pass filters in microstrip line, named by the method for coupling energy into the resonators. The first form, an end-coupled filter, has capacitive coupling from each end of the resonator into the end of the adjacent line. On the other hand, edge- or parallel-coupled filters have resonators that are coupled through the even- and odd-mode fields along the edges of the lines. Parallel-coupled filters are generally preferred, because they lead to as much as a 50% reduction in length. Furthermore, larger gaps between lines are permitted for parallel-coupled filters, easing the tolerances and permitting a broader bandwidth for a given dimensional tolerance. However, perhaps of greatest significance is the fact that the first spurious response for a parallel-coupled filter occurs at three times the center frequency of the filter, whereas for the end-coupled filter it occurs at only twice the center frequency.

9.4.1 The coupling mechanism

Consider the simplest form of quarter-wave parallel-coupled filter illustrated in Figure 9.16. This is a zero-order filter as it has no half-wavelength resonator elements. However, it does illustrate the coupling that takes place between the adjacent sections of a parallel-coupled band-pass filter. When it is compared with Figure 8.1, Figure 9.16a is seen to represent a single-section directional coupler with an open-circuit termination on both ports 2 and 3. The same circuit in a more practical form as the basic element of a filter is shown in Figure 9.16b. The fringing capacitances at both the open circuits and the step transitions in line width will have to be taken into account by appropriate reductions in the line lengths.

Consider the case illustrated in Figure 9.16b. For a unit voltage incident at port 1 and a matched load terminating port 4, from (8.26) for a parallel-coupled directional coupler

$$V_1 = 1$$
$$V_2 = -j\sqrt{1 - c^2}$$

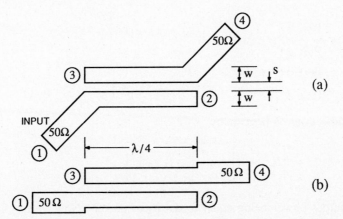

Figure 9.16 A quarter-wave parallel-coupled filter section drawn (a) as a basic directional coupler, with the coupled and direct ports left open-circuited, and (b) in more practical form as the basic element of a filter. Line length corrections for fringing capacitance are not shown.

$$V_3 = c$$
$$V_4 = 0 \tag{9.20}$$

V_2 and V_3 are the voltages of the waves incident onto the open-circuit terminations at their respective ports. Now V_2 is reflected from the open circuit with a phase change of $0°$, giving

$$V_2 = -j\sqrt{1 - c^2}$$

as an incident wave at port 2. Applying an equation of the form of (9.20) to obtain the resultant voltages at the other ports gives

$$V_1 = -j\sqrt{1 - c^2}\, V_2$$
$$= -j\sqrt{1 - c^2}\left[-j\sqrt{1 - c^2}\right] = c^2 - 1$$
$$V_3 = 0, \qquad \text{at the new isolated port}$$

and
$$V_4 = c\,V_2 = -jc\sqrt{1 - c^2} \tag{9.21}$$

Equation (9.21) shows that there is no addition to the original voltage, V_3, in (9.20). Similarly, the V_3 in (9.20) is reflected from its open circuit with a phase change of $0°$, giving

$$V_3 = c, \qquad \text{incident at port 3}$$
$$V_1 = c\,V_3 = c^2$$
$$V_2 = 0, \qquad \text{at the new isolated port}$$

and
$$V_4 = -j\sqrt{1 - c^2}\,V_3 = -jc\sqrt{1 - c^2} \tag{9.22}$$

Again, because the V_2 in (9.22) is zero, no modification to the original V_2 in (9.20) or (9.21) is required. Because there is a matched load at port 4, V_4 does not produce any further incident wave into port 4. Thus, for the original unit incident voltage at port 1 and for simplicity replacing the 50Ω by 1Ω, the following net *outward* powers are obtained:

Port 1: $\left|(c^2 - 1) + c^2\right|^2$

Port 2: 0

Port 3: 0

Port 4: $\left|-2j\,c\sqrt{1 - c^2}\right|^2$ $\qquad\qquad$ (9.23)

The total power out, P, is given by

$$P = \left|2c^2 - 1\right|^2 + \left|2c\sqrt{1 - c^2}\right|^2$$

i.e. $P = 1$

the total output power being equal to the input power. Now, if $c = 0.707$ (a 3dB coupler design) then, from (9.23), it is seen that there is no reflected power at the original input port and all the power is coupled through the structure to port 4. As the input frequency is varied about the center frequency, the more general relationships for a directional coupler (8.25) must be used and it will be found that the input port is

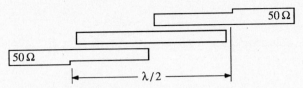

Figure 9.17 Two microstrip coupled sections forming a first-order band-pass filter. Note the reductions in line length to account for fringing capacitance.

no longer perfectly matched. Power reflected back to the input implies a less than unity transmission coefficient.

Hence, with correct dimensional choice, the structure of Figure 9.16b represents a band-pass filter where all the signal at the center frequency is transmitted with (ideally) zero insertion loss, while at other frequencies there will be varying insertion loss due to the loss of power by reflection at the input port.

If a 2-section filter, Figure 9.17, is constructed so that the output from port 4 becomes the input to the second section, then it is seen that a half-wavelength resonator section with quarter-wave coupling at each end is obtained. While it is possible to obtain a matched band-pass filter by cascading two identical 3 dB coupler sections in this manner, it turns out that other coupling ratios are also suitable, since the mismatch reflections will tend to cancel out at the center frequency. Even more generally, the condition $Z_{0e}Z_{0o} = Z_0^2$ for each filter section need not apply. In principle, it should be possible to design a band-pass filter with a desired 3 dB bandwidth by appropriately selecting the mismatch reflections for each section. An introduction to the design of band-pass filters will now follow.

9.4.2 Mapping functions

By a change of variables, it is possible to transform a band-pass response into a low-pass one and vice versa. For example, the band-pass filter response of Figure 9.18a is

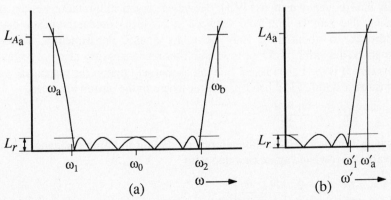

Figure 9.18 Mapping from (a) the band-pass filter response to that of (b) the low-pass prototype circuit

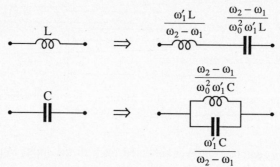

Figure 9.19 Lumped-element transformations from low-pass to band-pass circuits

exactly mapped to the low-pass prototype circuit, Figure 9.18b, by the variable transformation

$$\frac{\omega'}{\omega'_1} \leftrightarrow \frac{1}{\Omega}\left[\frac{\omega}{\omega_0} - \frac{\omega_0}{\omega}\right] \tag{9.24}$$

with $\omega_0 = \sqrt{\omega_2\,\omega_1}$ and where $\Omega = (\omega_2 - \omega_1)/\omega_0$ is the fractional bandwidth of the band-pass filter. When $\omega = \omega_1$ or ω_2 in (9.24) then $\omega'/\omega'_1 = 1$, i.e. the band-edge frequencies ω_1 and ω_2 map to the low-pass frequency ω'_1, where the attenuation is either 3 dB for the maximally-flat filter or the maximum pass-band ripple for a Chebyshev response. In terms of circuit components, the transformation in (9.24) is equivalent to transforming an inductance in the low-pass case to a series-resonant circuit in the band-pass case. Similarly, a low-pass capacitance becomes a parallel-resonant circuit, Figure 9.19.

The required series- and parallel-resonant sections in practical microstrip band-pass filters will be formed from transmission line resonators to simulate the performance of ideal resonators near resonance. Now it is a basic property of a transmission line resonator that multiple resonances exist, resulting in additional higher frequency pass-bands that are not predicted by the circuit of Figure 9.19a. For broadband filter design, it is necessary to use mapping functions that predict more accurately the frequency response [9.1]. However, practical limitations in the amount of coupling that can be achieved between microstrip resonators mean that only bandwidths up to about 15% may be realized and, consequently, narrowband approximations are valid. Very narrowband filter responses may also not be practical, but in this case it is the transmission line losses that will distort the response.

Consider a symmetrical band-pass filter where in the narrowband case

$$\omega_0 \approx \frac{\omega_1 + \omega_2}{2} \tag{9.25}$$

The mapping function (9.24), although not predicting multiple resonances, becomes an acceptable narrowband approximation with

$$\frac{\omega'}{\omega'_1} \approx \frac{2}{\Omega}\left[\frac{\omega - \omega_0}{\omega_0}\right] \tag{9.26}$$

This approximation is most accurate for small frequency deviations from ω_0 but

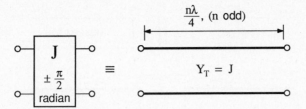

Figure 9.20 The equivalence between the immittance inverter and a transmission line that is an odd multiple of quarter-wavelengths long

will overestimate the stop-band attenuation of the transmission line filter, Figure 9.15, because the low-pass prototype does not model the higher frequency pass bands.

9.4.3 Resonators and immittance inverters

The elements of a prototype low-pass filter are alternating series inductance and shunt capacitance that transform to the series and parallel L-C resonant circuits for a band-pass filter, as illustrated in Figure 9.19. However, a practical network implementation with transmission line resonators favors the use of a tandem connection of resonators of the same type, namely open-circuited $\lambda/2$ resonant lengths of line. These resonators are joined together through networks that possess the properties of immittance inverters.

The ideal immittance inverter, characterized by the parameter, J, has the properties illustrated in Figure 9.20. The inverter has the properties of a transmission line, characteristic admittance $Y_T = J$ that is an odd-multiple of $\lambda/4$ long. As a $\lambda/4$ transformer, the transformation from the load to the input is given by

$$Z_{in} = \frac{1}{J^2} Y_L \quad \text{or} \quad Y_{in} = J^2 Z_L$$

(9.27)

The phase change for a wave transmitted through the transformer is $n(\pi/2)$ radian.

To understand the action of an immittance inverter in this context, consider the variation of susceptance with frequency for series- and parallel-resonant circuits, illustrated in Figure 9.21. From the Smith Chart, it is known that a quarter-wave

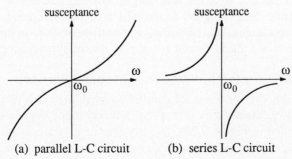

(a) parallel L-C circuit   (b) series L-C circuit

Figure 9.21 The susceptance variation of parallel- and series-resonant circuits near the resonant frequency

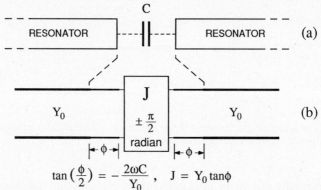

$$\tan\left(\frac{\phi}{2}\right) = -\frac{2\omega C}{Y_0} \,, \quad J = Y_0 \tan\phi$$

Figure 9.22 A series capacitance as an immittance inverter illustrated by (a) the capacitance between the ends of microstrip resonators, and (b) the immittance inverter equivalent circuit

length of transmission line will transform a normalized admittance to its reciprocal value and, with Figure 9.21a representing the frequency variation of admittance about the open-circuit value, the quarter-wave line will transform this admittance to that in Figure 9.21b, representing the admittance variation about the short-circuit value. Thus an immittance inverter, inserted in front of a series-resonant circuit, transforms it into a parallel-resonant circuit and vice versa. Unfortunately it turns out that the simple $\lambda/4$ transformer is not adequate in most cases as a practical immittance inverter, due to the limited range of characteristic impedances available in practical microstrip lines. However, other forms of inverter exist, e.g. the series capacitance of an end-gap coupled resonator, Figure 9.22, and two parallel-coupled lines that are to be described in the next section. Each of the above may be modeled by an inverter provided that additional line lengths are included to complete the model.

An end-coupled filter takes the form of several in-line $\lambda/2$ resonators with small capacitive coupling between adjacent resonators. The first and last resonators of the sequence are similarly coupled to the input and output lines. Each resonator essentially sees a high impedance at each end and may thus be modeled by a parallel-resonant equivalent circuit, §2.2.3. The equivalent circuit of the complete structure becomes an alternating sequence of immittance inverters, as in Figure 9.22, and parallel-resonant circuits with the line lengths ϕ being taken care of by resonator length adjustments. Edge-coupled filters will now be considered in greater detail because of the advantages that they offer, as described earlier. In this case, each $\lambda/2$ line functions both as a resonator and as a part of the immittance inverter.

9.4.4   Parallel-coupled lines as immittance inverters

In a microstrip line system, a pair of parallel-coupled lines with two of the ports left open circuit may be specified by their even- and odd-mode impedances and coupling length as illustrated in Figure 9.23a. As immittance inverters it is found, following Jones and Bolljahn [9.8] and Cohn [9.9], that the coupled lines are equivalent to an ideal inverter, J, together with line lengths ϕ, Figure 9.23b. The value of J is determined by Z_{0e} and Z_{0o} and may be considered substantially independent of ϕ for

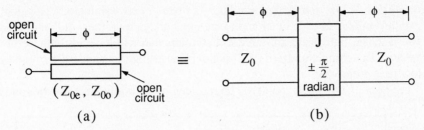

Figure 9.23 The equivalence between (a) parallel-coupled lines, and (b) an admittance inverter

$\phi \approx \pi/2$. Conversely, if a certain J is required then the Z_{0e} and Z_{0o} that are needed can be determined. With narrowband approximations that are valid if $\phi \approx \pi/2$ and useful for <10% bandwidths, J is assumed independent of frequency and the required coupled-line mode impedances are given by

$$Z_{0e} = Z_0 \left\{ 1 + J Z_0 + J^2 Z_0^2 \right\} \tag{9.28}$$

and

$$Z_{0o} = Z_0 \left\{ 1 - J Z_0 + J^2 Z_0^2 \right\} \tag{9.29}$$

where Z_0 is the characteristic impedance of the input and output connecting lines to the coupled section.

9.4.5 Band-pass filter design

Figure 9.24 illustrates the action of a parallel-coupled band-pass filter. The microstrip

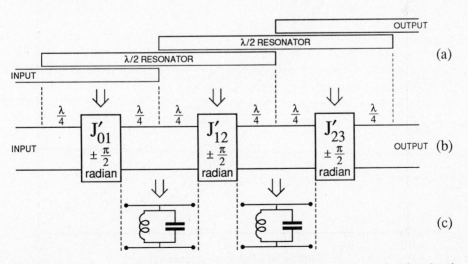

Figure 9.24 A 2-section band-pass filter with edge-coupled resonators showing (a) the microstrip layout, (b) the equivalent circuit of each $\lambda/4$-long coupler, and (c) each $\lambda/2$ line now represented by an equivalent parallel-resonant circuit

circuit, Figure 9.24a, shows two similar $\lambda/2$ open-circuit terminated resonators that are parallel coupled to each other and to the input and output lines. It is observed that for two resonant lines there are three $\lambda/4$ couplers that will have immittance inverter properties. The complete filter is a 2-section filter using the standard terminology of the low-pass prototype.

In Figure 9.24b, each coupler is represented by its immittance inverter and lines

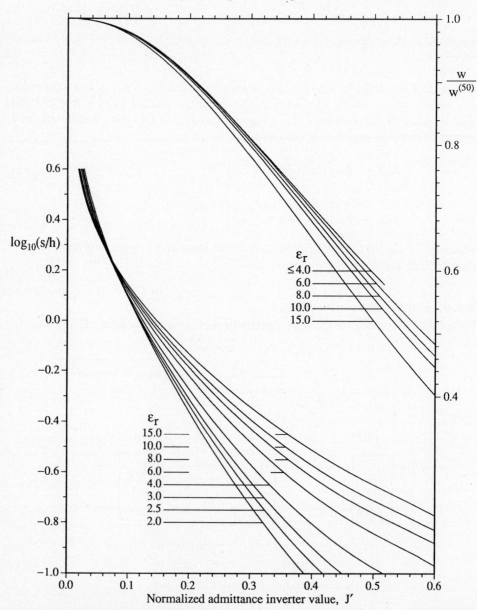

Figure 9.25 The line separation and width for a single section of an edge-coupled directional coupler matched to $50\,\Omega$ input and output lines

that are $\lambda/4$ long at the center frequency. Adjacent $\lambda/4$ lines form a $\lambda/2$ resonator. Now it is seen that the couplers not only provide the inverter function but, because of the $\lambda/4$ lines attached to the inverters, are integral parts of resonators as well. The line lengths at the input and output ports are indistinguishable from the connecting lines and are therefore of no further concern.

In a typical filter, the J values are such that the impedance levels through the filter are high with respect to Z_0. As a consequence, each $\lambda/2$ resonator in Figure 9.24b may be represented by a parallel-resonant circuit as in Figure 9.24c (see §2.2.3). The design problem is now the following: Given a low-pass prototype with element values κ_k, what should be the J values to realize it as an equivalent band-pass filter using coupled lines? Cohn [9.9, 9.10] and Matthaei et al. [9.1] have shown that the required normalized admittance inverter parameters $J' \equiv J/Y_0$ are

$$J'_{01} = \left[\frac{\pi \Omega}{2 \kappa_0 \kappa_1} \right]^{\frac{1}{2}} \qquad\qquad (9.30)$$

$$J'_{k,k+1} = \frac{\pi \Omega}{2} \times \frac{1}{\sqrt{\kappa_k \kappa_{k+1}}} \qquad k = 1, \cdots, n-1 \qquad\qquad (9.31)$$

$$J'_{n,n+1} = \left[\frac{\pi \Omega}{2 \kappa_n \kappa_{n+1}} \right]^{\frac{1}{2}} \qquad\qquad (9.32)$$

where κ_k are the low-pass prototype element values for an n-section filter and Ω is the fractional bandwidth.

The filter immittance inverters are symmetrical about the physical center of the filter for all orders of both maximally-flat and Chebyshev filters, there being $n+1$ inverters and n resonant $\lambda/2$ sections. Surprisingly, this is true even for a Chebyshev filter, with an *even* number of sections and $\kappa_0 \neq \kappa_{n+1}$, but where nevertheless $\kappa_0 \kappa_1 = \kappa_n \kappa_{n+1}$.

It now remains for the immittance inverters to be converted to the even- and odd-mode impedances for the coupled lines, and for these lines to be realized in a microstrip configuration. Normally, available equations [9.11] provide Z_{0e} and Z_{0o} as functions of s/h and w/h, whereas for design purposes the reverse is required. Using the quasi-static equations from [9.11], Figure 9.25 has been derived for this book to present s/h and w/h directly in terms of normalized immittance inverter values and with ε_r as a parameter.

Example 9.4

Design a microstrip band-pass filter with maximally-flat characteristics on a 1.5 mm thick substrate with $\varepsilon_r = 2.5$. Ignore the substrate and conductor losses. The filter specifications are (i) center frequency = 2.0 GHz, (ii) 0.1 dB passband > ±50 MHz, and (iii) attenuation at 2.5 GHz > 15.0 dB.

Solution:

The initial design will be carried out using the normalized frequency response for the low-pass equivalent with maximally-flat characteristics where, at the 3 dB point, $\omega_1' = 1$. Let ω_p' and ω_s' be the frequencies for the pass- and stop-band attenuations of 0.1 dB and 15.0 dB respectively. ω_p' depends on the number of sections in the filter. To satisfy the 0.1 dB pass-band attenuation requirement, then from (9.4) with $\omega_1' = 1$

$$0.1 \ = \ 10\,log_{10}\big(1 + (\omega_p')^{2n}\big)$$

i.e. $$(\omega_p')^{2n} \ = \ 0.02329$$

The stop-band attenuation specification must be satisfied at $\omega_s' = 10 \times \omega_p'$, at which frequency the attenuation is calculated and compared with the required minimum of 15 dB. The results are given in Table 9.6.

Table 9.6　Results for Example 9.4

n	ω_p'	ω_s'	Attenuation at ω_s'	Upper f_{3dB} GHz
1	0.153	1.53	5.2	2.328
2	0.391	3.91	23.7	2.128
3	0.625	6.25	47.8	2.080
2	-	3.15	20.0	2.159

　　A minimum of two sections is required for the filter to achieve the stop-band specification. For n = 2, the design just meets the specification within the pass band while it is significantly better than the specification for the stop band. Thus the stop-band attenuation may be reduced and is chosen as 20 dB. The results are given in the final row of the table. At the same time, the theoretical pass-band attenuation will also be reduced and is now less than 0.043 dB across the band.

　　The microwave band-pass filter design requires knowledge of the fractional 3 dB bandwidth. Thus the upper 3 dB frequency is also evaluated and is given in the final column of the table. The low-pass prototype values for an n = 2 maximally-flat filter are $\kappa_0 = 1.0$, $\kappa_1 = 1.414$, $\kappa_2 = 1.414$, and $\kappa_3 = 1.0$. The fractional 3 dB bandwidth, $\Omega = (2.159 - 1.841)/2.00$, i.e. $\Omega = 0.159$.

From (9.30), $$J_{01}' \ = \ J_{23}' \ = \ \left\{ \frac{\pi \times 0.159}{2 \times 1 \times 1.414} \right\}^{\frac{1}{2}}$$

$$= \ 0.420$$

From (9.31), $$J_{12}' \ = \ \frac{\pi \times 0.159}{2} \times \frac{1}{\sqrt{1.414 \times 1.414}}$$

$$= \ 0.177$$

From Appendix 3, for a 50 Ω line, $w^{(50)} = 1.5 \times 2.837 = 4.26$ mm. From Figure 9.25, for $J' = 0.420$, $log_{10}(s/h) = -1.00$ and $w/w^{(50)} = 0.676$ giving s = 0.15 mm and w = 2.88 mm. Likewise, for $J' = 0.177$, it is found that s = 0.85 mm and w = 3.95 mm. The filter layout is shown (not to scale) in Figure 9.26. The

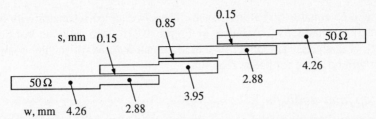

Figure 9.26 The line width (w) and separation (s) for a 2-section band-pass filter on $\varepsilon_r = 2.5$ substrate, 1.5 mm thick. Line length adjustments for fringing capacitance at the open circuits and step changes in line width have not been incorporated. *(Not to scale.)*

effective permittivity, and thus the length, for each coupler is found as the geometric mean of $\varepsilon_{eff}^{(e)}$ and $\varepsilon_{eff}^{(o)}$, as derived from Appendix 4. Adjustments to the resonator lengths for the fringing capacitance at the open circuits and step changes in the line width will complete the design.

9.5 BAND-STOP FILTERS

An efficient band-stop filter allows most signals to pass through while only attenuating a narrow band of frequencies. Design procedures for several types of band-stop filter structure are given by Schiffman and Matthaei [9.12]. An introduction to band-stop filters which builds on knowledge of parallel-coupled lines from the previous chapter is presented here. Consider the $\lambda/2$ long microstrip resonator which is terminated at each end by open circuits. From (8.25), there will be no coupling if the complete $\lambda/2$ length is parallel-coupled to the main transmission line. Maximum coupling to the resonator is achieved when only a $\lambda/4$ length of the resonator is parallel-coupled to the main line in a similar manner to a directional coupler. The remaining $\lambda/4$ length, at right angles here to the main line to prevent further coupling across to it, is required if the band-stop filter is to have an open-circuit termination at each end of the resonator. This is illustrated in Figure 9.27,

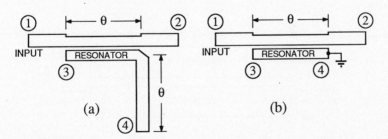

Figure 9.27 Microstrip band-stop filters showing (a) the practical realization with open-circuit terminations to the resonator, and (b) the coupled section as a resonator with one end short-circuited

where θ will be equal to 90° at resonance. The $\lambda/4$ coupled resonator with open- and short-circuit terminations, as shown in Figure 9.27b, will now be considered for simplicity in analysis. The effect of the additional $\lambda/4$ length for the complete $\lambda/2$ resonator will be noted for use as required.

Band-stop filter analysis

The band-stop filter characteristics of an ideal pair of parallel-coupled lines, Figure 9.28, that satisfy the requirements for a directional coupler in §8.2, namely

$$Z_{0e}Z_{0o} = Z_0^2 \qquad (9.33)$$

are considered. In general, the electrical length of the coupled section is θ. A unit incident voltage wave at port 1 sets up voltage waves at ports 2 and 3, with port 4 isolated. The coupling at port 3 from (8.25), including the θ dependence, is

$$d = \frac{j\mathbf{c}\sin\theta}{D} \qquad (9.34)$$

where $\qquad D = \sqrt{1 - \mathbf{c}^2}\cos\theta + j\sin\theta \qquad (9.35)$

The transmission out of port 2

$$t = \frac{\sqrt{1 - \mathbf{c}^2}}{D} \qquad (9.36)$$

These waves may be visualized as being in Z_0 lines of zero length, assumed attached to the ports of the coupler just in front of the actual termination at that port. Port 2 is matched and will not have any input wave. Ports 3 and 4 are terminated by an open circuit and a short circuit respectively and will both reflect waves back into the coupling region.

Let the total input waves at ports 3 and 4 be u and v respectively. The values of u and v are still to be calculated from the boundary conditions. These input waves and their transmitted and coupled components are illustrated in Figure 9.28. Applying the boundary condition for a short circuit at port 4

$$v + ut = 0 \quad \Rightarrow \quad v = -ut \qquad (9.37)$$

and for an open circuit at port 3

$$u = d + vt = d - ut^2 \qquad (9.38)$$

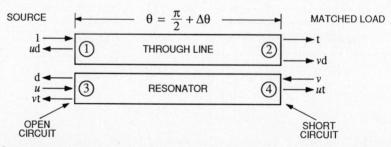

Figure 9.28   Voltage waves in zero-length Z_0 lines attached to the ends of the coupled lines of the band-stop filter of Figure 9.27b

Therefore $\quad u = \dfrac{d}{1+t^2}, \quad v = \dfrac{-dt}{1+t^2}$ $\qquad\qquad$ (9.39)

The voltage reflection coefficient at port 1

$$\Gamma = ud = \frac{d^2}{1+t^2} \qquad\qquad (9.40)$$

The voltage transmission coefficient through to port 2

$$T = t + vd = t\left\{1 - \frac{d^2}{1+t^2}\right\} \qquad\qquad (9.41)$$

When $\theta = \pi/2$ and the coupled length is $\lambda/4$, then

$$d = c \quad \text{and} \quad t = -j\sqrt{1-c^2} \qquad\qquad (9.42)$$

giving, from (9.39)

$$u = \frac{1}{c} \quad \text{and} \quad v = +j\frac{\sqrt{1-c^2}}{c} \qquad\qquad (9.43)$$

Now from (9.40) and (9.41), the reflection and transmission coefficients are $\Gamma = 1$ and $T = 0$ as required at the center frequency of an ideal band-stop filter.

The filter bandwidth

Consider the case where $\theta = \dfrac{\pi}{2} + \Delta\theta$, so that $\sin\theta = \cos\Delta\theta$ and $\cos\theta = -\sin\Delta\theta$. From (9.34) to (9.36)

$$D = -\sqrt{1-c^2}\sin\Delta\theta + j\cos\Delta\theta$$

and $\qquad d = \dfrac{jc\cos\Delta\theta}{D}, \quad t = \dfrac{\sqrt{1-c^2}}{D} \qquad\qquad$ (9.44)

The 3 dB bandwidth will be determined by the conditions $|\Gamma|^2 = |T|^2 = \frac{1}{2}$, i.e. half of the incident power is reflected and half is transmitted to the load. Substituting for d and t from (9.44) into (9.40), it follows (see Exercise 9.11) that

$$|\Gamma|^2 = \frac{c^4\cos^4\Delta\theta}{4(1-c^2)^2\sin^2\Delta\theta + c^4\cos^4\Delta\theta} \qquad\qquad (9.45)$$

For $|\Gamma|^2 = \frac{1}{2}$, then

$$4(1-c^2)^2\sin^2\Delta\theta = c^4\cos^4\Delta\theta \qquad\qquad (9.46)$$

This equation may be solved exactly for $\sin^2\Delta\theta$. However, for narrowband filters, from (9.46), $\Delta\theta \ll 1 \Rightarrow c \ll 1$ and

$$\sin^2\Delta\theta \approx (\Delta\theta)^2 \approx \frac{c^4}{4} \qquad\qquad (9.47)$$

For the full bandwidth, $2\Delta\theta = c^2$. Hence the quality factor of the filter, defined as the ratio of the resonant frequency to the 3 dB bandwidth for a filter exhibiting a single resonance, is given by

$$Q = \frac{f_0}{\Delta f} \approx \frac{(\pi/2)}{2\Delta\theta} = \frac{\pi}{2c^2} \qquad\qquad (9.48)$$

The analysis for the structure in Figure 9.27a follows the same steps, except that (9.37) is rewritten by applying the boundary condition at the open circuit at port 4 and transferring the terms back to the port 4 end of the coupled section. Thus (9.37) now becomes

$$v = ut(\cos 2\theta - j\sin 2\theta) \qquad (9.49)$$

This equation reduces to (9.37) at resonance when $\theta = \pi/2$.

Example 9.5

The band-stop filter in Figure 9.27b is constructed with line dimensions as for a 13 dB directional coupler. Calculate the 3 dB bandwidth from an exact solution of (9.46), if the center frequency is 2.0 GHz. Compare with the approximate result given by (9.48).

Solution:

For a 13 dB coupler, $c = 0.2239$. Thus (9.46) becomes

$$3.61 \sin^2\Delta\theta = 0.002512 (1 - \sin^2\Delta\theta)^2$$

With $s \equiv \sin^2\Delta\theta$

$$0.002512\, s^2 - 3.6091\, s + 0.002512 = 0$$

Solving for s, taking the solution where $s \leq 1$ gives $s = 0.000695$, i.e. $\sin\Delta\theta \approx \Delta\theta = \sqrt{s} = 0.0264$. This compares favorably with the approximate value of 0.0251 given from (9.47). The 3 dB bandwidth at 2.0 GHz

$$\Delta f = \frac{2\Delta\theta}{(\pi/2)} \times f_0 = \frac{2 \times 0.0264}{(\pi/2)} \times 2 \ \text{GHz}$$

i.e. $\qquad \Delta f = 67\,\text{MHz}$

Resonator losses in a practical filter include conductor and dielectric losses, as well as radiation losses from the open-circuit terminations. These losses have most effect on the performance of narrowband filters where they will (i) increase the filter 3 dB bandwidth, (ii) reduce the magnitude of the reflected signal at resonance, and (iii) reduce the band-stop attenuation at resonance.

EXERCISES

9.1   i)   Calculate the 3 dB frequency for a 5-section low-pass filter with maximally-flat characteristics that has a 0.2 dB attenuation at 1.0 GHz.

   ii) Repeat for a Chebyshev response with a 0.2 dB passband ripple and 30 dB attenuation at 1.0 GHz.

9.2   Repeat Exercise 9.1 for a high-pass filter.

9.3   Repeat Exercise 9.1 for a band-pass filter with 1.0 GHz replaced by 10.0 ± 1.0 GHz.

9.4   Sketch the microstrip layout for Exercise 9.1(i), using a 1.5 mm thick, $\varepsilon_r = 2.5$ substrate and 20 Ω and 150 Ω lines.

9.5   Calculate the reflection coefficient at low frequencies for a Chebyshev low-pass filter with a 0.6 dB ripple, having (i) two sections and (ii) three sections.

9.6 Consider the circuit in Figure 9.16. Describe its behavior:
  i) when port 3 has a short-circuit termination and port 2 remains open circuit,
  ii) when both ports 2 and 3 have short-circuit terminations.

9.7 For the circuit illustrated in Figure 9.16 with the line parameters giving 3 dB coupling, calculate the filter response as a function of frequency. In particular, determine the bandwidth of the filter.

9.8 Repeat Exercise 9.7 with the central coupling section being as for a 15 dB directional coupler.

9.9 Determine the scattering parameters at resonance of the circuit shown below, in which two sections with 15 dB coupling as in Exercise 9.8 are cascaded.

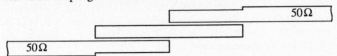

9.10 Assuming lossless and zero thickness lines, calculate the resonant frequency and 3 dB bandwidth for the circuit illustrated below, with the parameters w/h as for a 50 Ω line, $\varepsilon_r = 2.5$, s/w = 0.1, h = 1.5 mm, and $l = 50$ mm $\approx \lambda/2$ at the center frequency.

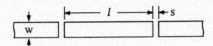

9.11 i) For the band-stop filter illustrated in Figure 9.27b, derive the expression for $|\Gamma|^2$ in (9.45).
  ii) Derive the corresponding expression for $|T|^2$ and check that

$$|\Gamma|^2 + |T|^2 = 1$$

9.12 i) Show how a band-stop filter may be constructed using a pair of identical band-pass filters, two matched loads and a 3 dB quadrature coupler.
  ii) Repeat part (i), but with a 3 dB, 180° hybrid coupler.

9.13 Compute all the critical dimensions for the band-stop filter shown in Figure 9.27b. The filter has 3 dB frequencies of 2.0 GHz ± 20 MHz and 50 Ω connecting lines. It is to be constructed on a substrate with $\varepsilon_r = 2.5$, h = 1.5 mm.

REFERENCES

[9.1] Matthaei, G. L., Young, L. and Jones, E. M. T., *Microwave Filters, Impedance-matching Networks, and Coupling Structures*, McGraw-Hill, New York, 1964.

[9.2] Malherbe, J. A. G., *Microwave Transmission Line Filters*, Artech House, Norwood, MA, 1979.

[9.3] Saal, R., *Der Entwurf von Filtern mit Hilfe des Kataloges Normierten Tielpasse*, Telefunken GMBH, Backnang, West Germany, 1963.

[9.4] Zverev, A. I., *Handbook of Filter Synthesis*, Wiley, New York, 1961.

[9.5] Williams, A. B., *Electronic Filter Design Handbook*, McGraw-Hill, New York, 1961.

[9.6] Ingalls, M. and Kent, G., "Monolithic capacitors as transmission lines", *IEEE Trans. Microwave Theory and Techniques*, Vol. MTT-35, No. 11, November 1987, pp. 964-70.

[9.7] Howe, H. Jr., *Stripline Circuit Design*, Artech House, Dedham, MA, 1974.

[9.8] Jones, E. M. T. and Bolljahn, J. T., "Coupled-strip-transmission-line filters and directional couplers", *IRE Trans. Microwave Theory and Techniques*, Vol. MTT-4, No. 2, April 1956, pp. 75-81.

[9.9]   Cohn, S. B., "Parallel-coupled transmission-line-resonator filters", *IRE Trans. Microwave Theory and Techniques*, Vol. MTT-6, No. 2, April 1958, pp. 223-31.

[9.10]   Cohn, S. B., "Direct-coupled-resonator filters", *Proc. IRE*, Vol. 45, No. 2, February 1957, pp. 187-96.

[9.11]   Kirschning, M. and Jansen, R. H., "Accurate wide-range design equations for the frequency-dependent characteristic of parallel coupled microstrip lines", *IEEE Trans. Microwave Theory and Techniques*, Vol. MTT-32, No. 1, January 1984, pp. 83-90. Corrections: *IEEE Trans. Microwave Theory and Techniques*, Vol. MTT-33, No. 3, March 1985, p. 288.

[9.12]   Schiffman, B. M. and Matthaei, G. L., "Exact design of band-stop microwave filters", *IEEE Trans. Microwave Theory and Techniques*, Vol. MTT-12, No. 1, January 1964, pp. 6-15. Addendum: *IEEE Trans. Microwave Theory and Techniques*, Vol. MTT-12, No. 3, May 1964, pp. 369-82. Corrections: *IEEE Trans. Microwave Theory and Techniques*, Vol. MTT-13, No. 5, September 1965, p. 703.

10 Miscellaneous components

10.1 INTRODUCTION

A selection of circuit elements and components that do not fit readily into the contents of other chapters is presented here. In §10.2, transitions between microstrip and other types of transmission lines are described. Not only are these transitions used at an interface between microstrip lines and the outside environment but also, in the case of slot lines, as a part of the circuit in the microstrip structure. Microstrip antennas are also basically a transition between propagation media. Here, only the important resonant patch antenna, from a choice of many antenna designs, is discussed in §10.7.

Lumped components, L, C and R, in §10.3 have a wide variety of uses, e.g. in matching techniques, as feedback elements in amplifiers and oscillators and in bias networks and filters. In §10.4, we consider power dividers for use as power splitting elements in antenna arrays and for the inputs to parallel high-power amplifiers, as well as their use as power combiners at the outputs of such amplifiers.

Microstrip resonators, fabricated as lines forming a part of the microstrip circuit, are a fundamental part of Chapter 9. Another type of resonator is the dielectric resonator, §10.6, generally implemented as a loosely coupled element above the microstrip line. This type has applications in high-Q filters and stable oscillator design. Other resonant modes, this time in a ferrite disc that replaces a part of the substrate itself, are important in the understanding of the non-reciprocal performance of circulators considered in §10.5.

10.2 LAUNCHING TECHNIQUES

With microwave subsystems being constructed using microstrip lines, it may be necessary to interface to other forms of transmission line, either for measurement purposes or for overall system design, as for example, for connection to an antenna. It is typical to use coaxial connections below about 5 GHz, waveguide connections above 20 GHz and either type of line in the intermediate frequency range. Basic transitions to coplanar lines are also included for completeness, in particular to slot line and coplanar waveguide, see e.g. Gupta et al. [10.1], even though these lines are otherwise beyond the scope of this book. The detailed design of any transition to a microstrip line will necessarily depend on such parameters as frequency, bandwidth, line impedances and substrate properties, leading to a variety of designs. In the following sections, typical structures are illustrated.

231

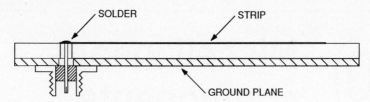

Figure 10.1   A coaxial to microstrip connection through the ground plane

10.2.1   Coaxial line to microstrip transition

The design illustrated in Figure 10.1 is a low frequency design that is only suitable below about 2 GHz using low permittivity ($\varepsilon_r \leq 2.6$) substrate materials. It may also be used as a coaxial feed to a microstrip antenna element, described in §10.7. The coaxial line is fed through the ground plane and substrate material and the inner conductor is soldered directly to the microstrip line. With the coaxial connector firmly attached to the ground plane, good electrical contact is maintained with the ground conductor.

For operating frequencies above 1 GHz, it is preferable to use the method of England [10.2] illustrated in Figure 10.2. The field variations through the transition are smoother than for the previous design. This is particularly so when the microstrip line is fabricated on 0.5 mm sapphire substrate. Coaxial connectors that have small tags, or continuations of the circular center conductor beyond the face of the body of the connector, are attached to the microstrip line with either a pressure contact or a solder joint. Specialized equipment may be required to determine the precise dimensions associated with this termination if reflection coefficients less than 0.01 are to be achieved. However, the following general precautions should be noted:

i)   Avoid a mechanical design that may destroy the extension of the coaxial line center conductor.

ii) Maintain good electrical contact at the points where the microstrip ground plane contacts the inner circumference of the outer conductor of the coaxial line.

iii) Reduce the capacitance of the center conductor and microstrip line in the vicinity of the transition, since it is in this region that there is generally excess capacitance due to the fringing fields from the coaxial line center conductor to the microstrip ground plane and from the microstrip line to the body of the coaxial connector. The capacitance is reduced by tapering the width of the microstrip line so that it is no wider at the edge of the substrate than the width of the pin from the coaxial line.

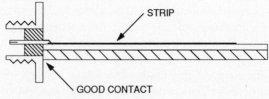

Figure 10.2   An edge-feed coaxial to microstrip connection, from England [10.2] (© 1976, IEEE)

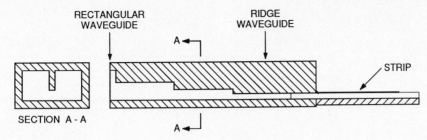

RECTANGULAR WAVEGUIDE

RIDGE WAVEGUIDE

A

STRIP

SECTION A - A

A

Figure 10.3 A rectangular waveguide to microstrip transition, from Van Heuven [10.4] (© 1976, IEEE)

10.2.2 Rectangular waveguide to microstrip transition

Some waveguide to microstrip transitions may be regarded as a combination of a waveguide to coaxial transition followed by coaxial to microstrip transition, with the coaxial section often approaching a zero length. Such designs are not illustrated here. However, there are two designs which make use of the actual field distributions in the waveguide and microstrip lines. In the first design by Schneider et al. [10.3], illustrated in Figure 10.3, several dimensions are critical for good performance.

The rectangular waveguide is transformed through a series of quarter-wave transformers to a ridge waveguide, where the final ridge section is separated from the opposite plane face by the microstrip substrate material. In this design, it is necessary to maintain good electrical contact between the ground plane and waveguide wall, as well as between the ridge and the microstrip line. There is still an abrupt discontinuity at the end of the waveguide, even though the transition has effectively modified the waveguide fields to match those of the microstrip line under the strip itself.

A second design for 18-26 GHz by Van Heuven [10.4] transforms the waveguide through a parallel plate structure and balun to the microstrip line. This design has the advantage that the transition structure may be fabricated as a part of the microstrip circuit.

10.2.3 Microstrip to slot-line transition

The slot line for microwave transmission [10.5] may be used in conjunction with microstrip lines to give a greater flexibility in component and system design (e.g. see [10.6]). Only the basic transition design is presented here.

Refer to Figure 10.4a and consider a microstrip line with a short circuit termination through the substrate to the ground plane. Further, consider a narrow slot in the ground plane, passing at right angles under the microstrip line and adjacent to the short circuit. For an input signal along the microstrip line, an electric field will be excited at right angles to the direction of the slot, setting up a guided wave that will propagate in both directions along the slot line. If the impedance looking in one direction along the slot line appears as an open circuit at the plane of the microstrip transition, then there will be a net signal propagating in the other direction. For low reflections from the transition, the characteristic impedances of the slot line and microstrip line must be equal. On any given substrate, the effective permittivity for a

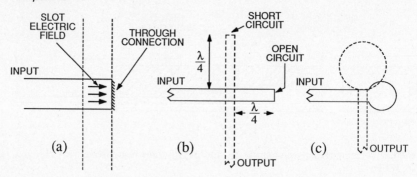

Figure 10.4 The microstrip to slot-line transitions, where solid lines represent the microstrip circuits and the dashed lines show the extent of the slots in the ground plane

microstrip line will be greater than for the slot line, since with the slot line a greater percentage of the electromagnetic energy flows outside the substrate.

The basic transition in Figure 10.4b is formed using $\lambda/4$ lines in conjunction with terminations that are easiest to fabricate, namely a microstrip open circuit and a slot-line short circuit. The difference in physical lengths of the $\lambda/4$ lines is due to the difference in effective permittivities. Broader bandwidth designs that continue to be easy to fabricate, Figure 10.4c, use circular or radial elements to give the required junction impedances [10.7].

10.2.4   Microstrip to coplanar waveguide transition

A coplanar waveguide has three parallel conductors in the one plane, with the two outer conductors as a ground plane that ideally extends out to infinity. The TEM-mode of propagation in a coplanar waveguide has electric field lines from the center conductor to ground and magnetic field lines around the center conductor.

In the transition by Riaziat et al. [10.8], illustrated in Figure 10.5, the strip of the microstrip line continues through to become the central conductor of the coplanar

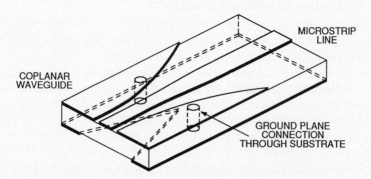

Figure 10.5 A microstrip to coplanar waveguide transition, from Riaziat et al. [10.8] (Reprinted with permission of *Microwave Journal*, from the June 1987 issue, © 1987 Horizon House, Inc.)

waveguide. Ground conductor symmetry must be maintained at all planes along the transition to avoid setting up parallel-plate modes and unwanted higher-order modes in the coplanar waveguide. Suppression of the microstrip mode beyond the transition region is achieved by using plated-through holes between the upper and lower ground planes. These connections effectively short out any normal component of electric field and will aid the current flow between the two configurations of the ground plane. The microstrip ground plane may also be tapered away from the center line to leave a true coplanar waveguide.

10.3 LUMPED COMPONENTS

Circuits built with microstrip components can be extremely bulky at the lower microwave frequencies. To reduce the size of the complete circuit, it may be preferable to use lumped L and C components. An alternative view is to observe that for a given physical area, a greater range of component values is possible with lumped components. A similar physical size constraint manifests itself also at higher microwave frequencies in monolithic microwave integrated circuits (MMICs) where, with microstrip structures, too great an area of GaAs substrate would be required and lumped components could be preferred.

Whether a structure may be represented by a lumped equivalent circuit is determined by its physical size in relation to the wavelength. The key equation to consider in this context is again (3.26). In Chapter 3 it was shown that (3.26) reduces to Laplace's Equation when the physical dimensions of the component are very much less than a wavelength. The behavior of the component may then be properly described by a lumped L, C or R, as the case may be. In practice, the dimensions should be less than $\lambda/20$ [10.9].

An advantage of lumped elements is that their use facilitates the achievement of broadband designs, since they present equivalent circuits that are not frequency dependent, in contrast with fixed length transmission line sections. On the other hand, when lumped elements are used at frequencies approaching the limit of their applicability, the simple L, C or R equivalent has to be modified by the inclusion of parasitic elements. Pettenpaul et al. [10.10] found that several formulae were needed for an accurate component description that included all parasitic effects. Another disadvantage of lumped elements is that they may require more processing steps in manufacture: e.g. capacitors may require dielectric overlays, spiral inductors may require cross-over paths. The lumped elements may come in film form, deposited on the dielectric substrate, or as chip components [10.11] to be attached to the circuit, as in non-monolithic (i.e. hybrid) microwave integrated circuits (MICs).

Lumped capacitors are illustrated in Figure 10.6. The interdigital capacitor, Figure 10.6a, relies on the strip-to-strip capacitance of parallel conducting fingers on a substrate, with the capacitance of a number of finger pairs combined in parallel. Larger capacitances can be obtained with the overlay structure in Figure 10.6b, but at the expense of additional processing steps. For by-pass and coupling capacitors, even larger capacitance values are often required. These can be achieved with the chip capacitor shown in Figure 10.6c. There is the possibility of two ways of mounting

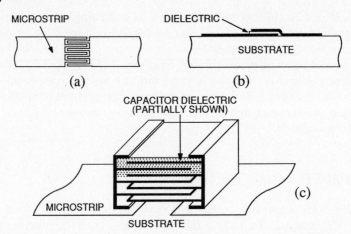

Figure 10.6 A capacitor as a discrete circuit element with (a) a plan view of an interdigital capacitor, (b) a cross-section of an overlay capacitor, and (c) a chip capacitor

chip capacitors: either as shown in Figure 10.6c or with the chip rotated by 90°, so as to have the capacitor plates at right angles to the substrate. As shown in Figure 10.6c, a chip capacitor can be analyzed [10.12] as a folded transmission line with periodic discontinuities at the folds, leading to multiple periodic resonances at higher frequencies. The latter orientation mentioned above eliminates some of the resonances [10.12].

A simple ribbon inductor is identical to a short length of high impedance microstrip line which has been discussed in earlier chapters. For such an inductor, the inductance is proportional to length and larger inductances are obtained by bending the strip in a meander line or a loop as illustrated in Figure 10.7a and b. Unfortunately, mutual coupling has now to be taken into account and behavior as a simple inductor is lost, especially if the strips are relatively close together. Much larger inductances are obtained by inductive loops, either square as in Figure 10.7c, or circular, but parasitic effects are now much more pronounced. Spiral inductors with

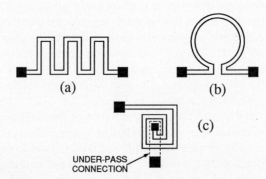

Figure 10.7 Inductors as discrete components constructed with (a) a meander line, (b) a single loop, and (c) a (square) spiral

the turns on top of each other and separated by dielectric spacers also come in the chip form [10.11].

Lumped resistors are required for biasing active circuits and as terminations for certain ports in multi-port devices, such as directional couplers and the power splitters of §10.4.

Several approximate formulae for inductance and capacitance values of lumped components are available, e.g. in [10.13, 10.14], while more elaborate formulations are quoted, e.g. in [10.9, 10.15], together with experimental results [10.9]. A very detailed treatment that is suitable for computer-aided design is to be found in [10.10].

10.4 POWER DIVIDERS AND COMBINERS

An n-way hybrid power divider with in-phase outputs was originally described by Wilkinson [10.16]. The basic 2-way divider, and some of the techniques for increasing its useful bandwidth, will be described here. Being a reciprocal device, it may also be used as a low-loss power combiner of identical signals, such as may result in an antenna array or from the outputs of a pair of parallel-connected transistor power amplifiers.

Consider the circuit configuration shown in Figure 10.8. The circuit is a four-port network where one of the ports is permanently terminated by a resistor, R. With the correct value for this resistor, the network can be matched for incident waves at all the other ports. In the circuit design, the two Z_T lines are kept apart so that there is negligible edge coupling between them. However, they must be brought close together again at plane B where port 4 is terminated with a resistor.

At the junction A, the two output lines appear in parallel with a common voltage applied to each of them. There will be a good input match and equal power division if the input impedance of the individual output networks at A is 100Ω. The lines from A to B are $\lambda/4$ transformers that transform the 50Ω load impedances at ports 2 and 3 to the 100Ω required at the input. Thus $Z_T = 70.7\,\Omega$. With identical potentials at all times on each terminal of the resistor at B, there will be no power loss within the resistor. The resistor value does not influence the input match as seen at port 1. However, its value is important if minimal reflections are also desired for any signal that may enter the divider from ports 2 or 3.

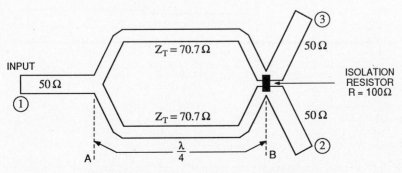

Figure 10.8 An equal-split power divider

Consider feeding in an input signal, balanced with respect to the ground plane, at the resistor port (port 4), but without the resistor present. By symmetry, there will be no output voltage at A and thus at port 1. This zero voltage at A will transform via the $\lambda/4$ transformers to zero currents at B, Exercise 1.10, so that the two lines to the left of B present open circuits to port 4. Ports 2 and 3 appear in series across port 4 and, since the ground plane is at the center point of the series combination, the two output signals will be out-of-phase. The input impedance for a signal applied to port 4 is the series combination of $2\times50\Omega$, i.e. 100Ω. Thus port 4 must be terminated by this impedance if, by reciprocity, there are to be no reflections for two equal-magnitude out-of-phase signals when they are fed into ports 2 and 3. Hence, $R = 100\Omega$. Since arbitrary inputs into ports 2 and 3 may be regarded as the sum of pairs of in-phase and out-of-phase signals, and since the in-phase signals will see perfect matches at ports 2 and 3 for any value of R, $R = 100\Omega$ will thus provide matching for any signals into ports 2 and 3.

The resistive termination on port 4 serves also to isolate the two output ports so that, for example, an input at port 2 will not produce any output at port 3. Again, this is easiest to see in terms of the reciprocity principle. If two equal power input signals are fed into ports 1 and 4 with a phase relationship that gives a sum output at port 2 and zero output at port 3 then, by reciprocity, an input at port 2 will give equal outputs at port 1 and into the resistor at port 4, leaving port 3 isolated. This has also been referred to, in terms of scattering parameters, in Exercise 2.7. When the circuit in Figure 10.8 is used as a power combiner, the in-phase components of two input signals at ports 2 and 3 appear at port 1, while the out-of-phase components are dissipated in the isolation resistor.

It is seen in the foregoing discussion that there are even- or odd-mode outputs depending on whether port 1 or 4 is used for the input. Parad and Moynihan [10.17] used an even- and odd-mode analysis to determine the line impedances that are required for an arbitrary power division and with an increased bandwidth compared with the basic design in Figure 10.8. Their results are summarized in Figure 10.9,

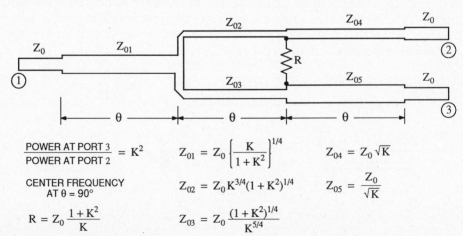

$$\frac{\text{POWER AT PORT 3}}{\text{POWER AT PORT 2}} = K^2 \qquad Z_{01} = Z_0\left[\frac{K}{1+K^2}\right]^{1/4} \qquad Z_{04} = Z_0\sqrt{K}$$

$$\text{CENTER FREQUENCY} \atop \text{AT }\theta = 90° \qquad Z_{02} = Z_0 K^{3/4}(1+K^2)^{1/4} \qquad Z_{05} = \frac{Z_0}{\sqrt{K}}$$

$$R = Z_0\frac{1+K^2}{K} \qquad Z_{03} = Z_0\frac{(1+K^2)^{1/4}}{K^{5/4}}$$

Figure 10.9 Design equations for an improved split-tee power divider, from Parad and Moynihan [10.17] (© 1965, IEEE)

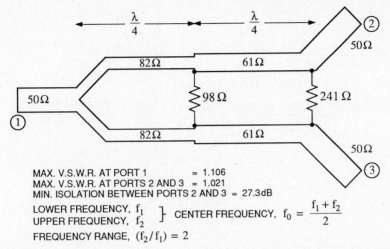

Figure 10.10 A broadband equal-split power divider, from Cohn [10.18] (© 1968, IEEE)

where it is seen that in the special case of equal power split, the two output transformers, Z_{04} and Z_{05} are not required (being $50\,\Omega$), $Z_{02} = Z_{03} = 59.5\,\Omega$, but a transformer on the input line is required with $Z_{01} = 42.0\,\Omega$.

Cohn [10.18], with an even- and odd-mode analysis that allows for more than one fixed resistor, shows how the network for equal-power division can be improved further. The analysis uses Chebyshev polynomials to give an equi-ripple V.S.W.R. for each port across the pass band. The circuit design for a 2-section divider that has an octave bandwidth is shown in Figure 10.10.

10.5 CIRCULATORS

A microstrip Y-junction circulator, illustrated in Figure 10.11, has three identical ports. At the junction where the lines meet, there is a non-reciprocal element formed from ferrite material biased by a d.c. magnetic field. Power through the circulator flows in accordance with the port sequence: $1 \rightarrow 2 \rightarrow 3 \rightarrow 1$. Thus the scattering parameter matrix (magnitudes only) for the ideal circulator is

$$[S] = \begin{bmatrix} 0 & 1 & 0 \\ 0 & 0 & 1 \\ 1 & 0 & 0 \end{bmatrix}$$

(10.1)

Typical performance figures for a practical circulator are input VSWR < 1.2, insertion loss < 0.5dB and isolation > 20dB over a wide frequency range, say an octave bandwidth. The circulator may be used as an isolator when one of the ports, say port 3, is terminated with a match load. As an isolator, it may be used as a constant impedance load for output impedance sensitive devices since the input impedance to the circulator is now insensitive to the impedance attached to the through port (see Exercise 10.2).

At the junction where the three symmetrical microstrip lines meet, a

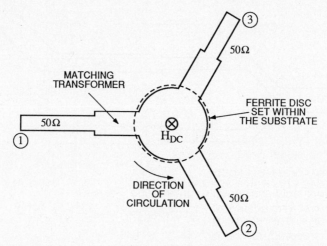

Figure 10.11   A symmetrical three-port junction circulator

transversely magnetized ferrite disc, with a d.c. magnetic field applied along the axis of the disc, replaces the normal substrate material. Fay and Comstock [10.19] describe in detail how a resonant field configuration may exist within the disc in the absence of an applied magnetic field. The fields of the resonant mode, shown in Figure 10.12a, may be pictured as the standing wave pattern generated by two contra-rotating waves. Each rotating wave is itself a resonant mode, with the wave being in-phase with itself after one complete revolution. For an input wave at port 1, part of the wave will be reflected at that port and there will be equal magnitude transmitted waves to ports 2 and 3.

A d.c. magnetic field is applied to the ferrite material which now exhibits a tensor permeability. The two contra-rotating waves will have different phase coefficients and thus the resonant frequencies of the two rotating modes are separated. The circulator is operated at a frequency between the two resonant modes, where the

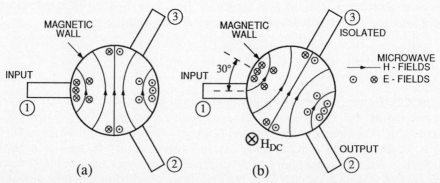

Figure 10.12   The dipolar mode of a ferrite disc, with (a) no d.c. applied magnetic field, and (b) with an applied d.c. magnetic field rotating the mode to isolate port 3, from Fay and Comstock [10.19] (© 1965, IEEE)

higher- (lower-) frequency resonant mode will have an inductive- (capacitive-) reactance component at the input. This results in a non-zero phase difference between the electric and magnetic fields for each rotating wave at the input, port 1. For one of the waves, the electric field leads the magnetic field (say by 30°) while for the other wave it lags by the same amount. Thus the electric field associated with the standing wave formed by the two rotating modes no longer coincides with the input port, but is moved 30° away. In effect, the standing wave pattern in the disc has been rotated, such that there is a null electric field component at (and no transverse magnetic field with respect to) the isolated port. A simplified view of this situation is shown in Figure 10.12b.

Effective coupling between the individual ports and the ferrite disc for a broadband, low loaded-Q, circulator involves a transformation from the 50Ω line to a lower impedance at the disc perimeter. This increases the coupling to the resonator and is achieved with a quarter-wave transformer as shown in Figure 10.11 or with a section of tapered line.

10.6 DIELECTRIC RESONATORS

Microstrip line filters with band-pass and band-stop characteristics were described in Chapter 9. In those filters, the basic resonator element was a $\lambda/2$ length of transmission line with an open-circuit termination at each end. In this section, the resonant lengths of microstrip line are replaced by high-permittivity, low-loss, dielectric resonators that improve the filter characteristics by:

i) reducing the physical dimensions of the filter,

ii) avoiding the problems associated with unequal even- and odd-mode phase velocities for coupled lines,

iii) improving the temperature stability with appropriate dielectric materials,

iv) achieving narrowband characteristics while maintaining a low insertion loss.

To achieve a narrowband characteristic for a band-pass filter, there must be a resonant circuit that is able to store electromagnetic fields with a minimum of loss of energy within the resonator, i.e. a cavity with a high unloaded quality factor, Q_U. By definition (see e.g. [10.20])

$$Q_U = \frac{\text{total energy stored}}{\text{energy dissipated per radian}} \bigg|_{\text{at resonance}} \tag{10.2}$$

This is the most general definition of Q. For the special case of a single resonance, Q becomes the reciprocal of the fractional bandwidth, as in (9.48). The energy dissipation is internal to the composite resonator structure for Q_U calculations and may include contributions to the loss from the electric fields in the resonator and lossy substrate and the current flow in the ground plane. The external Q, Q_E, is calculated in terms of the energy dissipation external to the composite resonator structure and is generally determined by all the coupling mechanisms from the feed lines into and out of the resonator. The loaded Q, Q_L, takes into account all causes of energy dissipation and is given by

$$\frac{1}{Q_L} = \frac{1}{Q_U} + \frac{1}{Q_E} \tag{10.3}$$

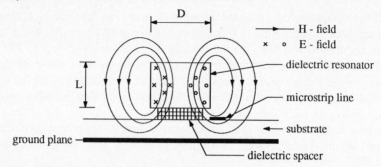

Figure 10.13 Magnetic field coupling between a cylindrical dielectric resonator in the circular symmetric TE_{011} mode and a microstrip line

With a low-loss dielectric material forming a resonator that contains the majority of the stored energy, $Q_U \approx (\tan\delta)^{-1}$. If Q_U is sufficiently high, then Q_L is almost entirely determined by the coupling factors between the outside environment and the resonator.

The general measurement of Q is described in microwave terms by Sucher [10.21] while the determination of Q, specific to the use of dielectric resonators with microstrip lines, is given by Khanna and Garault [10.22].

Dielectric resonators for microstrip circuits take the form of short lengths of circular or square cross-section dielectric waveguide, with the cross-section placed either on or parallel to the substrate. The situation where the resonator is separated from the substrate by a low-loss, low permittivity, spacer is illustrated in Figure 10.13. As with all transmission line and waveguide resonators there will be an infinite set of resonant modes, each having a resonant frequency that depends on the geometry. The lowest order circular-symmetric mode (TE_{011}) in a circular cross-section resonator is normally used. This mode is the one that is illustrated in Figure 10.13, where it is seen that the electric fields are parallel to the substrate surface and that the magnetic field pattern for the resonant mode has components that are suitable for coupling to the magnetic fields of the TEM wave on the microstrip line.

The resonant frequency is influenced by many parameters that include the geometry and permittivity of the resonator, the dielectric spacer (if present) and the substrate, as well as the proximity of the ground plane and shielding enclosures that may be required to prevent radiation from the high intensity fields in the unshielded resonator.

Measurements by Day [10.23] on various TiO_2, $\varepsilon_r = 85$, resonators on 0.635 mm alumina substrate are shown in the form of a mode chart in Figure 10.14. The usefulness of this mode chart is limited by the fact that normally $(D/L)^2 > 2.5$ for small height resonators, where D and L are as defined in Figure 10.13. It does, however, show that the TE_{011} mode of Figure 10.13 has the lowest resonant frequency when $(D/L)^2 > 1.86$. For any other resonator material, with a relative permittivity ε_r, the mode chart may be used in conjunction with

$$f'_r \approx f_r \left[\frac{85}{\varepsilon_r} \right]^{\frac{1}{2}}$$

(10.4)

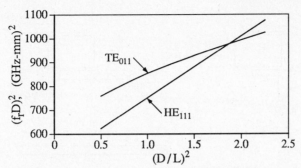

Figure 10.14 Cylindrical dielectric resonator design chart, based on results for $\varepsilon_r = 85$ discs on 0.635 mm alumina substrate, from Day [10.23] (© 1970, IEEE). The resonant frequency is f_r.

to give an initial estimate of the resonant frequency, f_r'.

Temperature stability

The selection of low-loss compound ceramic materials, typically $Ba_mTi_nO_p$, by Tsironis [10.24] has given extremely stable dielectric resonators. These materials exhibit either a near-zero temperature coefficient or a small negative value for f_r that compensates for any positive coefficient changes associated with the energy stored in the substrate. The unloaded Q may be increased for these resonators by separating them from the substrate with a low-loss dielectric spacer, such as quartz with $\varepsilon_r = 3.8$. The spacer causes a reduction in the magnitude of the fringing magnetic fields at the ground plane, resulting in lower attenuation from the induced currents that flow in the imperfect conductors.

Coupling mechanisms

Magnetic flux coupling between a microstrip line and a dielectric resonator has been illustrated in Figure 10.13. The coupling increases as the strip is brought closer to the resonator. Further coupling may be achieved by exciting the resonator with a pair of lines that are anti-phase driven, Figure 10.15a. This has been used by Iveland [10.25] as the coupling mechanism for end elements of a series of resonators that form a band-pass filter. Basic band-pass structures are given with the coupling methods in Figure 10.15, (b) and (c), the latter being analyzed in detail by Bonetti and Atia, [10.26]. Podcameni and Conrado [10.27] used the configuration in Figure 10.15d together with other variations in the design of band-pass and reflection (band-stop) filters. With this structure, the transmission coefficient between input and output may be varied with a change in the position, θ, of the dielectric resonator. Maximum transmission is achieved when $\theta = 90°$.

Tuning mechanisms

A conducting surface placed within the fringing fields above the dielectric resonator will increase the resonant frequency. The conducting surface may be in the form of a plate attached to a screw inserted through a shielding cover plate. The adverse thermal characteristics of this technique on the resonant frequency were reduced by

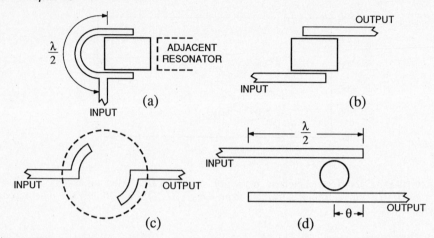

Figure 10.15 Coupling structures from microstrip lines to dielectric resonators. (a) Coupling to the end element of a series of resonators, from Iveland [10.25] (© 1971, IEEE), (b) and (c) single-section band-pass filters, e.g. from Bonetti and Atia [10.26] (© 1981, IEEE), and (d) variable coupling band-pass filter, from Podcameni and Conrado [10.27] (© 1985, IEEE).

Shimoda et al. [10.28] by attaching a further low permittivity dielectric spacer above the main resonator element. Only weak fringing fields will exist at the upper surface of this spacer, which is coated with a conducting film. As the conductor is etched away from the center of the covering disc, the resonant frequency of the composite structure is decreased.

10.7 MICROSTRIP ANTENNAS

The ability to construct antennas on the same substrate as other microstrip components provides for a simple low-profile structure for radiating elements. All forms of radiating structures fabricated on a substrate over a ground plane may be classed as microstrip antennas. However, the discussion here will focus on a particular type that is also known as a patch antenna. The importance of selecting the circuit dimensions and substrate parameters in such a way as to minimize radiation and all other unwanted modes of propagation was stressed in the design of microstrip circuits, in §4.5. For a microstrip antenna, on the other hand, these radiation effects are to be enhanced.

For radiation away from the antenna, there will be no shielding cover plate and hence no problems associated with waveguide cavity modes in an enclosed structure. For enhanced radiation, the following design approaches are employed:
i)   Broad low impedance lines are used.
ii) The substrate height is increased and low permittivity substrate materials are used. Unfortunately, these conditions tend to enhance the coupling to surface wave modes and thus care must be exercised to ensure that the design remains within the surface-wave mode limitations.

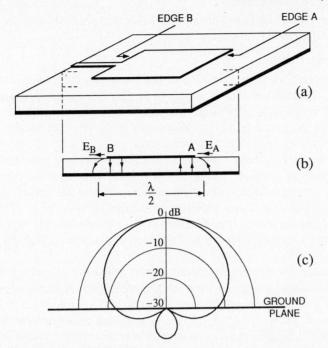

Figure 10.16 (a) The microstrip rectangular-patch antenna, (b) the electric field distribution under the patch, and (c) a typical radiation pattern for the patch near resonance

iii) Resonant elements are used to increase the field strengths at open-circuit planes from which radiation will occur.

A resonant length of very low impedance line, fed directly from a microstrip line, is illustrated in Figure 10.16a. This is an efficient structure in terms of the amount of power that may be radiated from a given area. The actual radiation emanates from the edges A and B. After making an allowance in the length of the line for the fringing fields from the open-circuit plane A, and from plane B where the large impedance mismatch also appears as an approximate open circuit, the overall resonator length is $\lambda/2$, where λ is the wavelength in the wide microstrip line.

Consider the typical values of $\varepsilon_{eff} \approx 2.0$ and $\lambda \approx 0.7\,\lambda_0$. The contributions from sources at A and B to the fields at a far-field point above the ground plane will never cancel completely. The radiation patterns may be described in two orthogonal planes, each one also orthogonal to the ground plane. Plane P–P is in the plane of the section, Figure 10.16b, and Q–Q orthogonal to it. A simplified far-field radiation pattern in the plane P–P is illustrated in Figure 10.16c. The finer detail of this pattern, especially below the ground plane, depends on circuit parameters such as the position of the input line as well as the size of the finite ground plane. Along the normal to and above the substrate, i.e. in the broadside direction, the electric field is polarized parallel to the length direction of the resonant section of line, i.e. as given by the electrical field components (E_A and E_B) at the edges of the resonator in Figure 10.16b. In the orthogonal plane Q–Q, a similar but slightly broader radiation pattern is given if the line width is less than $\lambda/3$. This pattern will depend on the length of the radiating edges at the open-circuit planes.

Circular polarization

There is considerable interest in the use of patch antennas for mobile communications via satellites. If radiation with a circular polarization is used, i.e. a wave where the electric field vector has a constant magnitude but rotates with time, the signal received by the patch can be made independent of the physical rotation of the patch about the line joining the source and the receiver.

The basic patch antenna in Figure 10.16a gives a linearly polarized wave. If the antenna is made as a square patch, it will support two resonant modes that have orthogonal polarizations for the far-field radiation. There are several techniques for exciting both modes, e.g. using a microstrip line feeding one corner of the square, by incorporating an asymmetry into the patch to couple between the two modes, or by excitation from a coaxial feed through the ground plane to a diagonal point on the patch.

A square patch that has degenerate modes at the one frequency does no more than provide a linear polarized wave as a combination of the original two linear components. For circular polarization, the individual plane polarized components must be in time phase quadrature. This is achieved by separating the degenerate modes into two uncoupled modes that have closely separated resonant frequencies. A rectangular, but almost square, patch will provide this requirement. Now, when the antenna is used at a frequency between the two resonances, one mode may be driven with the phase of the mode voltage leading the impressed current by 45°, while the other mode voltage lags the impressed current by the same amount, leading to the time phase quadrature between the field patterns of the two modes. A survey paper by Carver and Mink [10.29] and the book by Bahl and Bhartia [10.30] describe analytical and experimental design approaches for this and other microstrip antennas and are suggested for further reading on microstrip antenna technology.

EXERCISES

10.1 Determine the scattering parameter matrix (magnitudes only) for a circulator that, at each port, has an input VSWR = 1.2, transmission loss = 0.5 dB and isolation = 20 dB.

10.2 The circulator in the previous problem is used as an isolator with a matched load (VSWR = 1.05) connected to port 3. A load, with a VSWR = 3.0, is connected to port 2. Calculate the maximum input VSWR that can occur at port 1.

10.3 Estimate the dimensions of a cylindrical TiO_2 dielectric resonator, $\varepsilon_r = 85$, in the TE_{011} mode on an alumina substrate, if $D = 1.5 L$ and the resonant frequency is 10.0 GHz.

10.4 Design a broadband power divider with a power-split ratio of 1.0 dB.

10.5 Estimate the two lowest resonant frequencies of a rectangular patch antenna, 48 mm × 50 mm on a 1.5 mm thick substrate. For the substrate, $\varepsilon_r = 2.5$.

REFERENCES

[10.1] Gupta, K. C., Garg, R. and Bahl, I. J., *Microstrip Lines and Slotlines*, Artech House, Dedham, MA, 1979.

[10.2] England, E. H., "A coaxial to microstrip transition", *IEEE Trans. Microwave Theory and Techniques*, Vol. MTT-24, No. 1, January 1976, pp. 47-8.

[10.3] Schneider, M. V., Glance, B. and Bodtmann, W. F., "Microwave and millimetre wave hybrid integrated circuits for radio systems", *Bell System Technical Journal*, Vol. 48, No. 4, July 1969, pp. 1703-26.

[10.4] Van Heuven, J. H. C., "A new integrated waveguide-microstrip transition", *IEEE Trans. Microwave Theory and Techniques*, Vol. MTT-24, No. 3, March 1976, pp. 144-7.

[10.5] Cohn, S. B., "Slot line on a dielectric substrate", *IEEE Trans. Microwave Theory and Techniques*, Vol. MTT-17, No. 10, October 1969, pp. 768-78.

[10.6] Dickens, L. E. and Maki, D. W., "An integrated-circuit balanced mixer, image and sum enhanced", *IEEE Trans. Microwave Theory and Techniques*, Vol. MTT-23, No. 3, March 1975, pp. 276-81.

[10.7] Schüppert, B., "Microstrip/slotline transitions: modeling and experimental investigation", *IEEE Trans. Microwave Theory and Techniques*, Vol. MTT-36, No. 8, August 1988, pp. 1272-82.

[10.8] Riaziat, M., Bandy, S. and Zdasiuk, G., "Coplanar waveguides for MMICs", *Microwave Journal*, Vol. 30, No. 6, June 1987, pp. 125, 128, 130-1.

[10.9] Pengelly, R. S., *Microwave Field-effect Transistors — Theory, Design and Applications*, Wiley, New York, 2nd edn, 1986, Section 10.2.

[10.10] Pettenpaul, E., Kapusta, H., Weisgerber, A., Mampe, H., Luginsland, J. and Wolff, I., "CAD models of lumped elements on GaAs up to 18 GHz", *IEEE Trans. Microwave Theory and Techniques*, Vol. MTT-36, No. 2, February 1988, pp. 294-304.

[10.11] Laverghetta, T. S., *Practical microwaves*, Howard W. Sams and Co., Indianapolis, 1984, Section 4.9.

[10.12] Ingalls, M. and Kent, G., "Monolithic capacitor as transmission lines", *IEEE Trans. Microwave Theory and Techniques*, Vol. MTT-35, No. 11, November 1987, pp. 964-70.

[10.13] Combes, P. F.,Graffeuil, J. and Sautereau, J-F., *Microwave Components, Devices and Active Circuits*, Wiley, Chichester, 1987, Section 2.4.

[10.14] Liao, S. Y., *Microwave Circuit Analysis and Amplifier Design*, Prentice-Hall, Englewood Cliffs, NJ, 1987, Section 6.7.

[10.15] Arbie, P. L. D., *The Design of Impedance-matching Networks for Radio-frequency and Microwave Amplifiers*, Artech House, Dedham, MA, 1985.

[10.16] Wilkinson, E. J., "An n-way hybrid power divider", *IRE Trans. on Microwave Theory and Techniques*, Vol. 8, No. 1, January 1960, pp. 116-18.

[10.17] Parad, L. I. and Moynihan, R. L., "Split-tee power divider", *IEEE Trans. Microwave Theory and Techniques*, Vol. MTT-13, No. 1, January 1965, pp. 91-5.

[10.18] Cohn, S. B., "A class of broadband three-port TEM-mode hybrids", *IEEE Trans. Microwave Theory and Techniques*, Vol. MTT-16, No. 2, February 1968, pp. 110-16.

[10.19] Fay, C. E. and Comstock, R. L., "Operation of the ferrite junction circulator", *IEEE Trans. Microwave Theory and Techniques*, Vol. MTT-13, No. 1, January 1965, pp. 15-27.

[10.20] Altman, J. L., *Microwave Circuits*, D.Van Nostrand Co., Princeton, NJ, 1964.

[10.21] Sucher, M., "Measurement of Q" in *Handbook of Microwave Measurements*, Vol. 2, 3rd edn (eds Sucher, M. and Fox, J.), Polytechnic Press, New York, 1963.

[10.22] Khanna, A. and Garault, Y., "Determination of loaded, unloaded, and external quality factors of a dielectric resonator coupled to a microstrip line", *IEEE Trans. Microwave Theory and Techniques*, Vol. MTT-31, No. 3, March 1983, pp. 261-4.

[10.23] Day, W. R. Jr., "Dielectric resonators as microstrip-circuit elements", *IEEE Trans. Microwave Theory and Techniques*, Vol. MTT-18, No. 12, December 1970, pp. 1175-6.

[10.24] Tsironis, C., "Highly stable dielectric resonator FET oscillators", *IEEE Trans. Microwave Theory and Techniques*, Vol. MTT-33, No. 4, April 1985, pp. 310-14.

[10.25] Iveland, T. D., "Dielectric resonator filters for application in microwave integrated circuits", *IEEE Trans. Microwave Theory and Techniques*, Vol. MTT-19, No. 7, July 1971, pp. 643-52.

[10.26] Bonetti, R. R., and Atia, A. E., "Analysis of microstrip circuits coupled to dielectric resonators", *IEEE Trans. Microwave Theory and Techniques*, Vol. MTT-29, No. 12, December 1981, pp. 1333-7.

[10.27] Podcameni, A. and Conrado, L. F. M., "Design of microwave oscillators and filters using transmission-mode dielectric resonators coupled to microstrip lines." *IEEE Trans. Microwave Theory and Techniques*, Vol. MTT-33, No. 12, December 1985, pp. 1329-32.

[10.28] Shimoda, Y., Tomimuro, H. and Onuki, K., "A proposal of a new dielectric resonator construction for MIC's" *IEEE Trans. Microwave Theory and Techniques*, Vol. MTT-31, No. 7, July 1983, pp. 527-32.

[10.29] Carver, K. R. and Mink, J. W., "Microstrip antenna technology", *IEEE Trans. Antennas and Propagation*, Vol. AP-29, No. 1, January 1981, pp. 2-24.

[10.30] Bahl, I. J. and Bhartia, P., *Microstrip Antennas*, Artech House, Dedham, MA, 1980.

11 Active circuit characterization

11.1 INTRODUCTION

Active devices are embedded in microstrip lines to form functional circuits, each with their appropriate source and load. The aim of this chapter is to study the characterization of active two-port networks so as to establish the requirements that the design process imposes on the microstrip interface networks.

Very often the power gain has to maximized. However, when referring to power gain it is necessary to be more precise, as a number of different power gains may be defined and each has its own properties, significance and use. This is explored in §11.2. The case when the conjugately matched condition is satisfied simultaneously at the input and output is of considerable importance and is dealt with in §11.3. As the simultaneous conjugate matched condition is only meaningful if the two-port network is absolutely stable, stability is considered in §11.4. Power gain formulae for some important cases are given in §11.5, and §11.6 deals with two-port noise characterization. The chapter concludes with a discussion of design options.

Active two-port design involves many other considerations, e.g. non-linearities, bandwidth and parameter sensitivity, in addition to those included in this chapter. As a comprehensive exposition of amplifier and oscillator design is not the purpose here, the reader is presented with a number of important active two-port properties in a reasonably self-contained manner, showing how they reflect on microstrip circuit design. For a more complete treatment, the reader is referred to the references [11.1–11.5].

11.2 POWER GAINS

Consider the two-port network that is connected to a source and a load, as shown in Figure 11.1. The figure illustrates the power flow for the network. The actual power flowing into the input is P_{in}, while P_{out} is the power actually absorbed by the load. Two other powers are needed, namely P_{av_S}, the available power from the source, and P_{av_O}, the available power from the output of the two-port network. Naturally, $P_{av_S} \geq P_{in}$ and $P_{av_O} \geq P_{out}$.

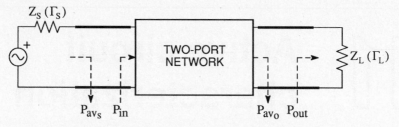

Figure 11.1   Power flow in a two-port network

Three power gains are now defined as follows:

$$G_p = \text{(Ordinary) Power Gain} = \frac{P_{out}}{P_{in}}$$

$$G_t = \text{Transducer Power Gain} = \frac{P_{out}}{P_{avs}}$$

$$G_a = \text{Available Power Gain} = \frac{P_{avo}}{P_{avs}} \tag{11.1}$$

The commonsense power gain is G_p, simply giving the ratio of power out to power in. In some texts, e.g. Gonzalez [11.1], this gain is referred to as the operating power gain. However, the transducer power gain is the most meaningful power gain from a physical point of view. It gives the ratio of the power that is actually delivered to the load to the power that could be delivered anyway to the load from the source by purely passive matching techniques. P_{avs} can always be extracted from a source by tuning out the source reactance and using a transformer to match the source resistance to the load. If the load is receiving less power than P_{avs}, the two-port network is not providing any useful amplification.

The available power gain is always used in the context of noise characterization. Many of the noise performance formulae are then considerably simplified, as the mismatch factors between a source and a load are taken care of automatically.

The power gains for any particular two-port network exhibit the following properties:

i) G_p depends on Z_L (Γ_L), but not on Z_S (Γ_S),

G_a depends on Z_S (Γ_S), but not on Z_L (Γ_L), while

G_t depends on both Z_S and Z_L

ii) $G_a \geq G_t, \quad G_p \geq G_t \tag{11.2}$

11.3 SIMULTANEOUS CONJUGATE MATCHING

The equality signs in (11.2) apply when there is conjugate matching at the output and input respectively. When there is conjugate matching at both the output and input, this is known as *simultaneous conjugate matching (s.c.m.)* and

$$G_a = G_t = G_p = G_{max} \tag{11.3}$$

where G_{max} is the largest value that any of the three power gains can possess. It is obvious that the three power gains, as they are defined in (11.1), are equal under s.c.m. However, a formal proof that they are also equal to the maximum gain that any of them can possess is required. Consider the differences in the way that maximum power flows are achieved at the input and output. At the output, it is the load that is adjusted to fit in with a fixed source as seen by the load. On the other hand, at the input it is the source that is adjusted for a fixed load as seen by the source. With a variable impedance source and a fixed load (Figure 11.2), maximum power into the load is *not* achieved when $Z_S = Z_L^*$, but is obtained when $Z_S = 0 - j Im(Z_L)$. Thus, it is not immediately apparent that simultaneous conjugate matching gives the maximum power gain, but surprisingly (11.3) is nevertheless true. This will now be proved by contradiction.

Proof

Let G' be the value of $G_a = G_p = G_t$ under s.c.m., assuming that G_{max} occurs at other than s.c.m. conditions. The largest of the maxima of G_a, G_t, G_p (denoted by G_{am}, G_{tm}, G_{pm}, respectively) must be greater than G' and, in view of the inequalities above, it is either G_{am} or G_{pm} that is the largest maximum. Let us assume that it is G_{am} that is in fact the largest maximum and thus that $G_{am} > G'$. Since $G_a = G'$ under s.c.m., $G_a = G_{am}$ must occur with a Z_S other than that required to give s.c.m. Let us denote this particular Z_S by Z_{Sm}. Let us now select Z_L to conjugately match the output when $Z_S = Z_{Sm}$. This will make $G_t = G_a$, but we will still have $G_a = G_{am}$, since G_a does not depend on Z_L. Thus we will have $G_t = G_{am}$. However, because the input is not conjugately matched, we will have $G_p > G_t$. That is, there is a value of $G_p > G_{am}$. This contradicts our assumption of G_{am} being the largest maximum.

A similar contradiction results if we assume G_{pm} to be the largest maximum. Hence, (11.3) has been proved.

Q.E.D.

Note 1
In view of the result in Exercise 11.1, simultaneous conjugate matching is implied by conjugate matching at either port for a lossless reciprocal network.

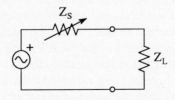

Figure 11.2   A variable source and a fixed load

Note 2

In the proof, we have implicitly assumed that the two-port network is absolutely stable. If the two-port network is potentially unstable, i.e. there exist passive source and load terminations that may cause the network to oscillate, none of these power gains achieves stationary values with passive source and load terminations and $G_{max} \rightarrow \infty$.

Note 3

When referring to the maximum power gain, it is superfluous to specify it as the transducer, available or ordinary power gain, as all of these maxima are the same.

11.3.1 Simultaneous conjugate matching in the general case

When the two-port network is absolutely stable, passive Γ_S and Γ_L exist to achieve simultaneous conjugate matching. Their values are denoted by $\Gamma_S(opt)$ and $\Gamma_L(opt)$ respectively and are given [11.1] in terms of the two-port scattering parameters by

$$\Gamma_S(opt) = \frac{B_1 - (B_1^2 - 4|C_1|^2)^{\frac{1}{2}}}{2C_1}$$

$$\Gamma_L(opt) = \frac{B_2 - (B_2^2 - 4|C_2|^2)^{\frac{1}{2}}}{2C_2} \tag{11.4}$$

where

$$B_1 = 1 + |s_i|^2 - |s_o|^2 - |\Delta|^2$$

$$B_2 = 1 - |s_i|^2 + |s_o|^2 - |\Delta|^2$$

$$C_1 = s_i - s_o^*\Delta, \quad C_2 = s_o - s_i^*\Delta \quad \text{and} \quad \Delta = s_i s_o - s_r s_f$$

11.4 STABILITY CONSIDERATIONS

In principle, the same techniques that are used to determine the stability of feedback amplifiers at lower frequencies could also be used at microwave frequencies, if an accurate equivalent circuit is available. Because of the necessity of including the package parasitic elements and various high frequency corrections to the equivalent circuit, the application of such techniques as the Nyquist and Bode criteria result in excessive complexity and it is preferable to determine stability in other ways.

If negative real parts in the input and output impedances are found when the network is terminated with real loads and sources respectively, then an oscillator can be constructed by tuning out the reactance at the appropriate port and terminating the port with a positive resistance, equal in magnitude to the relevant negative real part. If an oscillator can be constructed in this manner, the two-port network is potentially unstable. Potential instability does not mean that the two-port network is always unstable, but only that that there exist some combinations of passive load and source terminations for which it will oscillate.

Even though the presence of a negative real part implies instability, the converse is not true, as it is possible to have an oscillation even with positive real parts for the input and output impedances [11.6]. For example, such a situation can arise in multi-stage amplifiers with positive input/output resistances for the overall configuration, but with an oscillation produced by an inter-stage instability. However, the case of instability with positive real parts is not likely to be important in single-stage amplifier design and is frequently ignored. The reader is referred to Woods [11.6] for a thorough treatment of this point. The stability criteria that now follow in this section are all based on the requirement for a positive real part of the port immittance when there is a real termination on the other port.

Negative real parts in the input and output immittances imply that $|\Gamma_{in}| > 1$ and $|\Gamma_{out}| > 1$ respectively. It turns out that the regions of Γ_L values that produce $|\Gamma_{in}| > 1$ and $|\Gamma_{in}| < 1$ are separated by a circle in the Γ_L plane. Similarly, a circle separates the regions for Γ_S producing $|\Gamma_{out}| > 1$ and $|\Gamma_{out}| < 1$. Thus it is possible [11.1] to draw stability circles on the Smith Chart as follows:

For stability circles that *do not enclose* the Smith Chart origin, which is the case illustrated in Figure 11.3,

$$|\Gamma_{in}| < 1 \quad \text{if} \quad |s_i| < 1 \quad \textbf{and} \quad \Gamma_L \text{ falls } \textit{outside} \text{ the circle } (C_L, R_L)$$
$$\text{or if} \quad |s_i| > 1 \quad \textbf{and} \quad \Gamma_L \text{ falls } \textit{inside} \text{ the circle } (C_L, R_L)$$

$$|\Gamma_{out}| < 1 \quad \text{if} \quad |s_o| < 1 \quad \textbf{and} \quad \Gamma_S \text{ falls } \textit{outside} \text{ the circle } (C_S, R_S)$$
$$\text{or if} \quad |s_o| > 1 \quad \textbf{and} \quad \Gamma_S \text{ falls } \textit{inside} \text{ the circle } (C_S, R_S)$$

where
$$C_S = \frac{s_o \Delta^* - s_i^*}{|\Delta|^2 - |s_i|^2} \qquad R_S = \left| \frac{s_f s_r}{|\Delta|^2 - |s_i|^2} \right|$$

and
$$C_L = \frac{s_i \Delta^* - s_o^*}{|\Delta|^2 - |s_o|^2} \qquad R_L = \left| \frac{s_f s_r}{|\Delta|^2 - |s_o|^2} \right| \qquad (11.5)$$

Γ_L locations for stability/instability are illustrated in Figure 11.3. There is a

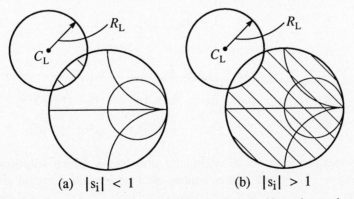

(a) $|s_i| < 1$ (b) $|s_i| > 1$

Figure 11.3 Stability circles in the Γ_L plane, showing the unstable region as the shaded area of the plane. Illustrated is the case of the stability circles not enclosing the origin.

corresponding set for Γ_S locations. For practical loads that exist within the unit circle, the hatched areas correspond to those loads that produce $|\Gamma_{in}| > 1$.

An easy way to remember whether the inside or outside of the stability circle is the stable region is to make use of Equations (2.24). For example, since a matched load ($\Gamma_L = 0$) at the center of the Smith Chart makes $\Gamma_{in} = s_i$, the center of the Smith Chart will correspond to a stable load if $|s_i| < 1$ and vice versa. This rule will also show which region will be the stable region in the case when the stability circles *do enclose* the origin.

11.4.1 Unconditional stability

Absolute, or unconditional, stability will clearly result if both input and output stability circles lie outside the unit circle on the Smith Chart. Remembering that the origin for the Γ_S and Γ_L planes is at the center of the Smith Chart, then for unconditional stability

$$|s_i| < 1 \quad \text{and} \quad |C_L| > R_L + 1$$

as well as $\quad |s_o| < 1 \quad \text{and} \quad |C_S| > R_S + 1 \quad\quad\quad\quad (11.6)$

11.4.2 Other formulations for stability

Other expressions that give necessary and sufficient conditions for unconditional stability, except in Case (2a) where the conditions are necessary but *not* sufficient, are given below. From [11.1, 11.6], they are written in terms of the stability factor, K, where

$$K = \frac{1 - |s_i|^2 - |s_o|^2 + |\Delta|^2}{2|s_f s_r|} \quad\quad\quad\quad (11.7)$$

The conditions are

1. $\quad\quad K > 1 \quad\quad \text{and} \quad |\Delta| < 1$

2. $\quad\quad K > 1 \quad\quad \text{and} \quad 1 - |s_i|^2 > |s_f s_r|$

$$1 - |s_o|^2 > |s_f s_r|$$

(2a) $\quad\quad K > 1 \quad\quad \text{and} \quad |s_i| < 1, \ |s_o| < 1$

3. $\quad\quad K > 1 \quad\quad \text{and} \quad B_1 = 1 + |s_i|^2 - |s_o|^2 - |\Delta|^2 > 0$

4. $\quad\quad K > 1 \quad\quad \text{and} \quad B_2 = 1 - |s_i|^2 + |s_o|^2 - |\Delta|^2 > 0$

When the two-port network is potentially unstable, a formal substitution of the s-parameters in (11.4) will lead to source and load impedances with negative real parts. Thus

$$\text{Potential instability} \ \Rightarrow \ |\Gamma_S(\text{opt})| > 1 \ \text{and/or} \ |\Gamma_L(\text{opt})| > 1$$

11.4.3 Stabilization techniques

In principle, an unstable two-port network may be stabilized with the same techniques that are used in feedback amplifiers at lower frequencies. However, the complexity of the feedback paths in a typical active device chip at microwave frequencies makes this approach impractical. It is preferable to tackle the stabilization problem by treating the two-port network as a "black-box", using only the information provided by its two-port parameters.

The only way that instability can arise in a unilateral two-port network, defined in §2.1.6 as one where $s_r = 0$, is if either $|s_i|$ or $|s_o|$ is greater than unity. On the other hand, if $|s_i|$ and $|s_o|$ are less than unity, instability can arise if there is feedback from output to input. Absence of this feedback with $s_r = 0$ can ensure that there is no instability. One way of stabilizing a two-port network is thus to deliberately unilateralize it with feedback elements, provided that $|s_i|$, $|s_o|$ remain less than unity. In general this cannot be achieved over a broad band of frequencies, in which case neutralization could be attempted with s_r being minimized as far as possible over the frequency band.

Another stabilization technique is that of loading the input and output. Intuition tells us that with sufficient loading any oscillation could be damped out, as long as the source of oscillation is accessible for loading. Mathematically this is best seen with the stability criterion expressed in terms of y-parameters as Llewellyn's stability criterion [11.7], namely

$$g_i > 0, \quad g_o > 0 \quad \text{and} \quad g_i g_o > \tfrac{1}{2}|y_f y_r|(1 + \cos\theta) \tag{11.8}$$

where $\quad g = Re(y) \quad \text{and} \quad \theta = \underline{/y_f y_r}$

A simple proof of (11.8) will be found in [11.8]. Parallel-resistance loading of the input or output effectively increases g_i or g_o respectively, as is clearly seen from the y-parameter equivalent circuit in Figure 2.19. With sufficient loading, which in the general case may require loading both the input and output, the stability requirement can be satisfied. The dual expression of (11.8) in terms of impedance or z-parameters shows that stability can also be achieved by series-resistance loading. However, one should not construe that in any particular case series or parallel loading are equally effective for stabilization. Among the reasons for this, one is related to the question of whether a negative resistance is effectively part of a series- or parallel-resonant circuit, as discussed further for oscillators in §11.7. Another, but in fact related, reason could be due to the possibility of instability, even with positive real parts of immittances. Finally, it should be noted that two-port networks could be envisaged where stability would not be achieved with any degree of loading.

By referring again to Figure 2.19, it can also be seen that parallel-resistance loading of the input or output does not alter y_f or y_r and thus, in view of the result

$$\frac{s_f}{s_r} = \frac{y_f}{y_r} \tag{11.9}$$

as presented in Exercise 2.10, s_f/s_r is not affected by parallel-resistance loading of the input or output. The significance of this result will become clear later in §11.5.2. Similar arguments show that s_f/s_r is also independent of series-resistance loading.

11.5 SOME POWER GAIN FORMULAE

Many power gain formulae are in general quite complicated. However, in some circumstances of practical importance they are either simple or can be presented in a simplified manner. Sample proofs of selected formulae are given as Examples 11.1 and 11.2 below. These two examples adopt a physical approach as preferred by the authors to the proofs found in other texts, which are based on the manipulation of equations, possibly with the help of signal flow graphs. Proofs of some of the other formulae are set out as exercises at the end of the chapter.

11.5.1 The matched condition

For the matched condition, $\Gamma_S = \Gamma_L = 0$, giving

$$G_t = |s_f|^2 \tag{11.10a}$$

$$G_p = \frac{|s_f|^2}{1 - |s_i|^2} \tag{11.10b}$$

and

$$G_a = \frac{|s_f|^2}{1 - |s_o|^2} \tag{11.10c}$$

Example 11.1

Prove that with $\Gamma_S = \Gamma_L = 0$, the transducer power gain $G_t = |s_f|^2$ as given in (11.10a).

Solution:

The source driving the two-port network is characterized by a_S, Γ_S as in §2.1.4 and §2.1.5. The incident and reflected waves at the input, a_1 and b_1, are seen from (2.17) to be related as follows:

$$a_1 = a_S + \Gamma_S b_1$$

When $\Gamma_S = 0$, then $a_S = a_1$ and, from (2.22)

$$P_{av_S} = \frac{|a_S|^2}{1 - |\Gamma_S|^2} = |a_1|^2$$

The incident and reflected waves at the output are a_2 and b_2 respectively. When $\Gamma_L = 0$, then $a_2 = 0$ and, consequently, P_{out} becomes $|b_2|^2$. Thus

$$G_t = \frac{|b_2|^2}{|a_1|^2}$$

Since $\Gamma_L = 0$, $b_2/a_1 = s_f$ by definition. Hence

$$G_t = |s_f|^2$$

Q.E.D.

11.5.2 The unmatched condition

For the unmatched condition, Γ_S and Γ_L have arbitrary values. Following the presentation of formulae by Gonzalez [11.1], the formulae for power gains may be written as follows:

$$G_t = \gamma_s |s_f|^2 \sigma_L \tag{11.11a}$$

or
$$G_t = \sigma_s |s_f|^2 \gamma_L \tag{11.11b}$$

$$G_p = \frac{|s_f|^2}{1 - |\Gamma_{in}|^2} \sigma_L \tag{11.11c}$$

$$G_a = \sigma_s \frac{|s_f|^2}{1 - |\Gamma_{out}|^2} \tag{11.11d}$$

where
$$\gamma_s = \frac{1 - |\Gamma_s|^2}{|1 - \Gamma_{in}\Gamma_s|^2}, \qquad \gamma_L = \frac{1 - |\Gamma_L|^2}{|1 - \Gamma_{out}\Gamma_L|^2} \tag{11.12a}$$

and
$$\sigma_s = \frac{1 - |\Gamma_s|^2}{|1 - s_i\Gamma_s|^2}, \qquad \sigma_L = \frac{1 - |\Gamma_L|^2}{|1 - s_o\Gamma_L|^2} \tag{11.12b}$$

Example 11.2

Prove that, for arbitrary Γ_S and Γ_L, the transducer power gain $G_t = \sigma_s |s_f|^2 \gamma_L$ as quoted in (11.11b).

Solution:

This result will be proved by starting with arbitrary (Γ_S, Γ_L) and then replacing Γ_L and Γ_S in turn by matched loads. It will be shown first that

$$G_t(\Gamma_S, \Gamma_L) = G_t(\Gamma_S, 0)\gamma_L \tag{11.13}$$

and then that

$$G_t(\Gamma_S, 0) = \sigma_s |s_f|^2 \tag{11.14}$$
$$\equiv \sigma_s G_t(0,0), \quad \text{from Example 11.1}$$

In the general case, $G_t(\Gamma_S, \Gamma_L)$ may be written as

$$G_t(\Gamma_S, \Gamma_L) = \frac{P_{out}}{P_{avs}} = \frac{|b_s|^2}{P_{avs}} \times \frac{P_{out}}{|b_s|^2} \tag{11.15}$$

where b_s is a traveling wave characterizing the two-port output in the same way that a_s characterizes the source, i.e. b_s is the traveling wave that would be launched by the two-port output into a matched line. It is denoted as a "b" wave because it is traveling in the reflected direction from the output of the two-port network.

Referring to the result

$$\frac{1 - |\Gamma_Y|^2}{|1 - \Gamma_X\Gamma_Y|^2} = \frac{P_L}{|a_s|^2} \tag{11.16}$$

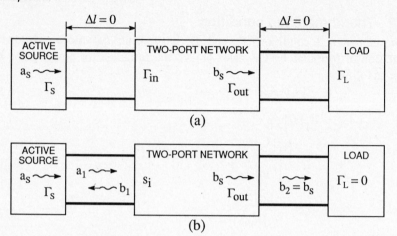

Figure 11.4 A two-port network, (a) with an arbitrary source and load, and (b) with the load in (a) replaced by a matched load

that is to be proved in Exercise 11.4, the quantities in (11.16) are now interpreted in terms of the quantities at the output in Figure 11.4a. Thus, letting $\Gamma_X = \Gamma_{out}$, $\Gamma_Y = \Gamma_L$, $a_S = b_S$ and $P_L = P_{out}$, makes

$$\frac{P_{out}}{|b_S|^2} = \frac{1 - |\Gamma_L|^2}{|1 - \Gamma_{out}\Gamma_L|^2} = \gamma_L \tag{11.17}$$

Now, moving from Figure 11.4a to Figure 11.4b by replacing Γ_L with a matched load (i.e. $\Gamma_L = 0$), b_S at the two-port output is unaltered, as the waves that would be launched into matched lines by the outputs of Figures 11.4a and 11.4b will be the same. However, because of the matched load in Figure 11.4b, $b_S = b_2$ and thus $P_{out} = |b_2|^2 = |b_S|^2$. The term $|b_S|^2/P_{avs}$ in (11.15) is thus seen as the G_t for the circuit in Figure 11.4b, i.e.

$$\frac{|b_S|^2}{P_{avs}} = G_t(\Gamma_S, 0) \tag{11.18}$$

Substituting (11.18) and (11.17) into (11.15) completes the first stage of the proof and gives (11.13), namely

$$G_t(\Gamma_S, \Gamma_L) = G_t(\Gamma_S, 0)\gamma_L$$

Remaining with Figure 11.4b

$$G_t(\Gamma_S, 0) = \frac{|b_2|^2}{P_{avs}} = \frac{|a_1|^2}{P_{avs}} \times \frac{|b_2|^2}{|a_1|^2} \tag{11.19}$$

Referring this time to the first equation in Exercise 11.4, the quantities are interpreted in terms of those at the input of the two-port network in Figure 11.4b. Remembering that the *output* in Figure 11.4b is a matched load makes $\Gamma_Y = s_i$. For the other terms, $\Gamma_X = \Gamma_S$, a_S is unaltered and P_{avs} is the available source power in both cases. Hence

$$\frac{|a_1|^2}{P_{av_s}} = \frac{1 - |\Gamma_s|^2}{|1 - s_i \Gamma_s|^2} = \sigma_s \tag{11.20}$$

Again, since $\Gamma_L = 0$, $b_2/a_1 = s_f$, so that (11.19) now becomes

$$G_t(\Gamma_s, 0) = \sigma_s |s_f|^2 \tag{11.21}$$

This completes the proof of the second stage, i.e. (11.14), and thus (11.11b) has been proved.

Q.E.D.

With unconditional stability, $G_t(max)$ may also be expressed [11.1] in terms of the stability factor as

$$G_t(max) = \left|\frac{s_f}{s_r}\right| \left\{K - \sqrt{K^2 - 1}\right\} \tag{11.22}$$

Setting $K = 1$ for the limiting case of unconditional stability gives

$$G_t(max) = \left|\frac{s_f}{s_r}\right| \tag{11.23}$$

Input/output resistive loading alters K without changing s_f/s_r, as explained in §11.4.3. Thus (11.22) shows that if a two-port network is stabilized by resistive loading of input/output, then the maximum power gain that can be achieved with no possibility of instability is given by (11.23). $|s_f/s_r|$ is thus called the *maximum stable gain* and is one of the figures of merit for an active two-port network.

11.5.3 The unilateral case

For the unilateral case, where there is no feedback from the output to input in the two-port network, $s_r = 0$ and the equations (11.11) reduce to

$$G_{t_u} = \sigma_s |s_f|^2 \sigma_L$$

$$G_{p_u} = \frac{|s_f|^2}{1 - |s_i|^2} \sigma_L$$

$$G_{a_u} = \sigma_s \frac{|s_f|^2}{1 - |s_o|^2} \tag{11.24}$$

In the unilateral case, simultaneous conjugate matching is achieved with $\Gamma_s = s_i^*$ and $\Gamma_L = s_o^*$, giving

$$G_{t_u}(max) = G_{p_u}(max) = G_{a_u}(max)$$

$$= \frac{|s_f|^2}{(1 - |s_i|^2)(1 - |s_o|^2)} \tag{11.25}$$

In terms of y-parameters, $G_{t_u}(max)$ is given by

$$G_{t_u}(max) = \frac{|y_f|^2}{4\, g_i\, g_o} \quad \text{where } g = Re(y).$$

(11.26)

as proved in Exercise 11.6.

If the unilateral case is assumed even when s_r is non-zero, as in the unilateral approximation, then the maximum error for the transducer gain will be bounded within the range given [11.9] by

$$\frac{1}{(1+u)^2} < \frac{G_t}{G_{t_u}} < \frac{1}{(1-u)^2}$$

(11.27)

where

$$u = \frac{|s_i\, s_r\, s_f\, s_o|}{(1 - |s_i|^2)(1 - |s_o|^2)}$$

(11.28)

provided that $|\Gamma_s| \le |s_i|$, $|\Gamma_L| \le |s_o|$ and $|s_i|$, $|s_o|$ are themselves < 1. In particular, (11.27) is valid when $G_{t_u}(max)$ is evaluated, as then $|\Gamma_s| = |s_i|$ and $|\Gamma_L| = |s_o|$.

11.5.4   Some further considerations

Mason's U-function is defined in terms of y-parameters as

$$U = \frac{|y_f - y_r|^2}{4\,(g_i\, g_o - g_f\, g_r)}$$

(11.29)

The activity or passivity of a network is determined by this function. For activity and the possibility of instability, in the case of g_i and $g_o > 0$, $U > 1$. A simple proof of this criterion is give by Jorsboe [11.10]. The condition $U = 1$ determines f_{max}, the maximum frequency of oscillation of a two-port network. f_{max} is a fundamental performance parameter of a high frequency transistor. For a unilateral network

$$U = \frac{|y_f|^2}{4\, g_i\, g_o}$$

(11.30)

which will be recognized as the $G_{t_u}(max)$ in (11.26).

U also has the property that it is *invariant under lossless reciprocal embedding* [11.7] so that, if a network is unilateralized with lossless reciprocal elements, its U-function is not altered. Thus, U is the maximum power gain of a network, after it has been unilateralized with lossless reciprocal elements. One practical implication of this property is that, in calculating U, it is not necessary to worry about the package parasitics of an active chip.

11.6   NOISE CHARACTERIZATION

In a typical amplifier cascade, the first stage is designed for a low noise performance, the middle stage for high gain and the final stage for the output power level. The lowest noise temperature for the first stage is achieved when it is driven by the appropriate source impedance. The source impedance that optimizes noise performance is in general quite different from that required for best gain but, unless the gain is exceedingly low, problems are not encountered.

The characterization of a noisy two-port network in terms of traveling wave quantities has been developed by Meys [11.11]. The noise temperature T_n of a two-port network as a function of the source reflection coefficient Γ_S is given [11.11] by

$$T_n = T_{n0} + 4 T_S \frac{R_n}{Z_0} \frac{|\Gamma_S - \Gamma_0|^2}{1 - |\Gamma_S|^2} \frac{1}{|1 + \Gamma_0|^2} \qquad (11.31)$$

where T_{n0}, R_n and Γ_0 are noise parameters that characterize the two-port network, Z_0 is the characteristic impedance with respect to which Γ_S and Γ_0 are defined and T_S is a standard temperature, typically 290K, and is part of the definition of R_n. R_n is always positive and thus the term added to T_{n0} is also always positive. This term achieves its minimum value of zero when $\Gamma_S = \Gamma_0$, in which case $T_n = T_{n0}$. Thus T_{n0} is the minimum noise temperature of the two-port network and is achieved when the source has the optimum source reflection coefficient equal to Γ_0. Starting with a given source impedance typically matched to the line, the designer must convert it to Γ_0 following the approach given in §6.11.

11.7 DESIGN OPTIONS

A two-port network that is absolutely stable is considered first. If gain is the primary consideration, a design for maximum power gain proceeds by choosing $\Gamma_S(\text{opt})$ and $\Gamma_L(\text{opt})$ from (11.4), to give simultaneous conjugate matching. The design calculations are simplified if it is assumed that $s_r = 0$. This is known as the *unilateral approximation*. The inequality (11.27) may be used to check the maximum error that might result from this approximation.

Given that matched loads are normal for the source and load terminations, the design problem is to derive appropriate matching networks that transform them to the desired terminations $\Gamma_S(\text{opt})$ and $\Gamma_L(\text{opt})$. The approach in §6.11 may be used for this purpose. Other values of Γ_S and Γ_L will be sought when the two-port network is potentially unstable, but again the given matched loads are transformed via matching networks to the Γ_S and Γ_L required. The same matching techniques may also be employed in low-noise design where a specific Γ_S is again required to achieve optimum noise performance.

If the two-port network is potentially unstable, simultaneous conjugate matching cannot be achieved and the maximum gain approaches infinity. Now, either an attempt may be made to stabilize the network, say as described in §11.4.3, or the potential instability is retained but the unstable regions are avoided by a suitable choice of Γ_S and Γ_L. Stability is assured if neither Γ_S nor Γ_L enters the instability-causing region as defined by the stability circles of §11.4, as then both $|\Gamma_{out}|$ and $|\Gamma_{in}|$ are less than unity.

That input and output loading can produce stabilization was seen in §11.4.3 from (11.8). Whether this loading is regarded as a part of the load or as a part of the two-port network itself does not alter the physical fact of the loading. If this loading is made a permanent part of the two-port network itself, then the resultant two-port network can be made *absolutely* stable, irrespective of what additional terminations the source and load may present. One important advantage that accrues if the two-port network is permanently stabilized is that the conjugately matched condition can be achieved at both the input and output. In view of the result in Exercise 11.1 and as also discussed in §2.1.5, this implies that with suitable lossless matching circuits a matched two-port network can be obtained. This is not possible if the two-port network is potentially unstable and the unstable regions are merely avoided. On the other hand, stabilization by loading the input or output will produce a reduction of gain and may be detrimental to other design aims.

The design process is helped by noting that terms of the form (11.12b) give circular loci on the Smith Chart. If

$$G = \frac{1 - |\Gamma|^2}{|1 - s\Gamma|^2} \tag{11.32}$$

then, in the complex Γ plane and with s constant, points that produce the same G lie on a circle with center C and radius R, given [11.1] by

$$C = \frac{Gs^*}{1 + G|s|^2} \tag{11.33}$$

and

$$R = \frac{\sqrt{(1 - G) + G|s|^2}}{1 + G|s|^2} \tag{11.34}$$

The loci for different values of G produce nested circles as illustrated in Figure 11.5 with the centers on the line between s^* and the origin. The maximum value of G

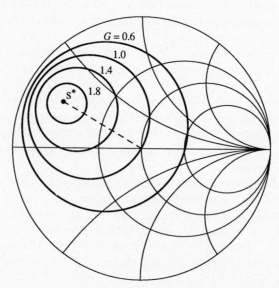

Figure 11.5 The loci of the normalized gain function, G, plotted in the Γ-plane for $s = 0.7\underline{/150°}$. When $\Gamma = s^*$, G is a maximum. In this case, $G(\text{max}) = 1.961$.

occurs when $\Gamma = s^*$. It is also interesting to note that the circle for $G = 1$ passes through the origin, i.e. $\Gamma = 0 \rightarrow G = 1$.

Nested circles in the Γ_S plane are also a feature of noise characterization. It will be noted that the term that contains the variable Γ_S in (11.31) can be re-expressed as follows:

$$N = \frac{|\Gamma_S - \Gamma_0|^2}{1 - |\Gamma_S|^2} = |\Gamma_0|^2 \left[\frac{|1 - (1/\Gamma_0)\Gamma_S|^2}{1 - |\Gamma_S|^2} \right] \tag{11.35}$$

This equation is the reciprocal of the term of similar form on the right hand side of (11.32). Thus loci of constant T_n are also circles in the Γ_S plane. Their centers and radii can be obtained after appropriate variable transformations from (11.33) and (11.34). With N defined in (11.35), the center C_N and radius R_N for constant N become [11.1]

$$C_N = \frac{\Gamma_0}{1 + N} \tag{11.36}$$

$$R_N = \frac{\sqrt{N^2 + N(1 - |\Gamma_0|^2)}}{1 + N} \tag{11.37}$$

Oscillator design at microwave frequencies is also considered from the two-port network point of view, rather than from an analysis of the positive feedback paths as at lower frequencies, because of the complexity of the equivalent circuits when all the required parasitic elements are included. The condition $\Gamma_S \Gamma_{in} = 1$ must hold if steady-state oscillations exist in a two-port network that is terminated by passive terminations, Γ_S and Γ_L. This condition follows from the fact that with steady-state oscillations present, the wave reflected from the input becomes the wave incident upon the source and the wave reflected from the source is the wave incident upon the input. As $|\Gamma_S| < 1$, it follows that $|\Gamma_{in}| > 1$. As is to be shown in Exercise 11.7, for $|s_i|$ and $|s_o|$ less than unity, $\Gamma_S \Gamma_{in} = 1$ also implies $\Gamma_L \Gamma_{out} = 1$ and vice versa. The condition $|\Gamma_{in}| > 1$ may be due either to the inherent instability of the two-port network or, if the two-port network is absolutely stable, to the potential instability deliberately introduced by incorporating feedback into the two-port network, for example through an inductance in the source lead of a FET in the common source configuration. Once $|\Gamma_{in}| > 1$ has been achieved, then an appropriate Γ_S must be presented to the two-port network.

While the condition for steady-state oscillations is $\Gamma_S \Gamma_{in} = 1$, the requirement for the growth of oscillations cannot be expressed in terms of these parameters. Some sources state that for oscillations to build up from noise disturbances, $|\Gamma_S \Gamma_{in}|$ must be greater than unity, but this is not correct. Even if there is a growth of oscillations, $|\Gamma_S \Gamma_{in}|$ can be made less than unity by an appropriate choice of characteristic impedance. For example, if Z_0 were chosen to equal a real Z_S, Γ_S would be zero, giving $|\Gamma_S \Gamma_{in}| = 0$ irrespective of whether or not Z_S allows growth of oscillations. The condition for growth of oscillation must be expressed in terms of immittances, either as

$$Re(Z_S) + Re(Z_{in}) < 0 \tag{11.38}$$

for a circuit that is effectively series resonant, or as

$$Re(Y_S) + Re(Y_{in}) < 0 \tag{11.39}$$

for the parallel-resonant case. Whether a series- or parallel-resonant condition exists cannot be determined from the two-port properties at a spot frequency. The behavior of the two-port network over a range of frequencies must be determined.

The reader is referred to the references at the end of this chapter, in particular [11.1, 11.4, 11.5], for further details on the design processes.

Example 11.3

Consider the following set of typical scattering parameters for a bipolar transistor at 1.0GHz:

$$s_i = 0.6\,\underline{/-100°} \qquad\qquad s_r = 0.04\,\underline{/33°}$$

$$s_f = 5.0\,\underline{/110°} \qquad\qquad s_o = 0.8\,\underline{/-30°}$$

Design a stable amplifier with a transducer gain that is greater than 14dB.

Solution:

The maximum transducer gain that may be achieved with the unilateral approximation is

$$G_{t_u}(max) = 108.5 \qquad (i.e.\ 20.4dB) \qquad\qquad \text{from (11.25)}$$

This will be achieved when

$$\Gamma_S = s_i^* = 0.6\,\underline{/+100°}$$

and $\qquad\quad \Gamma_L = s_o^* = 0.8\,\underline{/+30°}$

Provided that the stability conditions are met, this gain will be more than adequate as it gives a 6.4dB margin over the 14dB specification. To check for stability, the stability factor K is calculated as

$$K = 0.651 \qquad\qquad\qquad \text{from (11.7)}$$

Since $K < 1$, the two-port network is thus potentially unstable and the construction of stability circles is required. These circles are now plotted on the Smith Chart, on which it will be observed whether the source and load impedances are within or close to the unstable regions. In the Γ_S plane

$$C_S = 2.76\,\underline{/135°} \ \text{ and } R_S = 2.01 \qquad\qquad \text{from (11.5)}$$

Likewise in the Γ_L plane

$$C_L = 1.40\,\underline{/43°} \ \text{ and } R_L = 0.53$$

The stability circles are plotted on Figure 11.6, the figure being interpreted as either the Γ_S or the Γ_L plane as appropriate. Since $|s_i|$ and $|s_o|$ are both less than unity, the center of the Smith Chart corresponds to the stable Γ_S and Γ_L regions and the regions to be avoided in choosing Γ_S and Γ_L are contained *within* the stability circles.

It is seen that both s_i^* and s_o^* lie in stable regions, but somewhat close to the limits of stability. As the true transducer gain G_t approaches infinity when a

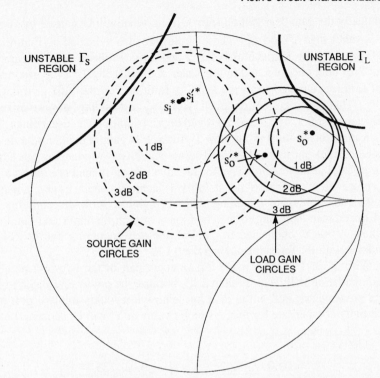

Figure 11.6 A single stage amplifier design from two-port parameters, showing the unstable region in the Γ_S plane together with the circles of gain degradation in dB about s_i^*. Corresponding plots are also given in the Γ_L plane.

stability circle is approached, $G_{t_u}(max)$, which is finite, would not be expected to be an accurate approximation to G_t when s_i^* and s_o^* are close to the stability boundary. To check the validity of the $G_{t_u}(max)$ approximation, u is now calculated to give

$$u = 0.42 \qquad \text{from (11.28)}$$

and $$\frac{1}{(1+u)^2} \Rightarrow -3.0\,\text{dB}, \quad \frac{1}{(1-u)^2} \Rightarrow 4.7\,\text{dB} \qquad \text{from (11.27)}$$

That is, at $\Gamma_S = s_i^*$ and $\Gamma_L = s_o^*$, $G_{t_u}(max)$ may differ from G_t by as much as 4.7 dB. If so desired, G_t may actually be calculated using (11.11a) or (11.11b).

It is not desirable to work close to the stability limits as the gain will then be very sensitive to parameter changes. It is better either to choose new values for Γ_S and Γ_L that are much farther from the stability limits or to make the transistor absolutely stable and design for the $G_{t_u}(max)$ of the new two-port network. In either case a reduction of gain will result, but there is a sufficient gain margin in the 6.4 dB value calculated earlier.

Gain circles are now constructed using (11.33) and (11.34) to indicate graphically the gain degradation from $G_{t_u}(max)$ that will be caused by choosing Γ_S and Γ_L other than s_i^* and s_o^*. These circles are also indicated on Figure 11.6 and are drawn for gain degradations of 1, 2 and 3dB. Probably a gain degradation of about 2dB could be tolerated in this case. A Γ_L of $0.35 \underline{/20°}$ is approximately at a point furthest removed from the stability limit and on the 2dB gain degradation circle. This Γ_L may be obtained from s_o^* by adding a normalized shunt conductance of 0.37 (i.e. $\approx 135\,\Omega$) together with some shunt susceptance.

Rather than choosing the new Γ_L for a 2dB gain degradation, a design will now be carried out that achieves absolute stability by connecting an admittance Y' in parallel at the output. The $135\,\Omega$ value just calculated is used as a guide to the choice of Y'. If absolute stability is achieved then, as explained earlier, simultaneous conjugate matching will be possible, with the consequence that good impedance matching to matched loads at both the input and output will be possible. With Y' connected in parallel, the new scattering matrix $[S']$ may be calculated with the formulae of Exercise 11.8.

With Y' of $1/(135\,\Omega)$ made a permanent part of the two-port network, the gain degradation will be more than 2dB, because the power dissipated in Y' is no longer power dissipated in the load but is now lost within the two-port network. To partially compensate for this, a smaller value of Y' will be chosen. Let us take $Y' = \dfrac{1}{200\,\Omega}$. This makes

$$s_i' = 0.609 \underline{/-98.2°} \qquad s_r' = 0.033 \underline{/35.4°}$$

$$s_f' = 4.123 \underline{/112.4°} \qquad s_o' = 0.491 \underline{/-33.7°}$$

and gives $K = 1.774$ (i.e. $K > 1$) together with $|\Delta| = 0.307$ (i.e. $|\Delta| < 1$). The new two-port network is now absolutely stable. The maximum unilateral transducer gain is now

$$G_{t_u}(max) = 35.59 \qquad \text{(i.e. 15.5dB)}$$

and is achieved with

$$\Gamma_S = s_i'^* = 0.609 \underline{/+98.2°} \quad \text{and} \quad \Gamma_L = s_o'^* = 0.491 \underline{/+33.7°}$$

The new $s_i'^*$ and $s_o'^*$ are also indicated in Figure 11.6.

The new $G_{t_u}(max)$ apparently meets the 14dB specification. However, it is necessary to check the errors introduced through the unilateral approximation by calculating the new u. With $u = 0.09$

$$\frac{1}{(1+u)^2} \Rightarrow -0.71\,\text{dB}, \qquad \frac{1}{(1-u)^2} \Rightarrow 0.77\,\text{dB}$$

It is seen that even with a possible downward error margin of 0.77dB, 14dB gain is still achieved.

Thus a $200\,\Omega$ resistor in parallel with the output adequately stabilizes the two-port network and choosing $\Gamma_S = 0.609 \underline{/98.2°}$ and $\Gamma_L = 0.491 \underline{/33.67°}$ will provide more than 14dB gain. At this point the design is complete, apart from calculating the required matching networks.

Comments

It is of interest to see how the true maximum transducer gain, $G_t(\text{max})$, compares with $G_{t_u}(\text{max})$. Using (11.22)

$$G_t(\text{max}) = 38.6 \qquad \text{(i.e. 15.9dB)}$$

Thus $G_t(\text{max})$ differs from $G_{t_u}(\text{max})$ by only 0.4dB. Of course, the Γ_S, Γ_L required to achieve $G_t(\text{max})$ will be somewhat different from those required for $G_{t_u}(\text{max})$. It is left to the reader to calculate the Γ_S, Γ_L for $G_t(\text{max})$ from Equations (11.4).

It is also of interest to calculate the maximum stable gain, given when $K = 1$, namely

$$\left| \frac{s_f}{s_r} \right| = \left| \frac{s_f'}{s_r'} \right| = 125 \quad \text{(i.e. 21.0dB)} \qquad \qquad \text{from (11.23)}$$

This shows that up to 21.0dB, or 5.1dB more than $G_t(\text{max})$, could have been obtained with a different stabilizing resistor arrangement. Thus, the present design is significantly overstabilized. If a higher gain were desired, different Y' values could be tried.

The reader will have noticed that, by simply choosing $\Gamma_S = \Gamma_L = 0$ (i.e. simple matched loads) a transducer gain $G_t = |s_f|^2$ giving 14.0dB could have been obtained and in the process satisfied the design requirements. Also, $\Gamma_S = \Gamma_L = 0$ would probably be far enough removed from the stability boundaries to provide a sufficiently stable design. However, the two-port network then would not be matched as seen from either the source or the load and there would be a significant V.S.W.R. on the connecting transmission lines. In particular, this means that the two-port network would provide a load or source impedance to the previous or following stages respectively that is dependent on the length of interconnecting transmission line.

EXERCISES

11.1 Prove that for a lossless reciprocal two-port network

$$\Gamma_S = \Gamma_{\text{in}}^* \iff \Gamma_L = \Gamma_{\text{out}}^*$$

11.2 This question relates to the stability properties of a two-port network.
   i) To what values do K, Δ and $G_t(\text{max})$ tend, as s_i and s_o tend towards zero?
   ii) What is the value of K for the balanced amplifier of Exercise 2.3(iii)? What are the implications of this result for stability?
   iii) When a two-port network is unilateral, verify that the formulae for $\Gamma_S(\text{opt})$, $\Gamma_L(\text{opt})$, and $G_t(\text{max})$ reduce to the correct values.
   iv) Show that the stability criterion is satisfied if $|s_i| < 1$, $|s_o| < 1$ and $s_r = 0$.
   v) If $|s_i| > 1$, show that $K < 1$ and/or $|\Delta| > 1$.

11.3 Derive an explicit expression for G_t in terms of s-parameters and Γ_S, Γ_L only.

11.4   A source a_s, Γ_X is connected through a zero length line of characteristic impedance Z_0 to a load Γ_Y. Show that

$$\frac{1-|\Gamma_X|^2}{|1-\Gamma_X\Gamma_Y|^2} = \frac{|a|^2}{P_{avs}} \quad \text{and} \quad \frac{1-|\Gamma_Y|^2}{|1-\Gamma_X\Gamma_Y|^2} = \frac{P_L}{|a_s|^2}$$

where "a" is the incident wave in the Z_0 line, and P_L is the power absorbed by the load. When $\Gamma_Y = 0$, the variable "a" becomes a_s.

11.5   Using the approach of Examples 11.1 and 11.2, derive the formulae for G_a and G_p:
   i)   in the matched case with $\Gamma_s = \Gamma_L = 0$
   ii) in the general case with arbitrary Γ_s and Γ_L.

11.6   Prove that in terms of y-parameters

$$G_{t_u}(\max) = \frac{|y_f|^2}{4\,g_i\,g_o} \quad \text{where } g = Re(y)$$

11.7   Prove that with $|s_i| < 1$ and $|s_o| < 1$

$$\Gamma_{in}\Gamma_s = 1 \iff \Gamma_{out}\Gamma_L = 1$$

11.8   Consider a two-port network with the scattering parameters $[S]$. Connect an admittance Y' in parallel across the output to produce a two-port network with the scattering parameters $[S']$. All parameters are normalized to Z_0. Show that

$$s_i' = s_i - \frac{s_r s_f \xi}{\xi'} \qquad\qquad s_r' = \frac{s_r}{\xi'}$$

$$s_f' = \frac{s_f}{\xi'} \qquad\qquad s_o' = \frac{1+s_o}{\xi'} - 1$$

where $\quad \xi = \dfrac{Y'Z_0}{2}, \quad \xi' = 1+\xi(1+s_o) \text{ and } \Delta = s_i s_o - s_r s_f$

11.9   Noting the relevant symmetries in the $[S']$ of Exercise 11.8, deduce the $[S']$ when Y' is connected in parallel at the input.

11.10 Consider a two-port network with the following scattering parameters:

$$s_i = 0.5\,\underline{/-90°} \qquad\qquad s_r = 0.05\,\underline{/33°}$$

$$s_f = 10.0\,\underline{/100°} \qquad\qquad s_o = 0.4\,\underline{/-20°}$$

   i)   Check for stability without and with a 100Ω load in parallel at the input and output in turn.
   ii)   Using the 100Ω loading at the input for stabilization, select Γ_s, Γ_L for maximum gain. Compare $G_{t_u}(\max)$ with and without stabilization.
   iii) Repeat part (ii) for 100Ω loading at the output.
   iv)   Return to the original two-port network without loading. Draw the appropriate gain degradation circles and choose Γ_s, Γ_L to obtain a G_{t_u} that is equal to the $G_{t_u}(\max)$ of the network with the 100Ω loading at the input.

11.11 The insertion gain of a two-port network is formally defined as P_{out}/P_{out}', where P_{out} is the output power as in Figure 11.1 and P_{out}' is the power that would flow to Z_L if the two-port network was removed and the source was connected directly to the load. Show that the insertion gain is identical to the transducer power gain if $Z_s = Z_L^*$.

REFERENCES

[11.1]   Gonzalez, G., *Microwave Transistor Amplifiers*, Prentice-Hall, Englewood Cliffs, NJ, 1984.

[11.2]   Haykin, S. S., *Active Network Theory*, Addison Wesley, Reading, MA, 1970.

[11.3]   Ha, T. T., *Solid-state Microwave Amplifier Design*, Wiley, New York, 1981.

[11.4]   Pengelly, R. S., *Microwave Field-effect Transistors Theory, Design and Applications*, Wiley, New York, 1986.

[11.5]   Vendelin, G. D., *Design of Amplifiers and Oscillators by the s-parameter Method*, Wiley, New York, 1982.

[11.6]   Woods, D., "Reappraisal of the unconditional stability criteria for active 2-port networks in terms of s-parameters", *IEEE Trans. on Circuits and Systems*, Vol. CAS-23, No. 2, February 1976, pp. 73-81.

[11.7]   Spence, R., *Linear Active Networks*, Wiley, London, 1970.

[11.8]   Jorsboe, H., "Criterion for absolute stability of two-ports", *IEEE Trans. on Circuits and Systems*, Vol. CAS-17, No. 4, November 1970, pp. 639-40.

[11.9]   Bodway, G. E., "Two port power flow analysis using generalised scattering parameters", *Microwave Journal*, Vol. 10, No. 6, May 1967, pp. 61-9.

[11.10] Jorsboe, H., "A simple proof of the central activity criterion", *IEEE Trans. on Circuits and Systems*, Vol. CAS-15, No. 3, July 1968, pp. 148-9.

[11.11] Meys, R. P., "A wave approach to the noise properties of linear microwave devices", *IEEE Trans. Microwave Theory and Techniques*, Vol. MTT-26, No. 1, January 1978, pp. 34-7.

12 Microstrip circuits and subsystems

12.1 INTRODUCTION

This chapter introduces the reader to selected practical microstrip circuits or subsystems. It brings together some of the ideas of earlier chapters and shows how they may be combined to produce functioning self-contained building blocks, that in turn may be a part of a complete microwave system.

The circuits have been adapted with our interpretations from those described in the literature and have been chosen for the simplicity and clarity with which they illustrate certain key circuit arrangements. The circuits described are a low-noise amplifier in §12.2, a balanced mixer in §12.3 and a selection of switching circuits in §12.4.

12.2 A LOW-NOISE AMPLIFIER

A two-stage low-noise amplifier circuit is illustrated in Figure 12.1, adapted from Fulton [12.1]. The input and output strips, ① and ⑭ respectively, are 50Ω lines for connection to the coaxial/microstrip transitions. The two FETs (e.g.⑤) have their sources connected with minimum path lengths to the ground plane and thus present the common source configuration to the matching circuits. The matching circuits use microstrip lines of the same characteristic impedance (50Ω) in series and shunt, except for lines ③ and ⑬ that are quarter-wave transformers. Three chip capacitors (e.g.②) provide microwave coupling between circuits at different d.c. bias potentials, as well as d.c. isolation of externally connected microwave circuits.

As this is a low-noise amplifier, the first stage input matching network, ③ and ④, converts the matched source to the source reflection coefficient, Γ_0, for which the transistor has its minimum noise figure. The second stage contribution to the noise figure is small, since the noise contribution of the second stage to the overall noise figure is reduced by the gain of the first stage. Nevertheless, when very low noise figures are to be achieved it is desirable also to optimize the second stage for low noise. Thus the interstage network (⑥, ⑦ and ⑧) transforms the Γ_{out} of the first stage to the noise-optimum source impedance Γ_0 to drive the input of the second stage.

The output matching network, ⑨ to ⑬, optimizes the power gain. It thus transforms the external 50Ω load to produce a conjugate match to the Γ_{out} of the

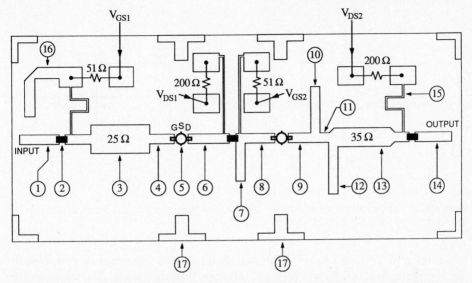

Figure 12.1 A low-noise two-stage amplifier circuit, adapted from Fulton [12.1] (Reprinted with permission of *Microwave Journal*, from the November 1984 issue, © 1984 Horizon House, Inc.)

second stage. This matching network is also designed to produce a band-pass filter characteristic in the gain at frequencies away from the center frequency.

Four separate bias voltages are provided to the gate and drain terminals of the two transistors. The series resistors in the bias supply connections can be used to monitor bias currents. The bias circuits are decoupled from the microwave circuits by high impedance (120Ω) lines (e.g.⑮) that are approximately λ/4 long. The large patches at the end of these lines, as well as the points where the d.c. bias potentials are connected, provide low impedance paths to ground. The rejection of spurious signals is most critical at the input to the first stage, where an even lower impedance to ground is achieved for the V_{GS1} supply by the open-circuit terminated 50Ω line, ⑯, that is approximately λ/4 long. Space is conserved by folding this line with a mitered right-angle corner.

The pads, ⑰, are connected to the ground plane and may provide points of attachment for interstage shields that pass over the center of each transistor and minimize unwanted feedback in the circuit.

12.3 MIXERS

12.3.1 Balanced mixers

Mixer circuits may be used whenever there is a need to translate signals between frequency bands. When two signals of differing frequencies are fed into a non-linear element such as a diode, numerous intermodulation products are produced, including the sum and difference frequencies of the signals. Mixer circuits are often configured

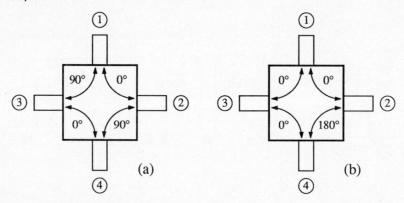

Figure 12.2 The phase relationships for hybrid networks, showing (a) the quadrature or 90° hybrid, and (b) the 180° hybrid

in the form of balanced mixers with two mixer diodes connected to two mutually isolated ports of a 3 dB hybrid network. Hybrids that are used in this connection are invariably one of two types: the 90° or quadrature hybrid and the 180° hybrid. Each type is illustrated diagrammatically in Figure 12.2. An input signal into any port of such a hybrid will deliver half the input power into two other ports and zero power into the remaining fourth port, i.e the isolated port. For both types of hybrid, there is mutual isolation between ports 1 and 4 and between ports 2 and 3. Each set of mutually isolated ports becomes the output ports when an input signal excites either port in the other set. The difference between the two types of hybrid lies in the phase differences between the outputs.

As illustrated in Figure 12.2, the two outputs are always 90° apart in the quadrature hybrid, while in the 180° hybrid they are either in-phase or 180° out-of-phase. For example, in Figure 12.2a, an input to port 1 will give outputs at ports 2 and 3 that are 90° apart from each other. However, it is important to note that the phase values shown in Figure 12.2 correctly give only the *phase differences* between outputs. There may be additional phase shifts from an input to both outputs, not affecting the phase difference between the outputs. These common phase shifts, if any, between the input and each output are not shown here.

The 3 dB edge-coupled directional coupler and the 3 dB Lange coupler, together with the branch-line coupler of earlier chapters, are examples of quadrature hybrids. The hybrid-ring coupler is an example of the 180° hybrid.

In a balanced mixer as illustrated in Figure 12.3, the signal (microwave or radio frequency, r.f.) and the local oscillator (l.o.) are applied to one pair of mutually isolated ports, say 1 and 4, and a pair of mixer diodes that are closely matched in their characteristics are connected to the other pair of ports. When the two diodes have the same orientation with respect to ground, as in Figures 12.3a, b and c, the intermediate frequency (i.f.) outputs across the two diodes are out-of-phase and may be combined by subtracting one output from the other. If the two diodes are connected with opposing polarities to ground, the i.f. outputs are in-phase and are combined with a summing circuit, as in Figure 12.3d. The subtraction and summing circuits may themselves be hybrid networks at the intermediate frequency.

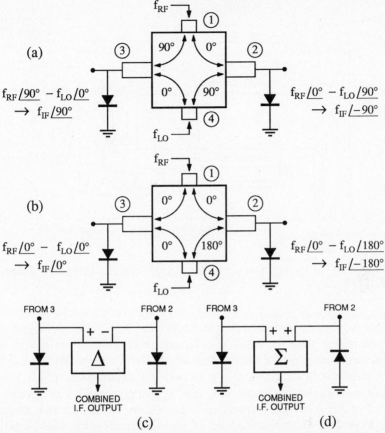

Figure 12.3 Balanced mixers with a quadrature hybrid in (a) and a 180° hybrid in (b). The diode outputs are combined as in (c) if they have the same orientation with respect to ground and as in (d) if they are connected anti-phase with respect to ground.

There are a number of reasons why balanced mixers are used. It turns out that the local oscillator amplitude modulation noise components, when mixed down to the i.f. band, appear across the two diodes in-phase when the i.f. signals are out-of-phase and vice versa, so that when the signal i.f. outputs are combined these noise components cancel. Considerable cancellation of local oscillator amplitude modulation noise can be achieved in this way. For similar reasons balanced mixers can reject many spurious intermodulation products. Further, because the r.f. and l.o. signals are applied to mutually isolated ports of the hybrid, there can be good isolation between these signals, although not in every case [12.2]. For further study in this area, the reader is referred to Maas [12.2].

12.3.2 A balanced mixer example

A balanced mixer using a 90° hybrid of the branch-line type is shown in Figure 12.4, based on Johnson [12.3]. The structure is quite symmetrical as far as the r.f. and l.o.

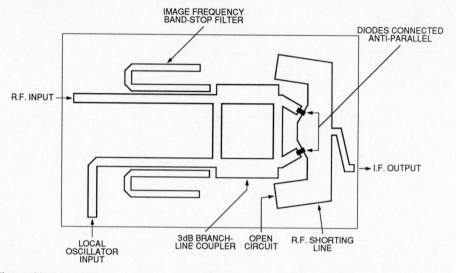

Figure 12.4 An image-rejection balanced mixer circuit, based on Johnson [12.3] (© 1968, IEEE)

inputs are concerned and the r.f. and l.o. inputs may be interchanged. The l.o. frequency is less than the input signal frequency, with the difference being the intermediate frequency. The image frequency is that frequency that also differs from the local oscillator by the same amount, i.e. the image frequency $= (f_{LO} - f_{IF})$.

The two diodes are connected with opposing polarities to ground and, as explained earlier, their i.f. outputs are to be added. At i.f., the two diodes appear simply in parallel, which results in the i.f. signal currents being added together. The i.f. ground return is not shown in Figure 12.4, but may take the form of a high impedance $\lambda_{LO}/4$ line with an i.f. short-circuit termination that is connected in parallel with the l.o. input line. The r.f. and l.o. frequencies are decoupled from the i.f. output by r.f. short circuits at the two diodes, presented here by the two open-circuit terminated low impedance $\lambda/4$ lines. The degree of isolation between the r.f. and l.o. circuits will depend on how well the diodes are matched to $50\,\Omega$ at these frequencies [12.2].

There are band-stop filters at the image frequency in the r.f. and l.o. lines. Their purpose is to present a reactive termination at the image frequency at each diode position. This improves the diode conversion loss and consequently reduces the noise of the mixer−i.f. preamplifier combination.

The band-stop filters are of the type discussed in §9.5, where it was shown that at the center of the band-stop band they have a reflection coefficient $\Gamma = +1$, i.e. they present an open circuit at their input. Thus, in this balanced mixer, there are open circuits at the image frequency at the r.f. and l.o. junctions to the hybrid network. These open circuits reflect certain impedances at the planes of the diodes. It was shown in Exercise 7.7 that these resultant impedances depend on the excitation mode (even or odd).

For analysis to determine the impedances seen by the diodes at the image frequency, the diodes are taken as the sources that are connected to the inputs of the hybrid. In fact the diodes are in front of the hybrid by lines that are $\sim \lambda/8$ long. The hybrid is now also taken as being terminated with open circuits at the other two ports for this frequency. It may be shown, using the results of Exercise 7.7 and noting the presence of the $\lambda/8$ lines, that the impedances seen by the diodes are open and short circuits for the even- and odd-mode excitation respectively. As a consequence, there is no power absorbed by the circuits external to the diodes at the image frequency.

12.4 SWITCHING CIRCUITS

Circuits with two or more discrete states of operation require switching elements to switch between the states. The ideal switch is one that can instantly change between high and low impedance conditions. Electronic devices that can be externally switched with d.c. voltages include the field effect transistor (FET) and the p-n, p-i-n and Schottky diodes.

12.4.1 Switching elements

The field effect transistor

The FET switch, as illustrated in Figure 12.5, is a three-terminal device with the switched states controlled by the gate voltage. It is preferably operated in shunt between the microstrip line and ground, an arrangement that is facilitated by the drain connections forming a through-line, which can be readily connected as a part of a microstrip line as in Figure 12.5b. The FET should be operated in the linear resistance region of its characteristics with $V_{DS} = 0$. At this operating point, shown in Figure 12.5a, there will also be a minimum power dissipation with only negligible reverse bias currents being supplied to the gate. With $V_{DS} = 0$, it will be seen that the microstrip line may be maintained throughout at d.c. ground potential for all the FETs. Further, interstage decoupling in a multi-FET circuit will not be required,

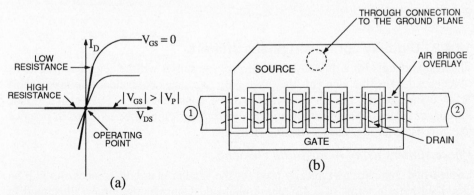

Figure 12.5 A field effect transistor, showing (a) the linear operating regions for a switch, and (b) the structure for a monolithic distributed switch approach, from Ayasli [12.4] (Reprinted with permission of *Microwave Journal*, from the November 1982 issue, © 1982 Horizon House, Inc.)

since the control voltages are provided to the third or gate terminal of each individual FET and thus will not affect the d.c. potentials of the microstrip lines.

In the ON-state with $V_{GS} = 0$, the microwave equivalent circuit between the drain and source will be a low resistance, typically of the order of 3Ω. In the OFF-state, when $|V_{GS}| > |V_P|$, there is a much higher resistance of the order of $3k\Omega$. V_P is the pinch-off voltage for the reverse biased gate/channel. However, there are also significant shunt capacitances in the OFF-state. It is for this reason that the FET is parallel-connected between the microstrip line (drain) and the ground plane (source) and the low impedance ON-state is used to provide the circuit isolation. This configuration is the form described by Ayasli [12.4] and shown as an integral part of a monolithic microwave integrated circuit in Figure 12.5b. Discrete FET components for a hybrid microwave circuit will be connected to the circuit in a similar manner but, of course, will have additional series inductances associated with the connecting leads. In the OFF-state, the capacitance components may be allowed for as a part of the matching circuits and the microwave signal may travel past the plane of the shunt FET with minimal reflections or absorption of power in the high impedance parallel equivalent circuit.

The p-i-n diode

The p-i-n diode, described in detail with circuit applications by White [12.5], is a 3-layer silicon diode with the heavily doped p and n regions separated by a high resistivity intrinsic layer.

The diode exhibits a low dynamic resistance when a forward bias current is applied and carriers are injected and maintained in the intrinsic region. This resistance, typically less than 1Ω at 1.0 GHz, applies even at high microwave current amplitudes as the waveform is not readily rectified, the reverse half-cycle of the waveform being too short to remove the charge that is stored in the intrinsic region. When a reverse bias voltage is applied, there are no injected charges in the intrinsic region and the diode is prevented from conducting, even under the application of large peak microwave voltages. With reverse bias, the microwave equivalent circuit of the diode is a few ohms in series with the junction capacitance that is of the order of 1 pF.

12.4.2   Digitally controlled phase shifters

The design details for a digitally controlled phase shifter will be considered in terms of idealized switching elements that have only two states equivalent to either a short or an open circuit. The phase shifter will be considered in a 50Ω characteristic impedance system. Modifications to the idealized design may be required in practice to compensate for non-ideal switching elements.

Phase shifters using loaded-line elements

Loaded-line elements are typically used for the smaller phase differences and will be described here in terms of a 45° phase element. Consider the circuit illustrated in Figure 12.6. The two stub lines are terminated with identical switching elements that, with their open- or short-circuit impedances, make the circuit identical to Figure 7.7 for the even- or odd-mode analysis of hybrid-line couplers. The two-port network

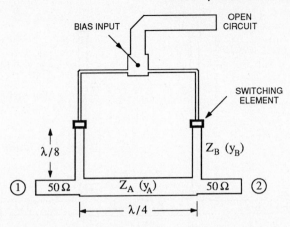

Figure 12.6 The loaded-line digital phase shifter. The microstrip line may be at d.c. ground, in common with any adjacent elements.

phase relationship is shown in Figure 12.7 and should be compared directly with Figure 7.8. The $-90°$ reference phase, with any additional phase shifts due to input and output line lengths, is a constant for both states and of no further concern in a differential phase shifter. The two states are symmetrical about the reference phase. For a 45° phase shifter, $\theta = 22.5°$. Based on (7.17)

$$\theta = \tan^{-1}(y_B) \tag{12.1}$$

giving $y_B = 0.4142$ and $Z_B = 121\Omega$. The through line between the two stubs is $\lambda/4$ long and, from (7.15), namely

$$y_A^2 = 1 + y_B^2 \tag{12.2}$$

has a normalized characteristic admittance that is always greater than unity. Thus the characteristic impedance is always less than 50Ω. With $y_B = 0.4142$, then $y_A = 1.082$, i.e. $Z_A = 46.2\,\Omega$. Both line characteristic impedances are within the practical limitations for microstrip lines.

Turning now to a 22.5° phase shifter with $\theta = 11.25°$, the respective values for Z_A and Z_B become 49Ω and 251Ω, with the impedances presented to the through line being $\pm j251\Omega$. The through-line impedance is close enough to 50Ω, but the high

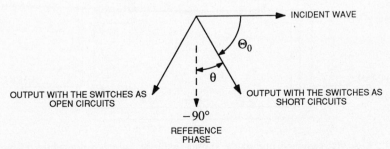

Figure 12.7 The phase relationships for a loaded-line digital phase shifter

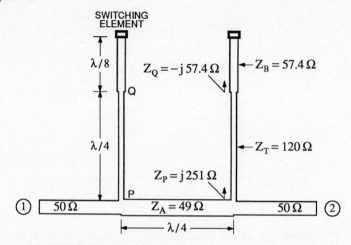

Figure 12.8 $3\lambda/8$ length stubs for small phase shift sections. The impedance values shown are for $22.5°$ phase shift with the switching elements in the open-circuit state.

value for the stub-line impedance may be outside the range of practical line values. To overcome this problem, a $3\lambda/8$ stub length may be used as illustrated in Figure 12.8, where the $\lambda/4$ section branching off the through line acts as a quarter-wave impedance transformer. Taking $Z_T = 120\Omega$ for the quarter-wave transformer, the $\pm j251\Omega$ impedance for the stub at P requires an impedance $Z_Q = \mp j57.4\Omega$ at Q. This is obtained with $Z_B = |Z_Q|$. Note that whereas short-circuit terminations on the $\lambda/8$ stubs gave the lesser total phase value, Θ_0, now the reverse is the case with $3\lambda/8$ stubs.

Phase shifters using switched-line elements

Switched-line phase shifters are used when the larger phase shifts are required. Between the input and output there are two separate paths, either one of which is selected by the switching elements. The case of a $90°$ element, illustrated in Figure

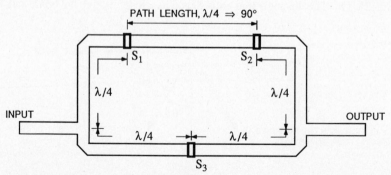

Figure 12.9 A $90°$ switched-line phase shifter, with the switches connected in parallel to the ground plane

12.9, will be considered. Here again, the switching elements are parallel-connected between the microstrip line and the ground plane.

In State 1, the parallel-connected elements S_1 and S_2 are low impedances to ground while S_3 is a high impedance. A direct path from input to output is provided across S_3. The input junction sees a $\lambda/4$ line terminated by the short circuit of S_1, which reflects an open circuit at the input junction. A similar effect occurs at the output because of S_2.

In State 2, the impedance of each switching element is changed and a direct path between the input and output junctions now passes across S_1 and S_2. S_3 is equidistant from each junction and its low impedance transforms through the $\lambda/4$ lines to appear as a high impedance at a junction. The through path in State 2 is $\lambda/4$ longer than for State 1, thus providing the 90° phase shift at the design frequency.

A switched-line circuit that maintains a constant difference between the switched phase states with frequency is described by Burns et al. [12.6]. The delayed or longer path has the additional line length to give a delay $\Delta\phi$. Two $\lambda/4$ short-circuit terminated lines, connected as parallel stubs with a separation of $\lambda/4$, are placed onto the reference or shorter path. Analysis shows that the phase variation with frequency of this circuit, as a part of the reference path, can be made to follow that of the additional matched length of line in the delayed path. Further, at the mid-band frequency, the stub lines appear as open circuits to the reference path and their separation is such that, for small frequency changes, there will be significant cancellation of the reflected waves from the planes of the stubs. As a consequence, a low V.S.W.R. will be maintained. To track the phase variation with frequency of the additional line length in the delayed path, the necessary characteristic impedance of each stub is given [12.6] by

$$Z_{stub} = \frac{\pi Z_0}{2\Delta\phi} \tag{12.3}$$

For the 90° element $Z_{stub} = Z_0$, while for a 180° element a lower impedance of $Z_0/2$ is required.

It is possible with switched-line phase shifters for the path that has apparently been blocked to become resonant. At resonance, a substantial signal may be transmitted through the path, affecting the transmission properties of the phase shifter. This occurs when a path is approximately $n\lambda/2$ long and is terminated at each end with either the high-series or low-shunt impedance of a switching element. The resonance may be damped out by switching in a parallel resistive load at a plane of voltage maximum along the line.

12.4.3 A transmit/receive switch

A transmit/receive (TR) switch is used to connect a transmitter and receiver to a common antenna. In either mode, when the switching circuits have been correctly set, a direct path with a low residual V.S.W.R. is required between the antenna port and either the transmitter or receiver port. This can be achieved with a large mismatch, e.g. a short circuit to ground, introduced into the unwanted path. The mismatch must be placed in such a way that the unwanted path appears as an effective

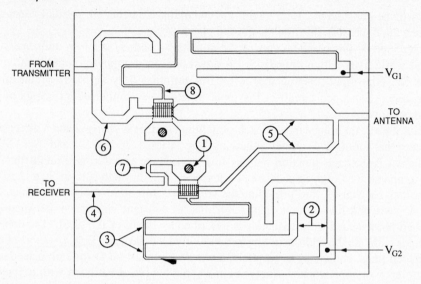

FROM
TRANSMITTER

V_{G1}

TO
ANTENNA

TO
RECEIVER

V_{G2}

Figure 12.10 A 10W TR switch chip fabricated on a GaAs substrate, based on Ayalsi et al. [12.7] (© 1982, IEEE)

open circuit at the combining junction. Thus, in the receiving mode, the path towards the transmitter must appear as an open circuit, so that as much received signal from the antenna is passed through to the receiver. For the transmit mode, it is also essential to prevent excessive power from entering and damaging the first stages of the receiver that will have been designed to operate at very low signal levels.

Ayasli et al. [12.7] describe a 10W TR switch chip that has been totally fabricated on a 0.1 mm GaAs substrate. For operation at 10GHz, the chip dimensions are 4.5×3.7 mm. Single-gate FETs with a construction similar to Figure 12.5 are used as switching elements. An interpretation of the circuit, adapted from [12.7], is illustrated in Figure 12.10. The insertion loss is of the order of 1.0dB with higher than 25dB isolation at any time between the transmitter and receiver ports. Assuming that 50Ω lines are brought out to the edge of the substrate, the other line impedances may be deduced from the approximate relative line widths on a GaAs substrate with $\varepsilon_r = 12.9$. Quasi-static approximations may be used, as it is found from (4.41) that the operating frequency is well below the frequency of 110GHz above which dispersion effects would become significant.

Interpreting the circuit of Figure 12.10

① A through connection to the ground plane.

② Folded open-circuit terminated $\lambda/4$ stubs with a low characteristic impedance ($\approx 43\Omega$) line. Each parallel-connected stub line gives a low microwave impedance at its junction to ③.

③ High characteristic impedance ($\approx 75\Omega$) $\lambda/4$ bias network lines. Each line transforms a low value of load impedance to a high value of input impedance.

④ 50Ω interconnecting lines.

⑤ $\lambda/4$ transformers ($\approx 32\Omega$ and 56Ω) to give an optimum performance for peak voltages and currents in the FETs when 10W is being transmitted (see [12.7]).

⑥ A double-stub matching network with $\approx \lambda/8$ between the stub lines.

⑦ The drain-source connection giving $V_{DS} = 0$ for both transistors. Note that, since the line length $\ll \lambda/4$, this element is also being used as a single stub, short-circuit terminated, matching element.

⑧ The gate bias voltage is fed through a low-pass filter, preventing microwave leakage from the gate circuit that may otherwise occur via the drain-gate capacitance. This low-pass filter is identical in function to that for the other FET, i.e. ③ and ②.

REFERENCES

[12.1] Fulton, R., "6GHz low noise amplifier using NE70083 GaAs FET", *Microwave Journal*, Vol. 27, No. 11, November 1984, pp. 189-91.

[12.2] Maas, S. A., *Microwave Mixers*, Artech House, Dedham, MA, 1986.

[12.3] Johnson, K. M., "X-band integrated circuit mixer with reactively terminated image", *IEEE Trans. Microwave Theory and Techniques*, Vol. MTT-16, No. 7, July 1968, pp. 388-97.

[12.4] Ayasli, Y., "Microwave switching with GaAs FETs", *Microwave Journal*, Vol. 25, No. 11, November 1982, pp. 61-4, 66, 68, 70-2, 74.

[12.5] White, J. F., "Diode phase shifters for array antennas", *IEEE Trans. Microwave Theory and Techniques*, Vol. MTT-22, No. 6, June 1974, pp. 658-74.

[12.6] Burns, R. W., Holden, R. L. and Tang, R., "Low cost design techniques for semiconductor phase shifters", *IEEE Trans. Microwave Theory and Techniques*, Vol. MTT-22, No. 6, June 1974, pp. 675-88.

[12.7] Ayasli, Y., Mozzi, R., Hanes, L. and Reynolds, L. D., "An X-band 10W monolithic transmit-receive GaAs FET switch", IEEE Microwave and Millimeter-wave Monolithic Circuits Symposium, Dallas, Texas, June 1982, pp. 42-6.

13 Microstrip line experiments

13.1 INTRODUCTION

The procedures that are described in this chapter provide an introduction to microstrip line experiments. The ease and precision of any measurement are necessarily related to the quality and sophistication of the equipment that is available for the purpose. However, it should be possible to perform experiments designed primarily for instructional purposes using any reasonable quality equipment. The emphasis will be placed on the measurement of fundamental microstrip parameters, as this will lead to a broader understanding for the design of microstrip circuits.

In §13.2, circuit requirements and procedures for the measurement of scattering parameters that were first introduced in Chapter 2 are discussed. A complete set of s-parameters gives a detailed specification of a two-port network. However, there are many instances where a reduced set of information may be all that is either attainable in view of equipment limitations or, indeed, required. The relationship between such measurements, e.g. V.S.W.R., insertion loss and s-parameters, is discussed.

The microwave literature abounds with experimental techniques for specific microwave measurements and books have been written specifically on the subject, e.g. [13.1–13.4]. Two experimental procedures have been selected because of their relevance to the basic properties of microstrip lines. The first experiment, §13.3, concerns the measurement of both the effective relative permittivity and the characteristic impedance of microstrip lines. This is followed in §13.4 by an experiment that uses resonant structures, through which a better understanding of some of the discontinuity effects as they apply to microstrip lines may be obtained.

The need for precision measurement has been considered to be of secondary importance here, when compared with the needs for simplicity and reliability in experimental design. In particular, some of the end-effect measurement techniques for discontinuity evaluation [13.5] that require further circuit etching between measurement sets have been omitted, because they do not fulfill the necessity for "non-destructive" testing in a teaching laboratory environment.

13.2 S-PARAMETER MEASUREMENTS

Any one of the microstrip circuits described in the earlier chapters may be characterized through their scattering parameters, from which most small-signal properties of the circuits may be inferred. Directional couplers, hybrids, filters, etc., as well as the sub-assemblies of Chapter 12, can have their small-signal properties verified through the measurement of their scattering parameters as functions of frequency. Scattering parameters may be measured either by the direct application of test incident waves, as was described in Chapter 2, or by using the six-port technique [13.6]. The six-port technique, where the magnitude and phase information for the scattering parameters may be deduced from power measurements only, is beyond the scope of this book and is being mentioned only for the sake of completeness.

To measure scattering parameters following the approach in Chapter 2, a test input is applied to one port and the other port is terminated in a matched load. The incident and reflected waves are sampled by means of directional couplers at the input and output ports. The output amplitudes of the directional couplers are compared and the ratios of each appropriate pair of outputs give the relevant scattering parameters. As only two waves are compared at any one time, two directional couplers are sufficient, if the measurement set-up is reconfigured for each individual s-parameter. The schematic diagram in Figure 13.1 shows three directional couplers, two at port 1 and one at port 2. The coupler outputs A and B are used to determine s_i and the coupler outputs A and C to determine s_f. Reversing the device under test (DUT) allows s_o and s_r to be similarly measured.

The phase difference between pairs of outputs is related to the phase of the appropriate s-parameter. The problem is that the phase differences of the appropriate waves at the input and output reference planes to the DUT are required, but it is the phase differences at the directional coupler outputs that are actually obtained. Variable path lengths must be inserted in the connecting transmission lines, so as to make the phase difference measured at the coupler outputs the same as at the reference planes of the DUT. Experimentally this is done by replacing the DUT with a device having known phase characteristics and adjusting the variable line until the correct phase output is measured. When measuring s_i or s_o, the DUT is replaced by a short circuit to produce a reflection coefficient with a constant 180° phase. The variable line is adjusted until *for all frequencies* the phase difference measured is also 180°. By insisting on the measured phase difference being 180° for all frequencies and not just one spot frequency, it is ensured that the transmission paths are exactly equal and do not just differ by $n\lambda/2$. In effect, the short circuit defines the position of each reference plane of the DUT.

When phase-calibrating for s_f and s_r, a suitable reference is a piece of through-

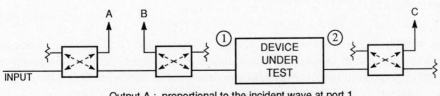

Output A : proportional to the incident wave at port 1
Output B : proportional to the reflected wave at port 1
Output C : proportional to the transmitted wave at port 2

Figure 13.1 The measurement configuration for s_i and s_f

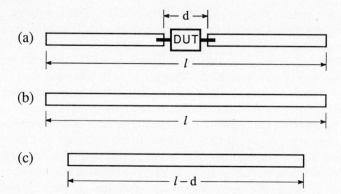

Figure 13.2 Replacing the device under test (DUT) by through-lines for calibration purposes. (a) DUT and its connecting lines. (b) Reference through-line of the same total length as the DUT and its connecting lines. (c) Reference through-line of the same length as the connecting lines.

line, either of the same total length as the DUT (including its connecting lines), Figure 13.2b, or of just the length of the connecting lines, Figure 13.2c. The use of the shorter line as a reference would give proper phases for s_f and s_r of the DUT. The use of the longer reference line would give the phases over and above the phases of s_f and s_r for an ideal line of length d.

It is important that reference planes established by the calibration procedure for reflection are the same as obtained by the calibration for transmission, as inconsistent measurements would otherwise result. In modern equipment only one calibration, say for reflection, is required [13.2]. The correct line lengths for the other measurements are then automatically obtained as a result of accurate line configurations within the equipment.

13.2.1 Related parameters

One often finds devices specified in terms of V.S.W.R., reflection coefficient, return loss, insertion loss or attenuation, isolation, etc., the latter quantities invariably being expressed in dB. These parameters are only fully meaningful if the source and load terminations are specified, almost invariably as matched loads. With matched conditions

$$|s_i| \;=\; \text{input reflection coefficient magnitude} \tag{13.1}$$

and $$\frac{1 + |s_i|}{1 - |s_i|} \;=\; \text{V.S.W.R. at the input} \tag{13.2}$$

with similar expressions for the output. Return loss is simply the reflection coefficient expressed in dB and, with matched terminations, the insertion losses are $|s_f|$ and $|s_r|$ expressed in dB. Attenuation normally refers to a symmetrical device such as an attenuator, for which $|s_f| = |s_r|$ and for which $|s_f| < 1$. When $|s_f| > 1$, one speaks of gain (as in Chapter 11). Isolation and coupling are also scattering parameters expressed in dB in the specification of a multi-port network such as a hybrid, a directional coupler or a circulator.

A nomogram relating many of these quantities in the context of a transmission line appears as Figure 1.5.

13.2.2 De-embedding considerations

It may not be possible to make measurements on the DUT directly. For example, it may be necessary to mount a transistor in a test circuit with microstrip connecting lines, with the lines themselves terminated in coaxial connectors. The transistor s-parameters are then to be deduced from the external measurements. For accurate measurements a very precisely constructed test jig may be used. The term de-embedding refers to the procedure of deducing the parameters of a DUT from the external measurements, as in [13.7].

If the connecting lines, including the microstrip/coaxial transition, from the DUT to the outside world are just straight-through lossless transmission lines (or close enough to that), then de-embedding is only required to recover the correct phase information, as the s-parameter magnitudes are correctly given by the external measurements. For accurate measurements, de-embedding requires knowledge of the properties of the test circuit or the test jig. These characteristics have to be measured by replacing the DUT with devices that have precisely known characteristics, say short-circuits or matched loads. De-embedding when precise measurements are required generally involves rather complicated formulae and is best done with the help of a computer, as shown in [13.8].

13.3 MICROSTRIP LINE PARAMETERS

In this experiment, the characteristic impedance of a line and the relative permittivity of a substrate are measured. The experiment uses a straight length of high impedance line that is terminated at each end with coaxial connectors as described in Figure 10.2. The line length should be at least 1.5λ long at the highest frequency that will be used. This naturally assumes that an estimate of the relative permittivity of the substrate is available. On the same basis, a line impedance of 80Ω, say, being greater than the 50Ω characteristic impedance of the measuring equipment, is selected. In order not to have significant changes in the line properties due to dispersion, §4.4, the maximum operating frequency should be chosen such that $(f \times h)$ is less than, say, $3\,\mathrm{GHz.mm}$. The basic equipment required to make these measurements is
- a variable or sweep frequency oscillator, say $1.0\text{--}2.0\,\mathrm{GHz}$,
- a standing wave detector in a 50Ω line and a V.S.W.R. meter,
- a frequency counter (alternatively the free space wavelength deduced from the standing wave pattern could be used),
- a matched load for the 50Ω line.

The microstrip test section is considered as an impedance transformer with a characteristic impedance Z_T and of fixed physical length l, but of a variable fraction of a wavelength when the frequency is varied. With a matched load attached to the output connector, i.e. a load impedance Z_0, the input impedance Z_{in} at the junction of the input coaxial line to the microstrip line is given from (1.54) by

$$Z_{in} = Z_T \frac{Z_0 \cos(\beta l) + j Z_T \sin(\beta l)}{Z_T \cos(\beta l) + j Z_0 \sin(\beta l)} \tag{13.3}$$

The phase coefficient, β, for propagation along the microstrip line is given by

$$\beta = \frac{2\pi}{\lambda} = \frac{2\pi\sqrt{\varepsilon_{eff}}}{\lambda_0} \qquad (13.4)$$

and

$$\lambda = \frac{\lambda_0}{\sqrt{\varepsilon_{eff}}} \qquad (13.5)$$

where λ and λ_0 are the guided and free space wavelengths respectively. Equation (13.3) demonstrates the special properties of the transmission line when βl takes the following values:

i) $\beta l = n\pi$, where n is an integer. This gives $Z_{in} = Z_0$.
ii) $\beta l = n\pi/2$, where n is an odd integer. This gives $Z_{in} Z_0 = Z_T^2$.

In the experiment, the frequency is varied until conditions (i) and (ii) are achieved in turn. When condition (i) is achieved, l becomes an integral number of $\lambda/2$, from which ε_{eff} may be deduced. When condition (ii) is achieved, Z_{in} is measured and Z_T deduced from $\sqrt{Z_0 Z_{in}}$.

The experiment will now be described in more detail using typical, *and possibly idealized*, data. Consider a line length $l = 225$ mm between the two coaxial connectors and let the effective line width, allowing for any finite thickness of the line, be $w = h = 1.5$ mm. A matched load (50Ω) is connected to one end of the test microstrip line, the other end of which is connected to the 50Ω standing wave detector line. As the oscillator frequency is varied, the input impedance normalized to 50Ω traces out the locus on the Smith Chart, Figure 13.3a. This is now interpreted, or directly plotted, as the variation of the input V.S.W.R. with frequency, Figure 13.3b. From the V.S.W.R. plot in Figure 13.3b, a minimum value that is close to unity is measured at 1.398 GHz and next at 1.864 GHz with increasing frequency. Line properties, determined from either wavelength or frequency, are found from the sharp minima in the V.S.W.R. response rather than the very much broader maxima. At the two frequencies for a V.S.W.R. minimum, condition (i) above is satisfied with

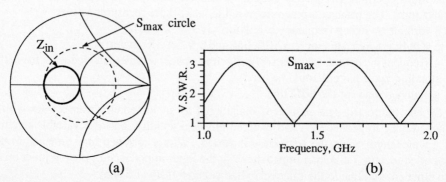

(a)   (b)

Figure 13.3 (a) The locus of the input impedance normalized to 50Ω of an unknown microstrip line terminated with a 50Ω load, and (b) the V.S.W.R., both as a function of frequency

a line that is $n\lambda/2$ long at 1.398 GHz and $(n+1)\lambda/2$ at 1.864 GHz. Thus, $n = 3$. At 1.864 GHz, $\lambda_0 = 160.8$ mm and since $(3 + 1) \times (\lambda/2) = l$ then, from (13.5)

$$\sqrt{\varepsilon_{eff}} = \frac{\lambda_0}{\lambda} = \frac{160.8}{112.5} \tag{13.6}$$

i.e. $\varepsilon_{eff} = 2.044$

The substrate relative permittivity may be derived from the effective permittivity value and w/h, if the effective filling factor, q, is known. The apparatus has been designed, it will be remembered, with the specific choice of $w/h = 1.0$. Kobayashi [13.9] gives q for values of ε_r and selected w/h, including $w/h = 1.0$. From this data, ε_{eff} for each value of ε_r may be derived and an algebraic expression $\varepsilon_{eff} = f(\varepsilon_r)$ obtained. Now the following calculations are made to determine ε_r:

i) ε_{eff}, following the procedure given in the example above,

ii) $B = 1 - \dfrac{1}{\varepsilon_{eff}}$,

iii) A, where $A = 1.506\,B - 0.649\,B^2 + 0.143\,B^3$ when $w/h = 1.0$,

iv) $\varepsilon_r = \dfrac{1}{1 - A}$.

Hence, with $\varepsilon_{eff} = 2.044$, the following values are obtained:

$$B = 0.5108, \quad A = 0.6190 \quad \text{and} \quad \varepsilon_r = 2.62.$$

Z_T *determination*

The peak V.S.W.R., $S_{max} = 3.1$, in Figure 13.3b occurs at those frequencies where the line length is an odd multiple of $\lambda/4$. From condition (ii), $Z_T = \sqrt{Z_{in} Z_0}$ and further $Z_{in}/Z_0 = S_{max}$. Note that it is also possible for $Z_{in}/Z_0 = 1/S_{max}$. However, this latter Z_{in} may be eliminated, since in the former case there will be a voltage maximum at the input plane to the line, but a voltage minimum for the latter. With $S_{max} = 3.1$ and a 50 Ω characteristic impedance for all other lines, $Z_T = 88.0\,\Omega$.

As a cross-check with the known line geometry of $w/h = 1.0$ and a measured substrate permittivity of 2.62, it is found from Table 3.2 that $Z_T = 88.5\,\Omega$.

Discussion

This experiment has been presented with an ideal set of results that may or may not be representative in practice. The reader is invited to consider the effects that may be observed in the results and possible solutions to improve the results as a consequence of

i) excess shunt capacitance at the coaxial line to microstrip junction,

ii) an imperfect matched load,

iii) the voltage minimum on the standing wave detector broadened by the background noise level,

iv) two voltage maxima along the standing wave detector having unequal magnitudes,

v) a close to, but non-integer, value for n,

vi) attenuation along the microstrip line,

vii) $Z_T = 50\,\Omega$.

Other methods are suitable for the accurate measurement of ε_{eff}. The measurement by Das et al. [13.10] of the difference in electrical path length between two identical lines, differing only in length, may be used to determine the frequency dependence of ε_{eff}. As 50Ω lines are desirable for the method, to some extent it is assumed that the permittivity is already known. Cavity methods [13.11, 13.12], that make use of a section of substrate material metallized to form a waveguide cavity, directly measure the relative permittivity of the bulk material at the resonant frequencies of the cavity.

13.4   DISCONTINUITY MEASUREMENTS

A collection of resonators that are coupled to their individual input lines through series gaps is fabricated on a substrate, as shown in Figure 13.4. Identical line widths giving 50Ω characteristic impedance and identical series capacitance gaps are used throughout. The size of the gaps should be about one-third of the substrate height. This will give a sharp resonance for an accurate measurement of the resonant

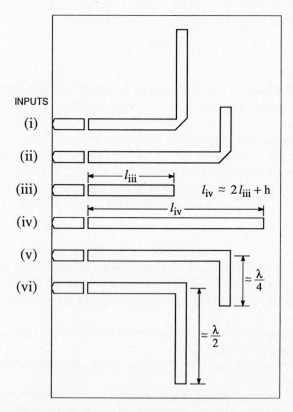

Figure 13.4 A schematic diagram showing full- and half-wavelength resonators that demonstrate the effective length of compensated corners in (i) and (ii), the effect of open-circuit and series-gap capacitances in (iii) and (iv) and the effect of uncompensated corners in (v) and (vi). (*Not to scale.*)

frequency. For convenience, each input line may be connected to its own coaxial connector. Frequencies are selected which, after discontinuities have been taken into account, make the resonators one wavelength long, except for resonator (iii) which is only $\lambda/2$ long.

The basic equipment required to make the measurements consists of
- a variable or sweep frequency oscillator, say 1.0–2.0GHz,
- a broadband directional coupler and a crystal detector to monitor the magnitude of the reflected wave,
- a frequency counter.

The average of two frequencies, at which equal magnitudes of the reflected signals close to but on each side of the resonant frequency are observed, should be used to determine the resonant frequency. This approach is more accurate than trying to determine the frequency for the minimum reflection that may either be quite broad or within the noise level.

The effective permittivity and open-circuit line extension

In Figure 5.2, it is observed that the open-circuit fringing capacitance for a 50Ω line on $\varepsilon_r = 2.5$ substrate is equivalent to a line extension of approximately $(0.5 \times h)$. The equivalent line extension for a very wide series gap, §5.6, will also tend towards the same value. Thus, as the fringing capacitances at the open circuit and the series gap for either resonator (iii) or (iv) will together be equivalent to a line extension Δl of approximately h, resonator (iv) should resonate at about the same frequency as (iii) if its length $l_{iv} + h \approx 2(l_{iii} + h)$, i.e. $l_{iv} \approx 2 l_{iii} + h$. At resonance, the input signal will be coupled into the resonant lines. Power will be absorbed by the losses within the resonator. This will be observed as a sharp dip in the reflected signal from the structure. The reflected signals from (iii) and (iv) are monitored in turn as frequency is varied and the resonant frequencies f_{iii} and f_{iv} are measured for each case. Comparing (iii) with (iv)

$$l_{iii} + \Delta l = \frac{\lambda_{iii}}{2} = \frac{c}{2 f_{iii} \sqrt{\varepsilon_{eff}}} \tag{13.7}$$

$$l_{iv} + \Delta l = \lambda_{iv} = \frac{c}{f_{iv} \sqrt{\varepsilon_{eff}}} \tag{13.8}$$

i.e. $$\varepsilon_{eff} = \left\{ \frac{c}{(l_{iv} - l_{iii})} \left[\frac{1}{f_{iv}} - \frac{1}{2 f_{iii}} \right] \right\}^2 \tag{13.9}$$

Equation (13.9) assumes that the end effects in (iii) and (iv) are unchanged by any small differences between f_{iii} and f_{iv}. The *sum* of the open-circuit and series-gap capacitances as an effective line extension, Δl, may now be determined through

$$\Delta l = \frac{c}{2 f_{iii} \sqrt{\varepsilon_{eff}}} - l_{iii} \tag{13.10}$$

90° corner reactances

Resonators (v) and (vi) are designed to measure the equivalent circuit inductance and capacitance components of a right-angle corner. The total lengths of each resonator,

l_v and l_{vi}, measured along the inside edge of the corner, are equal to l_{iv}. The corners will increase the overall length of each resonator and reduce the resonant frequency compared with f_{iv}. The equivalent electrical lengths are a wavelength at their respective resonant frequencies, f_v and f_{vi}. For (vi), with the corner at the center of the resonator where there is an open-circuit impedance plane giving maximum voltage and zero current, the increase over l_{iv} in its equivalent electrical length at f_{iv} determines the shunt capacitance of the corner.

Assuming identical end effects included in an equivalent length Δl and ε_{eff} independent of any small frequency changes, then in terms of the free space wavelengths

$$\sqrt{\varepsilon_{eff}}(l_{iv} + \Delta l) = \lambda_0^{(iv)} \tag{13.11}$$

and

$$\sqrt{\varepsilon_{eff}}(l_{vi} + \Delta l_{(C)} + \Delta l) = \lambda_0^{(vi)} = \frac{f_{iv}}{f_{vi}} \lambda_0^{(iv)} \tag{13.12}$$

where $\Delta l_{(C)}$ is the equivalent line length due to the shunt capacitance of the corner at an open-circuit plane. Thus

$$\Delta l_{(C)} = \left[\frac{f_{iv}}{f_{vi}} - 1 \right] \frac{\lambda_0^{(iv)}}{\sqrt{\varepsilon_{eff}}} - (l_{vi} - l_{iv}) \tag{13.13}$$

From (5.3), the shunt capacitance of the corner at an open-circuit plane is given by

$$C_C \approx \frac{\beta \Delta l_{(C)}}{\omega Z_0} = \frac{\sqrt{\varepsilon_{eff}} \Delta l_{(C)}}{c Z_0} \tag{13.14}$$

For (v), the corner is at a short-circuit impedance plane with maximum current and zero voltage. Similarly the increase from l_{iv} in the equivalent line length at f_{iv} determines the series inductance of the corner

$$L_C \approx \frac{\sqrt{\varepsilon_{eff}} Z_0 \Delta l_{(L)}}{c} \tag{13.15}$$

From these results, it will be found that the square corner reactances are not optimized for a 50Ω line, but that $\sqrt{L_C / C_C} < 50\Omega$. This is verified by the fact that the resonant frequency of the line that includes the corner capacitance is lower than when the corner inductance is included, i.e. $f_{vi} < f_v$.

The compensated corner

Resonators (i) and (ii) are identical to (vi) and (v) respectively, except that the corners have now been mitered to reduce the excess capacitance. Ideally this will make the corners behave as a short length of 50Ω line, irrespective of the impedance at the plane where they are situated. Thus the resonant frequencies f_i and f_{ii} should be equal and, since it is excess capacitance that has been mainly reduced by mitering, should be closer to f_v than f_{vi}. The equivalent length extension due to the corner, $\Delta l_{(C)}$, is given by

$$\Delta l_{(C)} = \Delta l_{(L)} = \left[\frac{f_{iv}}{f_i} - 1 \right] \frac{\lambda_0^{(iv)}}{\sqrt{\varepsilon_{eff}}} - (l_i - l_{iv}) \tag{13.16}$$

REFERENCES

[13.1] Sucher, M. and Fox, J. (eds) *Handbook of Microwave Measurements*, 3rd edn, Vols 1 & 2, Polytechnic Press, New York, 1963.

[13.2] Laverghetta, T. S., *Microwave Measurements and Techniques*, Artech House, Dedham, MA, 1976.

[13.3] Laverghetta, T. S., *Practical Microwaves*, H. W. Sams, Indianapolis, 1984.

[13.4] Ginzton, E. L., *Microwave measurements*, McGraw-Hill, New York, 1957.

[13.5] Gupta, C., Easter, B. and Gopinath, A., "Some results on the end effects of microstriplines", *IEEE Trans. Microwave Theory and Techniques*, Vol. MTT-26, No. 9, September 1978, pp. 649-52.

[13.6] Cronson, H. M. and Susman, L., "A six-port automatic network analyzer", *IEEE Trans. Microwave Theory and Techniques*, Vol. MTT-25, No. 12, December 1977, pp. 1086-91. *(Four other relevant papers also appear in this issue.)*

[13.7] Bauer. R. F. and Penfield, P. Jr., "De-embedding and terminating", *IEEE Trans. Microwave Theory and Techniques*, Vol. MTT-22, No. 3, March 1974, pp. 282-8.

[13.8] Benet, J. A., "The design and calibration of a universal MMIC test fixture", IEEE Microwave and Millimeter-Wave Monolithic Circuits Symposium, Dallas, Texas, June 1982, pp. 36-41.

[13.9] Kobayashi, M., "Analysis of the microstrip and the electrooptic light modulator", *IEEE Trans. Microwave Theory and Techniques*, Vol. MTT-26, No. 2, February 1978, pp. 119-26.

[13.10] Das, N. K., Voda, S. M. and Pozar, D. M., "Two methods for the measurement of substrate dielectric constant", *IEEE Trans. Microwave Theory and Techniques*, Vol. MTT-35, No. 7, July 1987, pp. 636-42.

[13.11] Howell, J. Q., "A quick accurate method to measure the dielectric constant of microwave integrated-circuit substrates", *IEEE Trans. Microwave Theory and Techniques*, Vol. MTT-21, No. 3, March 1973, pp. 142-3.

[13.12] Ladbrooke, P. H., Potok, M. H. N. and England, E. H., "Coupling errors in cavity-resonance measurements on MIC substrates", *IEEE Trans. Microwave Theory and Techniques*, Vol. MTT-21, No. 8, August 1973, pp. 560-2.

Appendix 1
The finite difference
method — applied to microstrip lines

There are many problems associated with microstrip lines, where a rigorous solution is considered too difficult and the fields are analyzed using numerical techniques. In this appendix, aspects of the finite difference method are studied in reasonable detail, so that the reader may be able to appreciate what has to be done to solve the problem of evaluating the capacitance of a microstrip line in a variety of situations. The capacitance values lead directly to the characteristic impedance and propagation coefficient for the line. The finite difference method is particularly useful if the microstrip line is situated in a shielding enclosure that limits the extent of the fields in the transverse plane. Other areas of use include the even- and odd-mode capacitances for parallel-coupled lines and the excess capacitance that is associated with transmission line discontinuities.

The numerical evaluation of the parameters of several transmission line configurations has been obtained using the finite difference method, e.g. [A1.1, A1.2]. The capacitance of any two-conductor transmission line may be found from a knowledge of the charge on the conductors and the potential difference between them. Let V(x,y) be the potential function throughout the cross-section of the transmission line, with the strip at a fixed potential above that of the ground plane. The continuous function V(x,y) must be a solution of Laplace's Equation in two dimensions

$$\nabla^2 V = \frac{\partial^2 V}{\partial x^2} + \frac{\partial^2 V}{\partial y^2} = 0 \tag{A1.1}$$

subject to the appropriate boundary conditions. In the finite difference method, a fine mesh is superimposed on the cross-section of the transmission line, Figure A1.1, and only the values ϕ_n of V(x,y) at the nodes of the mesh are considered.

The discrete form of partial differential equations

The partial differential equations are written in a finite difference form involving the potentials at the adjacent mesh nodes illustrated in Figure A1.1. Terms of the form $\partial^2 V/\partial x^2$ are replaced as functions of ϕ_n along the x-axis. Consider a function f(x) as shown in Figure A1.2, having discrete values ϕ_n at nodes along the x-axis. In this context, it will be more convenient to use the notation f_x for f(x), etc. The first derivative at a point x

$$f'_x = \frac{df_x}{dx} \tag{A1.2}$$

293

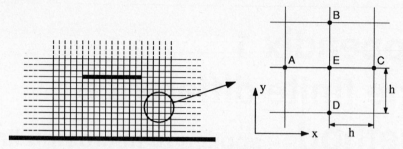

Figure A1.1   A typical mesh for the finite difference method

may be approximated by the forward difference formula

$$f'_x = \frac{\phi_{n+1} - \phi_n}{h}$$

(A1.3)

or by the backward difference formula

$$f'_x = \frac{\phi_n - \phi_{n-1}}{h}$$

(A1.4)

These two difference formulae generally give different results. Thus the central difference formula, which is an average of them and does not include the value of the function at the node at which the derivative is being found, will give a closer estimate of the derivative. The central difference formula is illustrated on Figure A1.2 and gives f'_x as

$$f'_x = \frac{\phi_{n+1} - \phi_{n-1}}{2h}$$

(A1.5)

A measure of the accuracy of these finite difference expressions may be obtained from a Taylor expansion of f_{x+h} and f_{x-h}.

$$f_{x+h} = f_x + h f'_x + \tfrac{1}{2}h^2 f''_x + \tfrac{1}{6}h^3 f'''_x + \cdots$$

(A1.6)

$$f_{x-h} = f_x - h f'_x + \tfrac{1}{2}h^2 f''_x - \tfrac{1}{6}h^3 f'''_x + \cdots$$

(A1.7)

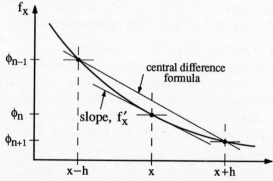

Figure A1.2   The potential function in one dimension

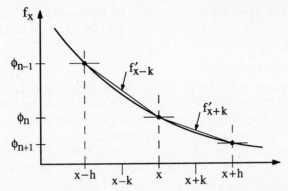

Figure A1.3 Inter-mesh points $(x \pm k)$ for evaluation of the second derivative at x

Using these expansions, it is seen that the forward and backward difference formulae will have errors of the order of h, while for the central difference formula the error is of the order of h^2. Thus, for an accurate representation of the problem, it is seen that there is an advantage in

i)   using the central difference formula, and
ii) using a mesh size, h, as small as possible.

In deriving an expression for the second derivative, consider two intermediate points along the x-axis at $(x-k)$ and $(x+k)$ as shown in Figure A1.3, where $k = h/2$. Using the central difference formula for the first derivative but with half the original interval gives

$$f'_{x+k} = \frac{\phi_{n+1} - \phi_n}{h} \tag{A1.8}$$

and
$$f'_{x-k} = \frac{\phi_n - \phi_{n-1}}{h} \tag{A1.9}$$

The second derivative at x now becomes

$$f''_x = \frac{f'_{x+k} - f'_{x-k}}{h} \tag{A1.10}$$

i.e.
$$f''_x = \frac{\phi_{n+1} - 2\phi_n + \phi_{n-1}}{h^2} \tag{A1.11}$$

This expression has an error of the order of h^2.

Consider the general node, E, as illustrated in Figure A1.1. At this node

$$f''_x + f''_y = 0 \tag{A1.12}$$

Therefore
$$\frac{\phi_C - 2\phi_E + \phi_A}{h^2} + \frac{\phi_B - 2\phi_E + \phi_D}{h^2} = 0 \tag{A1.13}$$

or
$$\phi_A + \phi_B + \phi_C + \phi_D - 4\phi_E = 0 \tag{A1.14}$$

In a two-dimensional finite difference solution of Laplace's Equation, (A1.14) is the equation that is generally used, except for special cases such as occur at boundaries and discontinuities.

Higher order representation

The three-node representation for the second order derivative (A1.11) is identical to the result that would be obtained if a quadratic equation were fitted through the three node potentials and differentiated twice with respect to the direction x at the center node. A quartic equation in terms of the node potentials at five points, namely $(x-2h)$, $(x-h)$, (x), $(x+h)$ and $(x+2h)$, when differentiated twice at $x=0$ gives an improved representation for the second derivative with an error in the second derivative of the order of h^4. This second derivative at x is

$$f''_x = \frac{-f_{x-2h} + 16f_{x-h} - 30f_x + 16f_{x+h} - f_{x+2h}}{12h^2} \tag{A1.15}$$

Unequal node spacings

Unequal node spacings commonly occur when the boundaries of the system do not coincide with the nodes of a regular mesh. With the final length of αh to a node that has been placed on the boundary, such that there are three nodes at $(x-h)$, (x) and $(x+\alpha h)$ as in Figure A1.4a

$$f''_x = \frac{2}{\alpha(\alpha+1)h^2} \left[\alpha f_{x-h} - (1+\alpha)f_x + f_{x+\alpha h} \right] \tag{A1.16}$$

It is usual in this equation for $0 < \alpha < 1$. However, if α is very small it may be advantageous to omit the node near the boundary, leaving the final mesh length to the boundary with α greater than but close to unity. This is illustrated in Figure A1.4b.

Boundary conditions

Two commonly used boundary conditions are the Dirichlet and Neumann boundary conditions. Consider a microstrip line where $\{\phi\}$ represents the discrete values of $V(x,y)$. The boundaries of a system impose constraints on the potentials and fields within the system. The Dirichlet boundary condition requires that the potential along the surface, $V(s)$, is a constant, i.e.

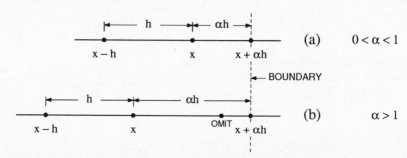

Figure A1.4 Node selections adjacent to a constant potential boundary, showing (a) the general selection of unequal node spacings near a boundary, and (b) a preferred selection with $\alpha > 1$, if α would be very small otherwise

$$V(s) = \kappa \tag{A1.17}$$

An example of the Dirichlet boundary condition would be a microstrip line where the discrete potentials on the ground plane must satisfy $\kappa = 0$, while for the conducting strip $\kappa = 1$ volt. Although in both cases there will be nodes placed on the boundaries, these node potentials will not appear as variables in the final matrix equation since they are constant.

The Neumann boundary condition is

$$\frac{\partial V}{\partial n}\bigg|_{surface} = 0 \tag{A1.18}$$

where n is the normal with respect to the surface. The elements of $\{\phi\}$ that lie on the boundary, being unknown potentials, will remain as variables in the final matrix equation. With electric potentials, this boundary condition represents either an open-circuit plane or a plane of symmetry at which the magnitude of the potential is a maximum. The second derivative at the boundary is given by (A1.11) when an image node has been introduced outside the boundary as illustrated in Figure A1.5. Equation (A1.18) is satisfied if

$$\phi_{n-1} = \phi_{n+1} \tag{A1.19}$$

Thus applying (A1.11) at the boundary

$$f_x'' = \frac{\phi_{n+1} - 2\phi_n + \phi_{n-1}}{h^2} = \frac{2}{h^2}\left\{\phi_{n+1} - \phi_n\right\} \tag{A1.20}$$

When there is a plane of symmetry at which (A1.18) applies, P(A1.20) is still used, thus omitting the potentials in the image region, in order that the problem may be formulated in terms of the potentials of one half of the system together with those potentials that lie on the plane of symmetry.

Boundary between two dielectric materials

The boundary between two dielectric materials that have relative permittivities ε_1 and ε_2 is illustrated in Figure A1.6.

A finite difference equation may be derived in terms of the potentials in the vicinity of E by considering the surface, S, around the node. From Gauss's Law, the total electric flux flowing outward through the surface is equal to the charge enclosed. For an ideal dielectric material with no conduction current flowing, there will be no

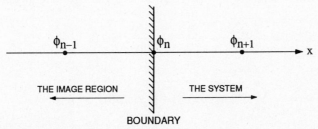

Figure A1.5 The image node representation for the Neumann boundary condition

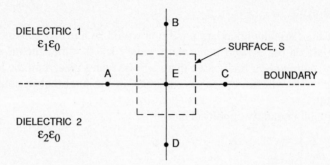

Figure A1.6 Nodes at the boundary between two dielectric materials

free charges on the dielectric interface. Thus, at the boundary of two isotropic dielectric media

$$\int D \cdot ds = 0 \quad \Rightarrow \quad \int_{S_1} \varepsilon_1 \nabla V \cdot ds_1 + \int_{S_2} \varepsilon_2 \nabla V \cdot ds_2 = 0 \tag{A1.21}$$

where S_1 and S_2 are the sections of the surface, S, in each dielectric region. Applying this equation at E in Figure A1.1 gives

$$\varepsilon_0 \left\{ \varepsilon_1 \frac{\phi_B - \phi_E}{h} + \varepsilon_2 \frac{\phi_D - \phi_E}{h} + \frac{(\varepsilon_1 + \varepsilon_2)}{2} \left[\frac{\phi_C - \phi_E}{h} + \frac{\phi_A - \phi_E}{h} \right] \right\} = 0 \tag{A1.22}$$

i.e. $(\varepsilon_1 + \varepsilon_2) \phi_A + 2\varepsilon_1 \phi_B + (\varepsilon_1 + \varepsilon_2) \phi_C + 2\varepsilon_2 \phi_D - 4(\varepsilon_1 + \varepsilon_2) \phi_E = 0$

$$\tag{A1.23}$$

The treatment of singularities

Equation (A1.14) results from the difference equations that represent Laplace's Equation in a Cartesian coordinate system. The principal error terms are those given by the 4th order terms of the Taylor Expansion. At a singularity and its neighboring points, the error terms do not converge [A1.3] and significant errors in the evaluation of the stored electrostatic energy and capacitance may occur.

However, if a branch type singularity, for example the edge of the strip of a microstrip line, is considered and taken as a local origin, it will be possible to solve Laplace's Equation uniquely in this region in terms of cylindrical coordinates, (r, θ). Thus

$$\nabla^2 V = r^2 \frac{\partial^2 V}{\partial r^2} + r \frac{\partial V}{\partial r} + \frac{\partial^2 V}{\partial \theta^2} = 0 \tag{A1.24}$$

with a solution

$$V = \sum_{k=-\infty}^{\infty} r^n \left(A_k \cos(n\theta) + B_k \sin(n\theta) \right), \qquad n = \frac{\pi k}{\alpha} \tag{A1.25}$$

where A_k, B_k are constants and α is the angle of the singularity that equals 2π for the

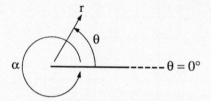

Figure A1.7 The geometry of an edge singularity

edge of a thin strip, as shown in Figure A1.7. Note that in (A1.25), n varies with k in the summation.

The strip is at a constant potential which may be taken as zero for the solution of (A1.25), with a constant added as necessary to all the potentials in the region. Thus, $A_k = 0$ because of the assumed zero potential, giving

$$V = V_0 + \sum_{k=-\infty}^{\infty} B_k r^n \sin(n\theta)$$
(A1.26)

The negative values of k in the summation are not required, as demonstrated by the following argument. The potential must be constant within a small enough region near the singularity. In particular, it must be constant, say V_{r_0}, on a small circle of radius r_0 that excludes the singularity as shown in Figure A1.8, requiring for $k > 0$

$$B_{-k} r_0^{-n} \sin(-n\theta) - B_k r_0^n \sin(n\theta) = 0$$
(A1.27)

i.e.
$$B_{-k} = -B_k r_0^{2n} \quad (k = 1, 2, \cdots)$$
(A1.28)

This equation must be true in the limit as $r_0 \rightarrow 0$, implying that $B_{-k} = 0$ for all positive k.

In [A1.3] the edge singularity was placed at the center of the mesh and was equidistant from the four surrounding nodes. Later work, [A1.4, A1.5], placed the singularity at a node with two advantages, namely: (i) equal mesh lengths were preserved to nodes elsewhere on the constant potential boundary, and (ii) a more detailed cylindrical representation of the potential variation could be derived.

For a thin strip in a uniform dielectric medium with even symmetry for the potentials about the plane of the conductor, Figure A1.9, the potential variation reduces [A1.5] to

$$V = \sum_{k=1}^{4} B_k r^{(2k-1)/2} \sin\left[\frac{2k-1}{2}\theta\right]$$
(A1.29)

Figure A1.8 Excluding the singularity by a small circle of radius r_0

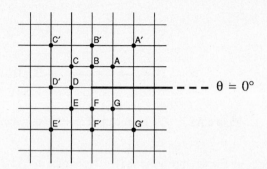

Figure A1.9   A strip singularity in a uniform dielectric medium

In this symmetrical situation, for example in a balanced strip transmission line, the seven nodes adjacent to the singularity have only four independent values which may be expressed by the four unknown node potentials at A, B, C and D. Taking the strip at zero potential, the infinite summation has been reduced to the four terms in (A1.29) for which the coefficients B_k may be evaluated. The factor $(2k - 1)/2$ in the sine term produces an even symmetry with $k = 1, \cdots, 4$, as an alternative to taking only the odd integers from $k = 1, \cdots, 7$. Four simultaneous equations may now be formed, expressing the potential at each of A', B', C' and D' in terms of B_k, $k = 1, \cdots, 4$. Solving for B_k expresses the B_k in terms of $\phi_{A'}, \cdots, \phi_{D'}$. Moving closer now to the singularity $\phi_A, \cdots, \phi_D$ can also be expressed in terms of B_k and thus in turn in terms of $\phi_{A'}, \cdots, \phi_{D'}$, giving

$$
\begin{bmatrix} \phi_A \\ \phi_B \\ \phi_C \\ \phi_D \end{bmatrix} = \begin{bmatrix} 0.2500 & 0.2608 & 0 & 0.0236 \\ 0.0653 & 0.4130 & 0.0769 & 0.0937 \\ 0 & 0.3080 & 0.2500 & 0.2844 \\ 0.0118 & 0.1876 & 0.1422 & 0.4129 \end{bmatrix} \begin{bmatrix} \phi_{A'} \\ \phi_{B'} \\ \phi_{C'} \\ \phi_{D'} \end{bmatrix}
$$

(A1.30)

This approach with the cylindrical coordinate form of Laplace's Equation provides a more accurate representation for the variation of the potential near the edge of the strip.

   In the case of a singularity associated with the strip of a microstrip transmission line, the situation is similar to that in Figure A1.9, except that now there is no symmetry with respect to the plane of the strip at the singularity because of the presence of only one ground plane. Now the infinite summation is carried out with seven terms as

$$
V = \sum_{k=1}^{7} B_k r^{k/2} \sin\left(\frac{k\theta}{2}\right)
$$

(A1.31)

and potentials at seven points have to be considered. Otherwise the procedure is as before. The following relation between $\phi_A, \cdots, \phi_G$ and $\phi_{A'}, \cdots, \phi_{G'}$ is obtained.

$$
\begin{Bmatrix} \phi_A \\ \phi_B \\ \phi_C \\ \phi_D \\ \phi_E \\ \phi_F \\ \phi_G \end{Bmatrix}
=
\begin{bmatrix}
0.2500 & 0.2554 & 0 & 0.0236 & 0 & 0.0054 & 0 \\
0.0639 & 0.3940 & 0.0697 & 0.0937 & 0.0072 & 0.0190 & 0.0014 \\
0 & 0.2790 & 0.2500 & 0.2844 & 0 & 0.0290 & 0 \\
0.0059 & 0.0938 & 0.0711 & 0.4129 & 0.0711 & 0.0938 & 0.0059 \\
0 & 0.0290 & 0 & 0.2844 & 0.2500 & 0.2790 & 0 \\
0.0014 & 0.0190 & 0.0072 & 0.0937 & 0.0697 & 0.3940 & 0.0639 \\
0 & 0.0054 & 0 & 0.0236 & 0 & 0.2554 & 0.2500
\end{bmatrix}
\begin{Bmatrix} \phi_{A'} \\ \phi_{B'} \\ \phi_{C'} \\ \phi_{D'} \\ \phi_{E'} \\ \phi_{F'} \\ \phi_{G'} \end{Bmatrix}
$$

(A1.32)

In this way the cylindrical coordinate solution of Laplace's Equation in the vicinity of a singularity may be linked with the Cartesian coordinate solution elsewhere in the system. The expense of having to introduce a least squares fit for the seven B_k coefficients in this example so as to include additional nodes, such as those between A' and B', is not warranted. Indeed, using only the radial variation of the $k = 1$ term of (A1.26) to the nodes B', D' and F' with an appropriate interpolation for B, D and F [A1.6], the percentage errors of microstrip line capacitance calculations were reduced by more than a factor of 10 compared with the results obtained ignoring the effects of the singularity. For the symmetrical case, where $\phi_{A'} = \phi_{G'}$, $\phi_{B'} = \phi_{F'}$ and $\phi_{C'} = \phi_{E'}$, (A1.32) reduces to (A1.30). For the microstrip transmission line with a dielectric substrate, relative permittivity ε_r, the boundary conditions at the air-dielectric interface must also be satisfied. From [A1.4], the potential in the air region is given by (A1.31) while in the dielectric region

$$
V = \sum_{k=1}^{4} B_{2k-1} r^{(2k-1)/2} \sin\left[\frac{2k-1}{2}\theta\right] + \frac{1}{\varepsilon_r} \sum_{k=1}^{3} B_{2k} r^k \sin(k\theta)
$$

(A1.33)

For each particular substrate material with a relative permittivity, ε_r, an equation similar to (A1.32) must be derived and used in the solution of the potential distribution for a microstrip transmission line.

The electric field strength across the ground plane

After the potential distribution at all the nodes throughout the cross-section of the microstrip line has been evaluated, to calculate the capacitance it is necessary to do a surface integration of total electric flux density around one of the conductors. Before this can be done, the electric field must be evaluated from the potential distribution just obtained. Any closed surface may be taken in principle. The actual surface of the strip conductor is a poor choice for the surface, because of the singularities and the rapidly changing electric field at the strip edges. Intermediate surfaces may be used, but it is probably easiest to form the surface across the complete ground plane.

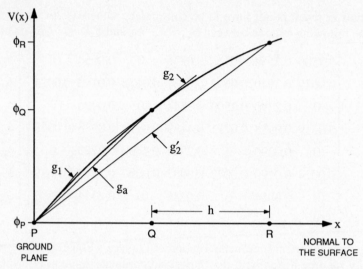

Figure A1.10 Nodes and their potentials on a normal to the ground plane

Consider the equidistant nodes that lie on a normal to the ground plane, Figure A1.10. At P, the electric field strength

$$\mathbf{E} = -\frac{\partial V}{\partial x} = -g_1 \tag{A1.34}$$

where g_1 is the gradient or tangent to $V(x)$ as a function of x at the ground plane. The simplest approximation would be to take $\mathbf{E}$ at P to equal g_a. A more accurate evaluation could be done as follows. Now

$$g_a = \frac{\phi_Q - \phi_P}{h} \tag{A1.35}$$

and as a first approximation

$$g_a \approx \frac{g_1 + g_2}{2} \rightarrow g_1 + g_2 = \frac{2(\phi_Q - \phi_P)}{h} \tag{A1.36}$$

However $g_2 \approx g_2' = \dfrac{\phi_R - \phi_P}{2h}$ \hfill (A1.37)

Therefore $g_1 \approx \dfrac{2(\phi_Q - \phi_P)}{h} - \dfrac{\phi_R - \phi_P}{2h}$ \hfill (A1.38)

i.e. $g_1 = \dfrac{-3\phi_P + 4\phi_Q - \phi_R}{2h}$ \hfill (A1.39)

This same equation may be found by fitting the polynomial

$$V(x) = a + bx + cx^2 \tag{A1.40}$$

to the node potentials at x = 0, h and 2h, differentiating with respect to x and then setting x = 0. Thus, from (A1.40), $g_1 = b$. This latter approach may be used with higher order polynomials if the node density along the normal is sufficiently high to warrant it. It may also be used whenever there are unequal node spacings along the normal.

Integration to calculate the charge

The electric field strength and the electric flux density are known now at discrete points across the ground plane. The total flux, Ψ, and therefore the enclosed charge, is evaluated numerically from the electric flux density. Let D_{1n}, D_{2n} and D_{3n} be the normal components of the electric flux density at adjacent nodes that are equally spaced with a separation, h, across the ground plane. For this 2h width of the ground plane, the trapezoidal rule for integration gives the flux and charge contributions

$$\Delta\Psi = \Delta Q = \frac{(D_{1n} + D_{2n})h}{2} + \frac{(D_{2n} + D_{3n})h}{2} \qquad (A1.41)$$

i.e.

$$\Delta Q = \frac{(D_{1n} + 2D_{2n} + D_{3n})h}{2} + O(h^3) \qquad (A1.42)$$

Using Simpson's Rule across the full 2h width gives

$$\Delta Q = \frac{(D_{1n} + 4D_{2n} + D_{3n})h}{3} + O(h^5) \qquad (A1.43)$$

This and other integration formulae are found by fitting the appropriate order polynomial to the data points and integrating over the region. The number of intervals for which a ΔQ is found using (A1.42) or (A1.43) is inversely proportional to the interval length, h, and the error terms in the calculation of the total charge on the ground plane are of the order of h^2 and h^4 respectively.

Example A1.1

The details of a complete two-dimensional problem are too long for presentation here. However, the techniques involved may be illustrated in this example where the capacitance of a parallel plate capacitor may be found.

Using the finite difference method, find the potential variation between the plates for the system illustrated in Figure A1.11. Calculate the capacitance per unit width and length. Compare the result with the true value for the parallel plate capacitor.

Solution:

For the width tending to infinity, the system is a one-dimensional one, as

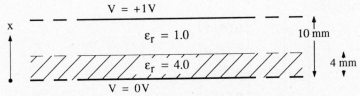

Figure A1.11 The geometry of a parallel plate capacitor

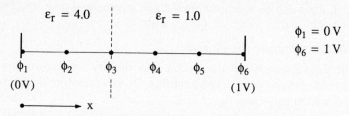

Figure A1.12 Nodes in one dimension for a parallel plate capacitor

illustrated in Figure A1.12. Equally spaced nodes are placed between the two plates. In one dimension, Laplace's Equation becomes

$$\frac{d^2V}{dx^2} = 0$$

If the distance between the nodes, h, is one unit ($= 2$ mm), then

$$\frac{d^2V}{dx^2} = 0 \implies \phi_{n+1} - 2\phi_n + \phi_{n-1} = 0$$

This is the equation that is applied at nodes 2, 4 and 5, i.e. for n = 2, 4, 5. Node 3 lies on the dielectric/air interface. At this boundary, the normal component of the electric flux density must be continuous, i.e.

$$\varepsilon_r^{(1)} E_{1n} = \varepsilon_r^{(2)} E_{2n}$$

Expressing this equation in finite difference form

$$\frac{\varepsilon_r (\phi_3 - \phi_2)}{h} = \frac{(\phi_4 - \phi_3)}{h}$$

giving $-\varepsilon_r \phi_2 + (\varepsilon_r + 1)\phi_3 - \phi_4 = 0$

Equations may now be formed in terms of the four unknown node potentials and expressed in matrix form as

$$\begin{bmatrix} -2 & 1 & 0 & 0 \\ -4 & 5 & -1 & 0 \\ 0 & 1 & -2 & 1 \\ 0 & 0 & 1 & -2 \end{bmatrix} \begin{bmatrix} \phi_2 \\ \phi_3 \\ \phi_4 \\ \phi_5 \end{bmatrix} = \begin{bmatrix} 0 \\ 0 \\ 0 \\ -1 \end{bmatrix} \quad \begin{matrix} \cdots \text{ at } \phi_2 \text{ node} \\ \cdots \text{ at } \phi_3 \text{ node} \\ \cdots \text{ at } \phi_4 \text{ node} \\ \cdots \text{ at } \phi_5 \text{ node} \end{matrix}$$

Solving gives

$$\begin{bmatrix} \phi_2 \\ \phi_3 \\ \phi_4 \\ \phi_5 \end{bmatrix} = \begin{bmatrix} 0.0714 \\ 0.1429 \\ 0.4286 \\ 0.7143 \end{bmatrix}$$

The capacitance is calculated from the charge per unit area on the plates for a known potential difference between them. The charge on either plate may be used since they are equal in magnitude. On the plate with a potential of 1.0 volt

$$Q = D_n = \varepsilon_0 E_n$$

i.e.

$$Q = \frac{\varepsilon_0 (1.0 - \phi_5)}{h}$$

With $h = 0.002$ m, the surface charge density

$$Q = \frac{8.854 \times 10^{-12} (1.0 - 0.7143)}{0.002}$$

$$= 1.265 \quad nC.m^{-2}$$

With a 1.0 volt potential difference between the plates, the capacitance of the plates is 1.265 nF.m^{-2}.

The exact solution for the capacitance is obtained by treating the system as two parallel plate capacitors C_1 and C_2, connected in series, with the air-dielectric interface representing an equipotential surface between the two capacitors. The capacitance per unit area in the dielectric region

$$C_1 = \frac{\varepsilon_r \varepsilon_0}{d} = \frac{4 \varepsilon_0}{0.004} = 1000 \varepsilon_0 \quad F.m^{-2}$$

Likewise

$$C_2 = 166.7 \varepsilon_0 \quad F.m^{-2}$$

giving

$$C = \frac{C_1 C_2}{C_1 + C_2} = 142.9 \varepsilon_0$$

i.e.

$$C = 1.265 \quad nF.m^{-2}$$

Note

In this example, the finite difference method has given an exact solution for both the potential distribution and the capacitance. This is because, being a one-dimensional problem and with the second derivative of the potential being zero, (A1.3) and (A1.4) represent the potential variation exactly. When the potential varies in a more complicated manner in two (or three) dimensions, the finite difference method will only lead to an approximate solution. However, the steps taken are similar to those presented in this example.

REFERENCES

[A1.1] Green, H. E., "The numerical solution of some important transmission-line problems", *IEEE Trans. Microwave Theory and Techniques*, Vol. MTT-13, No. 5, September 1965, pp. 676–92.

[A1.2] Wexler, A., "Computation of electromagnetic fields", *IEEE Trans. Microwave Theory and Techniques*, Vol. MTT-17, No. 8, August 1969, pp. 416–39.

[A1.3] Motz, H., "The treatment of singularities of partial differential equations by relaxation methods", *Quarterly of Applied Mathematics*, Vol. 4, No. 4, January 1947, pp. 371–7.

[A1.4] Woods, L. C., "The relaxation treatment of singular points in Poisson's Equation", *Quarterly Journal of Mechanics and Applied Mathematics*, Vol. 6, Pt. 2, 1953, pp. 163–85.

[A1.5] Whiting, K. B., "A treatment for boundary singularities in finite difference solutions of Laplace's Equation", *IEEE Trans. Microwave Theory and Techniques*, Vol. MTT-16, No. 10, October 1968, pp. 889–91.

[A1.6] Fooks, E. H. and Ladbrooke, P. H., "A co-ordinate transformation for the numerical analysis of microstrip transmission lines", *Proc. IREE Aust.*, Vol. 41, No. 2, June 1980, pp. 74–8.

Appendix 2
The method of sub-areas

This method, known also as the method of moments, is used to calculate the capacitance associated with a system of conductors. As described here, the quasi-static capacitance of a thin microstrip line will be evaluated. With minor enhancements, the approach may be used for capacitance calculations of thick microstrip lines [A2.1], lines with anisotropic substrates [A2.2], coupled lines [A2.3], open-circuit fringing capacitance [A2.4, A2.5], steps and gaps [A2.6, A2.7] as well as bends and junctions [A2.8].

Consider a uniform transmission line in free space that may be taken in cross-section as a two-dimensional problem. The electrostatic potential at the field point $P_j \equiv (x_j, y_j)$ due to a unit charge at the source point $P_i \equiv (x_i, y_i)$ is found from (3.38) to be

$$G(P_j : P_i) = -\frac{1}{2\pi\varepsilon_0} \ln\left[\sqrt{(x_j - x_i)^2 + (y_j - y_i)^2}\right] \qquad (A2.1)$$

if all the conductors are removed except for the filament upon which the charge is situated. The function (A2.1) is known as the Green's function for the region. In multi-dielectric problems, there will be different Green's functions depending on the relative locations in the dielectric regions of the source and field points. A superposition of all the potentials due to the individual contributions of the charges given by the variable charge density, $\rho(P_i)$, on the conductors leads to the integral equation

$$V(x_j, y_j) = \int G(P_j : P_i)\, \rho(P_i)\, dP_i \qquad (A2.2)$$

Of particular interest are the equations for the potentials at the conductors since these are specified in capacitance calculations. Thus, if the conductors are subdivided into a total of n sub-areas, it is possible to write (A2.2) for the potential at the geometric center of each area, giving the matrix equation

$$\mathbf{V} = [\,p\,]\, \mathbf{q} \qquad (A2.3)$$

where $\mathbf{V}$ is a column matrix of the voltages at the points (x_i, y_i) and $\mathbf{q}$ is a column matrix of the charges at (x_i, y_i). In this case, as it is $\mathbf{V}$ that is specified, (A2.3) is inverted to give

$$\mathbf{q} = [\,p\,]^{-1} \mathbf{V} \qquad (A2.4)$$

and the total charge as well as the charge distribution across the conductors may be found.

Before proceeding with an example that illustrates the power of the method as a numerical technique for capacitance calculation, there are two points that need to be noted.

Point 1   Planes of symmetry

Two strips with potentials $\pm V$ may be used to represent the microstrip conductor and its image that together give an equipotential surface of zero potential (the ground plane) between them. Allowing for the symmetry of charge between the conductor and its image, the complete system information is contained in the reduced size matrix equation for the potentials of the elements of one conductor in terms of the charges on that conductor and its image. Inverting the matrix and solving for the sub-area charges on the conductor with a potential of 1 volt with respect to the zero potential of the plane of symmetry leads directly to the capacitance of the line.

There is also a second plane of symmetry that passes through the center of the microstrip conductor, the ground plane and the conductor image. The potential at each point (x_i, y_i) on one half of the microstrip conductor may be expressed in terms of the four symmetrical charge components, $+q_i$, $+q_i$, $-q_i$ and $-q_i$, and their respective source to field point distances. In this way the matrix size for inversion is reduced by a factor of 4, compared with the complete system that does not account for symmetry.

Point 2   The handling of self-potential

The potential at the center of each sub-area is formed as a summation of the potential contributions due to a charge at the center of all sub-areas. Furthermore, this must include the potential at (x_i, y_i) due to the charge at (x_i, y_i), leading to a zero distance and apparent infinite potential. This is overcome by no longer considering the (x_i, y_i) charge as a point charge, but as a charge distribution over a small but finite dimension.

Consider one sub-area as illustrated in Figure A2.1, with a uniform charge density across its width. If the sub-area has a width ds_i and total charge q_i, then the potential at the center due to an element of length dr at a distance r is given by

$$V(x_i, y_i) = -\frac{1}{2\pi\varepsilon_0} \times \frac{q_i}{ds_i} \times ln(r)\,dr \tag{A2.5}$$

The integral over the right-hand half of the element is

$$V(x_i, y_i) = -\frac{q_i}{2\pi\varepsilon_0\,ds_i} \int_0^{ds_i/2} ln(r)\,dr \tag{A2.6}$$

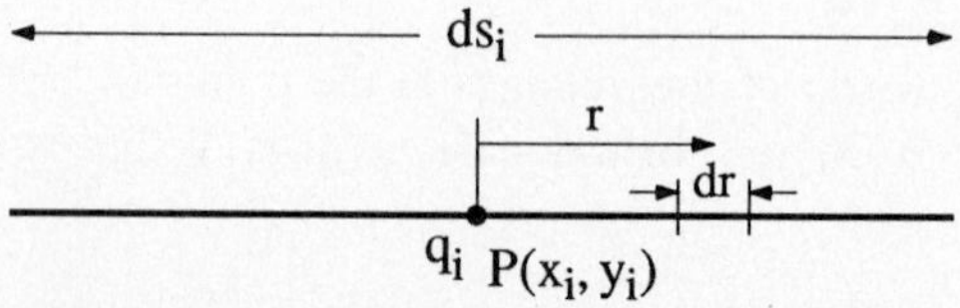

Figure A2.1   The geometry of one sub-area

Using $\int ln\,(r)\,dr = r\big(ln\,(r) - 1\big)$ in the integration of (A2.6) and doubling for the complete element gives the self-potential of a sub-area as

$$V(x_i, y_i) = -\frac{q_i}{2\pi\varepsilon_0}\left\{ ln\left[\frac{ds_i}{2}\right] - 1\right\} \tag{A2.7}$$

Example A2.1

Using the method of sub-areas, estimate the capacitances of microstrip lines in free space with w/h = 0.01, 0.1 and 1.0.

Solution:

Consider the microstrip line and its image as illustrated in Figure A2.2. The simplest situation is to represent each strip by one point charge. It is expected that this will lead to good results only for very narrow strips, but for wider strips more elements would be needed. The potential at P_1 due to the charges $+q$ and $-q$ is found by adding contributions as given by (A2.7) and (A2.1) respectively. Thus

$$V = -\frac{q}{2\pi\varepsilon_0}\Big[ln\,(w/2) - 1 - ln\,(2h)\Big] \tag{A2.8}$$

i.e.
$$C = \frac{2\pi\varepsilon_0}{1 + 2\,ln\,2 + ln\,h - ln\,w}\quad \text{F.m}^{-1} \tag{A2.9}$$

Results derived from (A2.9) are presented in Table A2.1.

For higher accuracy, the line and its image may each be represented by two equal point charges as illustrated in Figure A2.3. Because of the symmetry, the result is still simple enough to be expressed in one equation. The potential at P_1 is given by

$$V = -\frac{q}{2\pi\varepsilon_0}\Big[A' + B' - C' - D'\Big] \tag{A2.10}$$

where the four terms are

$\quad A' = ln\,(w/4) - 1 \qquad\qquad$ due to the charge at P_1
$\quad B' = ln\,(w/2) \qquad\qquad\quad$ due to the charge at P_2
$\quad C' = ln\,(2h) \qquad\qquad\quad\;\,$ due to the charge at P_3
$\quad D' = ln\left(\sqrt{4h^2 + (w/2)^2}\right) \quad$ due to the charge at P_4

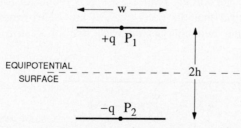

Figure A2.2 A single point charge representation for a microstrip line

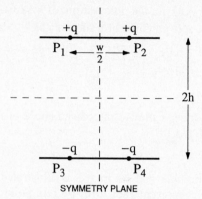

Figure A2.3 A two-point charge representation for a microstrip line

With a total charge of $\pm 2q$ on each plate, the normalized line capacitance

$$\frac{C}{\varepsilon_0} = \frac{2q}{\varepsilon_0 V} = \frac{4\pi}{-A' - B' + C' + D'} \tag{A2.11}$$

Results derived from (A2.11) and the answer for Exercise 3.8 are presented in Table A2.1.

Table A2.1 Normalized capacitance of a microstrip line in free space

| | Sub-areas per plate | | | |
| | 1 | 2 | 4 | |
	Equation (A2.9)	Equation (A2.11)	Exercise 3.8	Accurate value from [A2.2]
0.01	0.899	0.919	0.930	0.940
0.1	1.340	1.385	1.411	1.434
1.0	2.633	2.795	2.890	2.980

Note

In Appendix 5 of Wheeler's paper [A2.9], it is seen that a narrow microstrip line is equivalent to two quasi-circular wires with equal charges, having a line geometry as illustrated in Figure A2.4. This representation clearly shows that the charge is not equally distributed across the strip but must be increasing towards the edge of the strip. In solving Exercise 3.8, it is found for four charges equally spaced across the strip that the outer charges are 1.856 times greater than the inner charges. If the two charges on each half of a strip or its image are

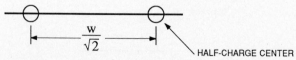

Figure A2.4 The two-wire line approximation to a microstrip line

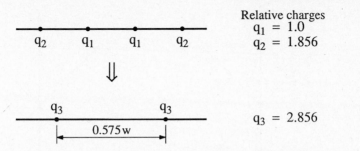

Relative charges
$$q_1 = 1.0$$
$$q_2 = 1.856$$

$$q_3 = 2.856$$

Figure A2.5 An equivalent charge distribution obtained by reducing four charges to two

combined together as one, maintaining the product of (charge) × (distance from strip center) to find the equivalent charge center, the result will be as illustrated in Figure A2.5. There has been a definite movement of the charge centers from 0.5w apart as in (A2.11) towards 0.707w as given in the accurate equivalence by Wheeler.

***

The use of images

In Figure A2.2, the single line-charge representation for a microstrip line, appearing as a point charge in the transverse plane, was used in conjunction with an image to solve for the capacitance per unit length of the line. The image charge was placed such that an equipotential surface coincided with the ground plane. To this stage, the microstrip line has been in a uniform dielectric medium, but now the mixed dielectric problem of a dielectric substrate must be accounted for in the solution. Charges in the vicinity of dielectric boundaries will be modeled in two stages:

i) A line charge in the vicinity of a flat and infinite boundary between two dielectric materials is shown in Figure A2.6. At the boundary, the incident flux, ψ, from region 2 is split into two parts with $K\psi$ being returned into the region and the remainder, $(1 - K)\psi$, continuing into region 1. Splitting up the flux leads to the image charge of Figure A2.6 which, together with the original charge, is used to determine the Green's function for each region.

ii) A line charge in the vicinity of a flat and infinite boundary of a uniform thickness slab of dielectric material is shown in Figure A2.7. From the model by Silvester [A2.1], the Green's function is derived for each of the three regions shown in the figure. The procedure of (i) is followed at each boundary and a Green's function is derived for each of the three regions shown in the figure. In general, there are an infinite number of images, each of which is known as a partial image.

Consider the filament of flux, as in Figure A2.6, emanating uniformly in all directions from a line source that is represented in the transverse plane by the point charge, q [A2.1]. At a point on the dielectric interface, some of the flux, $K\psi$, will be reflected and the remainder, $(1 - K)\psi$, will continue on into the adjacent dielectric region. Applying the boundary condition that the normal component of the electric

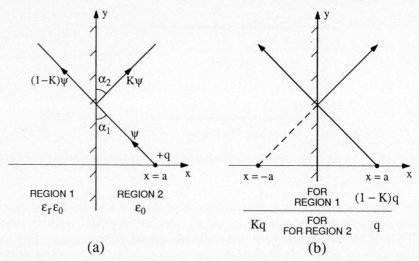

Figure A2.6 Flux from a charge near a dielectric boundary, showing (a) the flux lines, and (b) the partial image charges to be used for each half-plane, from Silvester [A2.1] (© 1968, IEE)

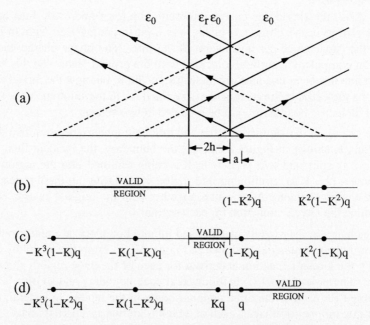

Figure A2.7 The flux and partial image charges for a line charge in the vicinity of an infinite dielectric slab of thickness 2h, showing (a) the flux lines and construction for the images, (b) the image charges for the potential evaluation in the region which does not contain the slab or the original charge, (c) the image charges for evaluation within the dielectric slab, and (d) the image charges for evaluation in the region containing the original charge, from Silvester [A2.1] (© 1968, IEE)

flux density must be continuous across the surface, then

$$(1 - K)\,\psi\sin(\alpha_1) \;=\; \psi\sin(\alpha_1) \,-\, K\,\psi\sin(\alpha_2) \tag{A2.12}$$

i.e. $\qquad \alpha_1 \;=\; \alpha_2 \tag{A2.13}$

Applying the boundary condition that the tangential component of the electric field must be the same on each side of the surface gives

$$\frac{1}{\varepsilon_r\varepsilon_0}(1 - K)\,\psi\cos(\alpha_1) \;=\; \frac{1}{\varepsilon_0}\Big\{\psi\cos(\alpha_1) \,+\, K\,\psi\cos(\alpha_1)\Big\} \tag{A2.14}$$

i.e. $\qquad (1 - K) \;=\; \varepsilon_r\,(1 + K) \tag{A2.15}$

or $\qquad K \;=\; -\,\dfrac{\varepsilon_r - 1}{\varepsilon_r + 1} \tag{A2.16}$

Thus for a charge in free space above a homogeneous dielectric half-space, this coefficient is independent of the angle of incidence and

$$-1 < K < 0$$

For the reflection of flux at an interface from the dielectric material to free space, K will now be positive and have the same magnitude as before. Thus, flux lines that have passed into the dielectric slab and out again have terms $(1 - K)(1 + K) = 1 - K^2$ in their magnitudes.

The potentials in region 1 of Figure A2.6a may be derived using a charge with a value $(1 - K)q$ at $x = a$ as shown in Figure A2.6b. In region 2 that includes the original charge, the potentials are derived using the original charge q at $x = a$ and the image charge at $x = -a$. Note that there are never any image charges within the region in which a potential is to be evaluated.

The Green's function (A2.1) for region 1 now becomes

$$G(P_j : P_i) \;=\; -\,\frac{(1 - K)}{4\pi\varepsilon_r\varepsilon_0}\,ln\big((x_j - a)^2 + (y_j - y_i)^2\big) \tag{A2.17}$$

while for region 2

$$G(P_j : P_i) \;=\; -\,\frac{1}{4\pi\varepsilon_0}\Big\{ln\big((x_j - a)^2 + (y_j - y_i)^2\big) + K\,ln\big((x_j + a)^2 + (y_j - y_i)^2\big)\Big\} \tag{A2.18}$$

A plot of the resultant electric fields in Figure A2.8 shows how the fields are drawn towards the dielectric region.

Multiple images

Silvester [A2.1] has shown how multiple images are required to derive the potentials associated with a charge in the vicinity of a slab of dielectric material, Figure A2.7. Although the parallel flux lines only apply for a field point at very large distances, they do locate the positions of the multiple images. The positions of these images, together with the values of the image coefficients, are independent of the angle of the flux lines to the slab surfaces. The potential at any field point may be derived from the charge, its images and their respective distances to the field point.

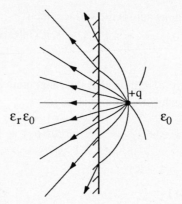

Figure A2.8   Electric fields from a line charge above a dielectric half-space

A simple use of the infinite summation of images will be illustrated in the following example. For further stages of refinement of the method, including the setting up of appropriate Green's functions, the representation of the charge variation across conductor surfaces and the method for integration of the resultant integrals within the summation, the reader is referred to the literature, e.g. [A2.1, A2.2].

Example A2.2

Estimate the capacitance per meter for a microstrip line with $w/h = 0.01$ on a substrate with $\varepsilon_r = 2.5$.

Solution:

In its simplest form, the microstrip line may be treated as a single sub-area with a charge +q. The ground plane equipotential surface is maintained by placing an image charge −q as illustrated in Figure A2.9. Each of these charges at A and B now has its own accompaniment of partial image charges as required for the dielectric slab boundaries. From Figure A2.7, it is seen that as the original charge approaches the surface and $a \to 0$, the potential at A due to the charge at A is given by the self-potential of the sum of the charge itself, q, and the first partial image, Kq, i.e. $(1 + K)q$. Thus, from (A2.7)

$$V = -\frac{(1+K)q}{2\pi\varepsilon_0}\left\{ ln\left[\frac{w}{2}\right] - 1 \right\} \tag{A2.19}$$

The remaining partial images for the charge at A sum as

$$V = -\frac{-K(1-K^2)q}{2\pi\varepsilon_0}\sum_{n=1}^{\infty}\left(K^{2(n-1)}\,ln(4nh)\right) \tag{A2.20}$$

As the final result depends on the ratio w/h and not on the individual values of w and h, for evaluation purposes take $h = 1$ and $w = 0.01$. The potential at A is evaluated in three stages. The first two stages evaluate the potential at A due to the charge at A and its partial images. The third stage evaluates the potential at A due to the partial image charges associated with the charge at B.

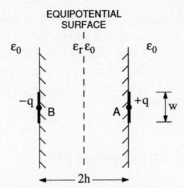

Figure A2.9 The geometry of a narrow line on a dielectric substrate

i) The self-potential of the charge and partial image that are placed at A is given from (A2.19), with $K = -0.4286$ as

$$V = -\frac{q}{\varepsilon_0} \times \frac{1 - 0.4286}{2\pi}\left\{ln(0.005) - 1\right\}$$

$$= \frac{0.5728\,q}{\varepsilon_0}$$

ii) The other partial images from the charge at A, from (A2.20) give the potential

$$V = -\frac{0.05568\,q}{\varepsilon_0}\left\{1.3863 + 0.3819 + 0.0838 + 0.0172 + 0.0034 + \cdots\right\}$$

$$= -\frac{0.1043\,q}{\varepsilon_0}$$

iii) Consider now the negative image charge at B together with its partial images. Their contributions towards the potential at A are given with a reinterpretation of case (b) of Figure A2.7 as

$$V = -\frac{-q}{\varepsilon_0} \times \frac{(1 - K^2)}{2\pi}\sum_{n=1}^{\infty}\left(K^{2(n-1)}\,ln((4n-2)\,h)\right)$$

$$= \frac{0.1299\,q}{\varepsilon_0}\left\{0.6931 + 0.3291 + 0.0777 + 0.0164 + 0.0033 + \cdots\right\}$$

$$= \frac{0.1454\,q}{\varepsilon_0}$$

Summing together the three contributions

$$V = \frac{0.6139\,q}{\varepsilon_0} = \frac{q}{C}$$

i.e.

$$C = 14.4 \text{ pF.m}^{-1}$$

This result may be compared with an accurate result that is derived from Kobayashi [A2.2] which gives $C = 0.940 \varepsilon_0$ for a line in free space with $w/h = 0.01$. The effective filling factor of 0.5432 for $\varepsilon_r = 2.5$ gives $\varepsilon_{eff} = 1.815$. These values give an accurate capacitance value for this line, against which the derived value of 14.4 pF.m^{-1} is to be compared, of $1.815 \times 0.940 \times \varepsilon_0 = 15.1 \text{ pF.m}^{-1}$.

REFERENCES

[A2.1] Silvester, P., "TEM wave properties of microstrip transmission lines", *Proc. IEE*, Vol. 15, No. 1, January 1968, pp. 43–8.

[A2.2] Kobayashi, M., "Analysis of the microstrip and the electrooptic light modulator", *IEEE Trans. Microwave Theory and Techniques*, Vol. MTT-26, No. 2, February 1978, pp. 119–26.

[A2.3] Bryant, T. G. and Weiss, J. A., "Parameters of microstrip transmission lines and of coupled pairs of microstrip lines", *IEEE Trans. Microwave Theory and Techniques*, Vol. MTT-16, No. 12, December 1968, pp. 1021–7.

[A2.4] Silvester, P. and Benedek, P., "Equivalent capacitances of microstrip open-circuits", *IEEE Trans. Microwave Theory and Techniques*, Vol. MTT-20, No. 8, August 1972, pp. 511–16.

[A2.5] Farrar, A. and Adams, A. T., "Computation of lumped microstrip capacities by matrix methods – rectangular sections and end effects", *IEEE Trans. Microwave Theory and Techniques*, Vol. MTT-19, No. 5, May 1971, pp. 495–7.

[A2.6] Benedek, P. and Silvester, P., "Equivalent capacitances for microstrip gaps and steps", *IEEE Trans. Microwave Theory and Techniques*, Vol. MTT-20, No. 11, November 1972, pp. 729–33.

[A2.7] Farrar, A. and Adams, A. T., "Matrix methods for microstrip three-dimensional problems", *IEEE Trans. Microwave Theory and Techniques*, Vol. MTT-20, No. 8, August 1972, pp. 497–504.

[A2.8] Silvester, P and Benedek, P., "Microstrip discontinuity capacitances for right-angled bends, T-junctions and crossings", *IEEE Trans. Microwave Theory and Techniques*, Vol. MTT-21, No. 5, May 1973, pp. 341–6.

[A2.9] Wheeler, H. A., "Transmission-line properties of a strip on a dielectric sheet on a plane", *IEEE Trans. Microwave Theory and Techniques*, Vol. MTT-25, No. 8, August 1977, pp. 631–47.

[A2.10] Wheeler, H. A., "Transmission-line properties of parallel strips separated by a dielectric sheet", *IEEE Trans. Microwave Theory and Techniques*, Vol. MTT-13, No. 2, March 1965, pp. 172–85.

[A2.11] Wheeler, H. A., "Transmission-line properties of parallel wide strips by a conformal-mapping approximation", *IEEE Trans. Microwave Theory and Techniques*, Vol. MTT-12, No. 3, May 1964, pp. 280–9.

Appendix 3
Microstrip line data

The data in this appendix are provided for use in the examples and problems that specify a substrate with $\varepsilon_r = 2.5$. Quasi-static approximations are assumed throughout. Specific w/h and Z_0 values are quoted in Table A3.1. These values are used for the simple curve fitting equations that are provided if calculations with intermediate line parameters are required.

Table A3.1 Microstrip line data for $\varepsilon_r = 2.5$

| RELATIVE PERMITTIVITY = 2.5 | | | | | |
| Analysis | | | Synthesis | | |
w/h	Z_0, Ω	ε_{eff}	Z_0, Ω	w/h	ε_{eff}
0.1	193.12	1.851	20	9.563	2.263
0.2	161.73	1.872	25	7.281	2.226
0.4	130.52	1.900	30	5.777	2.193
0.6	112.44	1.924	40	3.924	2.137
0.8	99.80	1.946	50	2.837	2.090
1.0	90.17	1.966	60	2.128	2.051
1.5	73.39	2.008	70	1.635	2.018
2.0	62.28	2.043	80	1.275	1.990
4.0	39.46	2.139	100	0.796	1.945
6.0	29.13	2.199	120	0.506	1.913
8.0	23.17	2.239	150	0.259	1.881
10.0	19.27	2.269	180	0.134	1.859

FOR ALL CURVE FITTING, $0.10 \leq w/h \leq 10.0$

ANALYSIS

Given w/h and with $\varepsilon_r = 2.5$, then the characteristic impedance is given by

$$Z_0 = e^x \tag{A3.1}$$

where the exponent

$$x = \sum_{i=0}^{4} A_i \left\{ ln(w/h) \right\}^i \tag{A3.2}$$

The effective permittivity

$$\varepsilon_{eff} = \sum_{i=0}^{4} B_i \left\{ ln(w/h) \right\}^i \tag{A3.3}$$

SYNTHESIS

Given the characteristic impedance, Z_0, then the ratio w/h is given by

$$\frac{w}{h} = e^y \tag{A3.4}$$

with

$$y = \sum_{i=0}^{4} C_i \left\{ ln(Z_0) - 4.0 \right\}^i \tag{A3.5}$$

Now, knowing w/h, ε_{eff} may be found from (A3.3).

Table A3.2 Coefficients for Equations A3.2, A3.3 and A3.5

i	A_i	B_i	C_i
0	4.5015	1.9657	0.9054
1	−0.4762	0.0950	−1.5627
2	−0.0832	0.0255	−0.3422
3	−0.0046	−0.0007	−0.2215
4	0.0018	−0.0014	−0.0943

Appendix 4 Formulae for parallel-coupled microstrip transmission lines

These low frequency expressions, as given by Kirschning and Jansen [A4.1] (© 1984, IEEE), make use of normalized values for the strip width and spacing

$$u = \frac{w}{h} \quad \text{and} \quad g = \frac{s}{h} \tag{A4.1}$$

They are modified here only to the extent that consistent terminology is maintained. A sample set of data with intermediate calculation values is provided in the table following the expressions, to assist in checking any computer implementation of the equations. Two equal width strips of negligible thickness are assumed. The two expressions, Z_0 and ε_{eff}, as they appear in this section relate to a single microstrip line of width w on the same substrate material and are derived from the analysis formulae of Table 3.2. For the following range of parameters:

$$0.1 \leq u \leq 10.0 \quad 0.1 \leq g \leq 10.0 \quad 1.0 \leq \varepsilon_r \leq 18.0$$

the errors quoted in [A4.1] are $< 0.7\%$ for $\varepsilon_{eff}^{(e)}$, $< 0.5\%$ for $\varepsilon_{eff}^{(e)}$, and $< 0.6\%$ for Z_{0e} and Z_{0o}.

THE EVEN-MODE EFFECTIVE PERMITTIVITY

$$\varepsilon_{eff}^{(e)} = \frac{\varepsilon_r + 1}{2} + \frac{\varepsilon_r - 1}{2} \left[1 + \frac{10}{v} \right]^{-a_e(v) \times b_e(\varepsilon_r)} \tag{A4.2}$$

with

$$v = \frac{u(20 + g^2)}{10 + g^2} + g \times exp(-g)$$

$$a_e(v) = 1 + \frac{1}{49} \times ln\left[\frac{v^4 + (v/52)^2}{v^4 + 0.432} \right] + \frac{1}{18.7} \times ln\left\{ 1 + \left[\frac{v}{18.1} \right]^3 \right\}$$

$$b_e(\varepsilon_r) = 0.564 \times \left[\frac{\varepsilon_r - 0.9}{\varepsilon_r + 3.0} \right]^{0.053}$$

THE ODD-MODE EFFECTIVE PERMITTIVITY

$$\varepsilon_{eff}^{(o)} = \varepsilon_{eff} + \left[\frac{\varepsilon_r + 1}{2} + a_o(u, \varepsilon_r) - \varepsilon_{eff} \right] \times exp(-c_o \times g^{d_o}) \tag{A4.3}$$

with $$a_0(u,\varepsilon_r) = 0.7287 \times \left\{ \varepsilon_{eff} - \frac{\varepsilon_r + 1}{2} \right\} \times (1 - exp(-0.179\,u))$$

$$b_0(\varepsilon_r) = \frac{0.747\,\varepsilon_r}{0.15 + \varepsilon_r}$$

$$c_0 = b_0(\varepsilon_r) - (b_0(\varepsilon_r) - 0.207) \times exp(-0.414\,u)$$

$$d_0 = 0.593 + 0.694 \times exp(-0.562\,u)$$

THE EVEN-MODE CHARACTERISTIC IMPEDANCE

$$Z_{0e} = Z_0 \times \left[\frac{\varepsilon_{eff}}{\varepsilon_{eff}^{(e)}} \right]^{\frac{1}{2}} \div \left\{ 1 - \frac{\sqrt{\varepsilon_{eff}}\,Z_0\,Q_4}{377} \right\} \qquad (A4.4)$$

with $$Q_1 = 0.8695 \times u^{0.194}$$

$$Q_2 = 1 + 0.7519\,g + 0.189 \times g^{2.31}$$

$$Q_3 = 0.1975 + \left\{ 16.6 + \left[\frac{8.4}{g} \right]^6 \right\}^{-0.387} + \frac{1}{241} \times ln \left\{ \frac{g^{10}}{1 + (g/3.4)^{10}} \right\}$$

$$Q_4 = \frac{2\,Q_1}{Q_2} \times \left\{ u^{Q_3} \times exp(-g) + u^{-Q_3}(2 - exp(-g)) \right\}^{-1}$$

THE ODD-MODE CHARACTERISTIC IMPEDANCE

$$Z_{0o} = Z_0 \times \left[\frac{\varepsilon_{eff}}{\varepsilon_{eff}^{(o)}} \right]^{\frac{1}{2}} \div \left\{ 1 - \frac{\sqrt{\varepsilon_{eff}}\,Z_0\,Q_{10}}{377} \right\} \qquad (A4.5)$$

with $$Q_5 = 1.794 + 1.14 \times ln \left\{ 1 + \frac{0.638}{g + 0.517 \times g^{2.43}} \right\}$$

$$Q_6 = 0.2305 + \frac{1}{281.3} \times ln \left\{ \frac{g^{10}}{1 + (g/5.8)^{10}} \right\} + \frac{ln(1 + 0.598 \times g^{1.154})}{5.1}$$

$$Q_7 = \frac{10 + 190 \times g^2}{1 + 82.3 \times g^3}$$

$$Q_8 = exp(-6.5 - 0.95 \times ln(g) - (g/0.15)^5)$$

$$Q_9 = \left\{ Q_8 + \frac{1}{16.5} \right\} \times ln(Q_7)$$

$$Q_{10} = Q_4 - \frac{Q_4 - Q_5 \times exp(ln(u) \times Q_6 \times u^{-Q_9})}{Q_2}$$

Table A4.1 Sample calculated values to verify the coupled-line equations

SAMPLE VALUES $u = 2.0 \quad g = 0.5 \quad \varepsilon_r = 2.5$			
EFFECTIVE PERMITTIVITIES			
$\varepsilon_{\text{eff}}^{(e)}$	2.1649	$\varepsilon_{\text{eff}}^{(o)}$	1.8870
v	4.8472	$a_0(u,\varepsilon_r)$	0.0765
$a_e(v)$	1.0010	$b_0(\varepsilon_r)$	0.7047
$b_e(\varepsilon_r)$	0.5283	c_0	0.5127
		d_0	0.7835
MODE CHARACTERISTIC IMPEDANCES			
Z_{0e}	66.84	Z_{0o}	45.10
Q_1	1.0220	Q_5	2.6237
Q_2	1.4141	Q_6	0.2525
Q_3	0.1702	Q_7	5.0941
Q_4	0.7575	Q_8	0.0000
		Q_9	0.0987
		Q_{10}	0.0000

REFERENCE

[A4.1] Kirschning, M. and Jansen, R. H., "Accurate wide-range design equations for the frequency-dependent characteristic of parallel coupled microstrip lines", *IEEE Trans. Microwave Theory and Techniques*, Vol. MTT-32, No. 1, January 1984, pp. 83–90. Corrections: *IEEE Trans. Microwave Theory and Techniques*, Vol. MTT-33, No. 3, March 1985, p. 288.

Answers to selected exercises

1.1 $33.3\,\Omega$

1.2   With $|\Gamma| = 0.1$,   i) 1.22,   ii) 1%,   iii) 20.0 dB,   iv) 99.0%,   v) 0.044 dB

1.3   ii) $1.164 \rightarrow 1.284$

1.4   0.087 dB,   V.S.W.R. = 60.5

1.6   ii) $(41.4 - j28.0)\,\Omega$

1.7 $\Gamma_L = 0.62\,\underline{/29.7°}$,   V.S.W.R. = 4.27, $l = 87.3$ mm

1.8   i) $(31.9 - j41.0)\,\Omega$

       ii) $(42.0 - j36.0)\,\Omega$

2.1   i) $\begin{bmatrix} 0 & 0.316 \\ 0.316 & 0 \end{bmatrix}$, magnitudes only

       ii) $\begin{bmatrix} 0 & e^{-j\theta} \\ e^{-j\theta} & 0 \end{bmatrix}$, where $\theta = 2n\pi$

         When $n\lambda = \dfrac{\lambda}{4}$, $\theta = \dfrac{\pi}{2}$, leading to $\begin{bmatrix} 0 & -j \\ -j & 0 \end{bmatrix}$

       iii) $\begin{bmatrix} \dfrac{n^2 - 1}{n^2 + 1} & \dfrac{2n}{n^2 + 1} \\ \dfrac{2n}{n^2 + 1} & \dfrac{1 - n^2}{n^2 + 1} \end{bmatrix}$

       iv)   a) $\begin{bmatrix} \frac{1}{3} & \frac{2}{3} \\ \frac{2}{3} & \frac{1}{3} \end{bmatrix}$ b) $\dfrac{1}{j\omega L + 2Z_0} \begin{bmatrix} j\omega L & 2Z_0 \\ 2Z_0 & j\omega L \end{bmatrix}$; [S] is unitary

       v)   a) $\dfrac{1}{1 + x^2} \begin{bmatrix} 1 - x^2 & -j2x \\ -j2x & 1 - x^2 \end{bmatrix}$, where $x = \dfrac{1}{k}$

         b) $\dfrac{1}{2x\cos\phi + j(1 + x^2)\sin\phi} \begin{bmatrix} j(1 - x^2)\sin\phi & 2x \\ 2x & j(1 - x^2)\sin\phi \end{bmatrix}$, where $\phi = 2n\pi$

       vi) $\dfrac{1}{\alpha^2 + Z_0^2} \begin{bmatrix} \alpha^2 - Z_0^2 & -2\alpha Z_0 \\ +2\alpha Z_0 & \alpha^2 - Z_0^2 \end{bmatrix}$; [S] is unitary

       vii) $\begin{bmatrix} 0 & 1 & 0 \\ 0 & 0 & 1 \\ 1 & 0 & 0 \end{bmatrix}$, magnitudes only; [S] is unitary

viii) $\begin{bmatrix} 0 & 0 \\ 1 & 0 \end{bmatrix}$, magnitudes only; [S] is NOT unitary

ix) $\begin{bmatrix} \xi & 0 \\ -(1+\xi)g_mZ_0 & 1 \end{bmatrix}$, where $\xi = \dfrac{1-j\omega CZ_0}{1+j\omega CZ_0}$

2.2 i) a) $\begin{bmatrix} -\frac{1}{3} & \frac{2}{3} & \frac{2}{3} \\ \frac{2}{3} & -\frac{1}{3} & \frac{2}{3} \\ \frac{2}{3} & \frac{2}{3} & -\frac{1}{3} \end{bmatrix}$ b) $\begin{bmatrix} 0 & \frac{1}{2} & \frac{1}{2} \\ \frac{1}{2} & -\frac{1}{4} & \frac{3}{4} \\ \frac{1}{2} & \frac{3}{4} & -\frac{1}{4} \end{bmatrix}$ c) $\begin{bmatrix} 0 & \frac{1}{2} & \frac{1}{2} \\ \frac{1}{2} & 0 & \frac{1}{2} \\ \frac{1}{2} & \frac{1}{2} & 0 \end{bmatrix}$

2.3 iii) $\begin{bmatrix} -\dfrac{\delta s_i}{2} & -j(s_r + \dfrac{\delta s_r}{2}) \\ -j(s_f + \dfrac{\delta s_f}{2}) & -\dfrac{\delta s_o}{2} \end{bmatrix}$

2.4 ii) $T_f = \dfrac{s_f}{1 - s_o\Gamma_L}$, $\qquad T_r = \dfrac{s_r}{1 - s_i\Gamma_s}$

iii) For $[S]_1$ followed by $[S]_2$

$$[S] = \frac{1}{1 - s_{o1}s_{i2}} \begin{bmatrix} s_{i1} - s_{i2}\Delta_1 & s_{r1}s_{r2} \\ s_{f1}s_{f2} & s_{o2} - s_{o1}\Delta_2 \end{bmatrix}, \qquad \Delta \equiv \begin{vmatrix} s_i & s_r \\ s_f & s_o \end{vmatrix}$$

3.2 i) 0.417 pF
3.3 i) 4.08 mm
  ii) 7.22 mm
3.4 (0.99 ± 0.04) mm
3.5 $24.0\,\Omega \rightarrow 136.8\,\Omega$
3.6 2.323
3.7 $33.9\,\Omega$, 0.105λ
4.1 $90.7\,\Omega$
4.2 0.073 mm
4.3 i) 0.044 dB.λ^{-1}
4.4 For copper at 1.0 GHz, $\delta = 2.09$ μm; at 10 GHz, $\delta = 0.66$ μm
4.5 ii) For copper at 1.0 GHz, $\Delta = 0.45$ μm
4.6 0.0028 dB.cm^{-1} $(0.28$ dB.m$^{-1})$
4.7 3.5 mm
4.9 0.7%
4.10 4.0 GHz
5.1 964 MHz
5.2 i) $31\,\Omega$, $16\,h/\lambda_0$ radian
5.3 i) 0.44
  ii) $0.44 + j0.01$
5.4 $(49.99 + j0.38)\,\Omega$, $(50.02 + j0.37)\,\Omega$
6.2 i) $z_L = 1.84 - j2.52$
6.3 i) $z = 0.70 - j0.51$
6.4 ii) 2.83 pF, 0.76 pF
6.6 $Z_T = 98.7\,\Omega$, $l = 0.106\lambda$
6.7 0.136λ (nearer to the load), 0.162λ

6.8 0.406λ

6.9   position $= 0.0155\lambda$,   length $= 0.348\lambda$, 11%

6.10 $l_1 = 0.07\lambda$, $l_2 = 0.042\lambda$

6.11 $l = 0.313\lambda$, $Z_T = 79\Omega$

6.12   i)   position $= 0.007\lambda$,   length $= 0.34\lambda$

      ii) 0.129λ (nearer to the load), 0.393λ

      iii) 0.079λ, 24.3Ω

      iv) 7.64 nH, 4.81 pF

      v) 0.053λ for the 100Ω line nearer the load, 0.182λ for the 25Ω line

6.13 $l_1 = l_2 = 0.078\lambda$

6.14   i) (31.6Ω), $(25.1\Omega, 39.8\Omega)$, $(21.2\Omega, 26.6\Omega, 37.6\Omega, 47.2\Omega)$

7.1 $C = 3.57$ dB

7.2   i) 4.77 dB, 70.7Ω and 40.8Ω

7.4   V.S.W.R. $= 1.125$

7.5 68.8Ω, 72.8Ω

7.6   at f_0,   i) V.S.W.R. $= 1.0$ ii) $C = 3.0$ dB

8.1 $C = 32$ dB at 1.04 GHz

8.2 $w = 8.08$ mm, $s = 0.86$ mm, $l = 35.1$ mm

8.4 11.7 dB

8.7   V.S.W.R. $= 19.0$, -10.0dB at port 3, -10.46dB at port 4

8.10 43%

9.1   i) 1.357 GHz

      ii) 641 MHz

9.2   i) 737 MHz

      ii) 1.56 GHz

9.3   i) 10.0 ± 1.3 GHz

      ii) 10.0 ± 0.6 GHz

9.7   Bandwidth $= 146\%$

9.10 $f_0 = 2.0$ GHz,   bandwidth $\approx \pm 3$ MHz

9.13   In coupling region, $w = 3.77$ mm, $s = 0.38$ mm, $l = 26.4$ mm

10.2 1.35

10.3 $D = 3.19$ mm, $L = 2.13$ mm

10.5 1.889 GHz, 1.964 GHz

11.2   i) $K \to \dfrac{1 + |s_r s_f|^2}{2\,|s_r s_f|}$, $\Delta \to s_r s_f$, $G_t(max) \to |s_f|^2$

11.3 $G_t = \dfrac{|s_f|^2(1 - |\Gamma_s|^2)(1 - |\Gamma_L|^2)}{\left|(1 - s_i\Gamma_s)(1 - s_0\Gamma_L) - s_r s_f \Gamma_s \Gamma_L\right|^2}$

11.10   i)   Without loading, $K = 0.97$ and $|\Delta| = 0.62$

       ii) $\Gamma_s = 0.36\,\underline{/118°}$, $\Gamma_L = 0.49\,\underline{/24°}$, 19.8dB, 22.0dB

       iii) $\Gamma_s = 0.57\,\underline{/83°}$, $\Gamma_L = 0.08\,\underline{/71°}$, 19.2dB, 22.0dB

Index